Lecture Notes in Electrical Engineering

Volume 1262

The book series *Lecture Notes in Electrical Engineering* (LNEE) publishes the latest developments in Electrical Engineering—quickly, informally and in high quality. While original research reported in proceedings and monographs has traditionally formed the core of LNEE, we also encourage authors to submit books devoted to supporting student education and professional training in the various fields and applications areas of electrical engineering. The series cover classical and emerging topics concerning:

- Communication Engineering, Information Theory and Networks
- Electronics Engineering and Microelectronics
- Signal, Image and Speech Processing
- Wireless and Mobile Communication
- Circuits and Systems
- Energy Systems, Power Electronics and Electrical Machines
- Electro-optical Engineering
- Instrumentation Engineering
- Avionics Engineering
- Control Systems
- Internet-of-Things and Cybersecurity
- Biomedical Devices, MEMS and NEMS

For general information about this book series, comments or suggestions, please contact leontina.dicecco@springer.com.

To submit a proposal or request further information, please contact the Publishing Editor in your country:

China

Jasmine Dou, Editor (jasmine.dou@springer.com)

India, Japan, Rest of Asia

Swati Meherishi, Editorial Director (Swati.Meherishi@springer.com)

Southeast Asia, Australia, New Zealand

Ramesh Nath Premnath, Editor (ramesh.premnath@springernature.com)

USA, Canada

Michael Luby, Senior Editor (michael.luby@springer.com)

All other Countries

Leontina Di Cecco, Senior Editor (leontina.dicecco@springer.com)

**** This series is indexed by EI Compendex and Scopus databases. ****

B. Annappa · Wei-Chiang Hong · Dasari Haritha · G. Lavanya Devi
Editors

High Performance Computing, Smart Devices and Networks

Select Proceedings of CHSN 2023

Springer

Editors
B. Annappa
Department of Computer Science and Engineering
National Institute of Technology Karnataka
Mangalore, Karnataka, India

Wei-Chiang Hong
Department of Information Management
Asia Eastern University of Science and Technology
Taipei, Taiwan

Dasari Haritha
Department of Computer Science and Engineering
Jawaharlal Nehru Technological University
Kakinada, Andhra Pradesh, India

G. Lavanya Devi
Department of Computer Science and Engineering
Andhra University
Visakhapatnam, Andhra Pradesh, India

ISSN 1876-1100 ISSN 1876-1119 (electronic)
Lecture Notes in Electrical Engineering
ISBN 978-981-97-7793-8 ISBN 978-981-97-7794-5 (eBook)
https://doi.org/10.1007/978-981-97-7794-5

This Springer imprint is published by the registered company Springer Nature Singapore Pte Ltd.
The registered company address is: 152 Beach Road, #21-01/04 Gateway East, Singapore 189721, Singapore

Dedicated to

Late Prof. Allam Appa Rao
Former Vice-Chancellor, JNTU Kakinada

Late Prof. Ch. Satyanarayana
Former Registrar, JNTU Kakinada

Foreword

The 4th International Conference on Computer Vision, High Performance Computing, Smart Devices and Networks (CHSN-2023) is aimed to bring researchers together working in this area to share their knowledge and experience. In this conference, topics of contemporary interest will be discussed to provide a holistic vision on the latest technologies for computer science and engineering. The scope includes data science, machine learning, computer vision, deep learning, artificial intelligence, artificial neural networks, mobile applications development, Internet of Things, etc. Conference participants are expected to gain relevant knowledge and better understanding of the applications of computer science in various fields.

CHSN-2023 would be both stimulating and informative with the active participation of galaxy of keynote speakers. We would like to thank all the authors who submitted the papers, because of which the conference became a story of success. We also would like to express our gratitude to the reviewers, for their contributions to enhance the quality of the papers. We are very grateful to the keynote speakers, reviewers, session chairs, and committee members who selflessly contributed to the success of CHSN-2023. We are very thankful to Jawaharlal Nehru Technological University Kakinada, for providing the basic requirements to host the CHSN-2023.

Last but not least, we are thankful for the enormous support of publishing partner, i.e., Springer for supporting us in every step of our journey toward success.

Dr. B. N. Jagadesh
Organizing Chair, CHSN-2023
Professor, VIT-AP
Amaravati, Andhra Pradesh, India

Program Committee

Dr. Ganta Chamundeswari, Sir C R Reddy College of Engineering, Eluru
Dr. Gurram Sunitha, Sree Vidyanikethan Engineering College
Dr. I. Gede Pasek Suta Wijaya, University of Mataram, Indonesia
Dr. J. Harikiran, VIT AP
Dr. J. Avanija, Sree Vidyanikethan Engineering College
Dr. Janmenjoy Nayak, Maharaja Sriram Chandra Bhanja Deo University, Baripada
Dr. Jayalaxmi G. Naragund, KLE Technological University
Dr. Juhi Jain, Amity School of Engineering and Technology
Dr. K Vivekanandan, Puducherry Technological University
Dr. K.Reddy Madhavi, Sree Vidyanikethan Engineering College
Dr. Kabir Kharade, Shivaji University
Dr. Karthikeyan Subramanian, University of Technology and Applied Sciences, Oman
Dr. Kumar Raja D. R, Reva University
Dr. L. Venkateswara Reddy, KG Reddy College of Engineering and Technology, Hyderabad
Dr. Laiphrakpam Dolendro Singh, National Institute of Technology Silchar
Dr. Latha Parthiban, Pondicherry University Community College
Dr. M. Brindha, National Institute of Technology, Tiruchirappalli
Dr. Malaya Dutta Borah, National Institute of Technology, Silchar
Dr. Megha Ummat, Sri Guru Gobind Singh College of Commerce, University of Delhi
Dr. Mohammed Abdul Qadeer, Aligarh Muslim University
Dr. Mohammed Nasser Al-Mhiqani, Keele University
Dr. Mohammed Pasha, Muffakham Jah College of Engineering and Technology
Dr. Mohd Fairuz Iskandar Othman, Universiti Teknikal Malaysia Melaka
Dr. Mohd Faizal Abdollah, Universiti Teknikal Malaysia Melaka
Dr. Mohd Zaki Mas'ud, Universiti Teknikal Malaysia Melaka
Dr. Naresh Babu Muppalaneni, National Institute of Technology Silchar
Dr. Naveen Palanichamy, Multimedia University, Malasiya
Dr. Nazrulazhar Bahaman, Universiti Teknikal Malaysia Melaka
Dr. Ng Kok Why, Multimedia University, Malasiya
Dr. Nitikarn Nimsuk, Thammasat University (Rangsit Campus), Thailand
Dr. Noor Hisham Bin Kamis, Multimedia University (MMU), Malaysia
Dr. P. Sateesh, MVGR College of Engineering
Dr. Partha Pakray, National Institute of Technology, Silchar
Dr. Pascal LORENZ, University of Haute Alsace, France
Dr. Prashant Giridhar Shambharkar, Delhi Technological University
Dr. Pratik Chattopadhyay, Indian Institute of Technology (BHU), Varanasi
Dr. R. Kanesaraj Ramasamy, Multimedia University, Malasiya
Dr. R. Murugan, National Institute of Technology, Silchar
Dr. R. Rajeshwara Rao, JNTU Vizianagram
Dr. Raihana Syahirah Abdullah, Universiti Teknikal Malaysia Melaka
Dr. Rajesh Prasad, American University of Nigeria
Dr. Rajib Kumar Jha, Indian Institute of Technology Patna

Dr. Ram Bilas Pachori, Indian Institute of Technology Indore
Dr. Ramanujam E, National Institute of Technology Silchar
Dr. Rashmi Saini, G. B. Pant Institute of Engineering and Technology
Dr. Ripon Patgiri, National Institute of Technology Silchar
Dr. Robiah Binti Yusof, Universiti Teknikal Malaysia Melaka
Dr. S. M. Warusia Yassin, Universiti Teknikal Malaysia Melaka
Dr. S. K. Chaya Devi, Vasavi College of Engineering
Dr. Sanjaya Kumar Panda, National Institute of Technology, Warangal
Dr. Sansanee Auephanwiriyakul, Chiang Mai University, Thailand
Dr. Santosh Singh Rathore, ABV-IIITM Gwalior
Dr. Sasikumar Gurumoorthy, Jerusalem College of Engineering
Dr. Sharifah Sakinah Syed Ahmad, Universiti Teknikal Malaysia Melaka
Dr. Sherali Zeadally, University of Kentucky
Dr. Siti Azirah Asmai, Universiti Teknikal Malaysia Melaka
Dr. Siti Rahayu Binti Selamat, Universiti Teknikal Malaysia Melaka
Dr. Sivasutha Thanjappan, Multimedia University, Malasiya
Dr. Sonika Dahiya, Delhi Technological University
Dr. Srinivasa Chakravarthi Lade, GITAM University
Dr. Surya Kameswari Uduga, Acharya Nagarjuna University
Dr. Surya S. R., College of Engineering Perumon, Kollam
Dr. Syed Muzamil Basha, REVA University
Dr. Syed Thouheed Ahmed, REVA University
Dr. T. G. Vasista, Pallavi College Engineering
Dr. Tanasanee Phienthrakul, Mahidol University, Thailand
Dr. Tomasz RAK, Rzeszow University of Technology, Poland
Dr. Uma N. Dulhare, Muffakham Jah College of Engineering and Technology
Dr. Umashankar Subramaniam, Prince Sultan University, Riyadh, Saudi Arabia
Dr. Veenu Mangat, Panjab University
Dr. Vishal Ramesh Satpute, Visvesvaraya National Institute of Technology, Nagpur
Dr. Vishal Saraswat, Bosch Engineering and Business Solutions
Dr. Wei-Chiang Hong, Asia Eastern University of Science and Technology, Taiwan
Dr. Yogan Jaya Kumar, Universiti Teknikal Malaysia Melaka
Dr. Zulkiflee Muslim, Universiti Teknikal Malaysia Melaka
Dr. Zuraida Abal Abas, Universiti Teknikal Malaysia Melaka
Mr. Ksh Robert Singh, Mizoram University
Mr. Nageshwar Nath Pandey, Siksha 'O' Anusandhan Deemed to be University
Mr. Rohit Pratap Singh, National Institute of Technology Silchar
Mr. Sahinur Rahman Laskar, National Institute of Technology Silchar
Mr. Sanjay Patidar, Delhi Technological University
Mr. Yonten Jamtsho, Royal University of Bhutan
Mr. Zakir Hussain, National Institute of Technology Silchar
Ms. Anjali Bansal, Delhi Technological University
Ms. Deepali Jain, National Institute of Technology Silchar
Ms. K. Bavani, Kalasalingam Academy of Research and Education
Ms. Priya Singh, Delhi Technological University

Ms. Sabuzima Nayak, National Institute of Technology Silchar
Ms. Shweta Meena, Delhi Technological University
Ms. Zeba Shamsi, National Institute of Technology Silchar

Keynote Speakers

Dr. R. Kanesaraj a/l ramasamy, Multimedia University, Cyberjaya, Selangor

Dr. Sarahani Binti Harun, Pensyarah Universiti, Malaysia

Dr. Mira kania Sabariah, Telkom University, Indonesia

Contents

Secure Data Sharing Using Enhanced Attribute Based Encryption for Blockchain Technology

V. Akshaya Keerthi and V. Vijayalakshmi

Abstract Sharing data safely and securely is the need of the hour in many industrial sectors where large volumes of data are generated and shared over the internet. Conventional data sharing schemes rely on centralized storage platform with definite third-party access control mechanism. Hence, it is prone to single point failure, malicious attacks, poor transaction transparency, etc. Advancements in cryptography and blockchain technology are revolutionizing the rapidly growing information systems sector. However, the real-time deployment of blockchain technology encounters many problems such as storage explosion, data leakage problem, etc. An additional layer of encryption can overcome the problem of data leakage. In order to address these issues, this paper presents an Enhanced Attribute Based Encryption (EABE) scheme with lightweight optimal key generation for blockchain technology to improve the security and traceability of shared data. The data is encrypted using Enhanced Attribute Based Encryption (EABE). The underlying key generation step is optimized using bio-inspired metaheuristic optimization algorithms such as Green Anaconda Optimization with Improved Black Widow Optimization (GAO-IBWO). Thus, the proposed EABE scheme forms an additional layer of security for blockchain technology with lightweight optimal key generation.

Keywords Enhanced attribute based encryption (EABE) · Green anaconda optimization (GAO) · Black widow optimization (BWO) · Key generation · Blockchain

V. Akshaya Keerthi (✉) · V. Vijayalakshmi
Department of Electronics and Communication Engineering, Puducherry Technological University, Puducherry, India
e-mail: akshayakeerthi@pec.edu

V. Vijayalakshmi
e-mail: vvijizai@pec.edu

B. Annappa et al. (eds.), *High Performance Computing, Smart Devices and Networks*, Lecture Notes in Electrical Engineering 1262,
https://doi.org/10.1007/978-981-97-7794-5_1

1 Introduction

Secure data transmission is the need of the hour in many sectors such as Industrial Internet of Things (IIoT), banking, healthcare application, Unmanned Aerial Vehicle (UAV), military applications, supply chain management, blue economy, etc. Cyber threats are increasing day by day due to increased rate of internet users. Large volumes of data are transmitted over the unprotected internet medium. Data collected from various sources are stored in centralized server which is prone to malicious attacks and single point failure. A single centralized authoritative organization is involved in access management. Thus, the digital data tracking becomes impossible and also encounters lack of security [1].

Development in cryptography and blockchain technologies has paved the way for more secure transaction of data. Cryptography deals with techniques and protocols that prevent unauthorized users from accessing and gaining knowledge of the confidential information. It plays a major role in today's rapidly growing information system. Many encryption techniques are available to ensure the security, accuracy, accountability, and confidentiality of data. Blockchain is a distributed ledger technology where group of data are stored in blocks. Many such blocks are created, and each block stores its own data, hash value of the current block and hash value of the previous block. All these connected blocks are referred as a ledger. The ledger information is known to all the participants in the network. It offers decentralization, immutability, security, and transparency. Thus, blockchain is considered as a potential idea for achieving security and privacy in many sectors [2].

Currently, blockchain-based data sharing models are implemented in many sectors. It overcomes the problem of centralized data storage. However, the classical blockchain proposal suffers from storage, resource consumption, throughput, delay, and computational complexity [3]. Blockchain encounters data leakage problems due to its open and transparent nature. An additional layer of transaction encryption is required. Encryption algorithms are the most appropriate solution to secure data on the blockchain. Most important criteria to be considered is the identification of the type of cryptography technique that well satisfies the problem of security and privacy of the data stored in blockchain. Attribute Based Encryption (ABE) technique is well suited for secure data transmission. It is an asymmetric encryption technique that comes under the broad category of identity-based encryption. It offers one-to-many encryption and decryption of data and thus minimizes the system complexity [4].

The organization of this paper is as follows. Section 2 outlines the related work. Section 3 explains the existing research article related to the key generation technique using bio-inspired algorithms. The proposed approach of Enhanced Attribute Based Encryption technique for blockchain technology is briefly described in Sect. 4. Section 5 elaborates the result analysis. Finally, the conclusion and future work is illustrated in Sect. 6.

2 Related Work

Nowadays, secure data sharing has become a challenge in various sectors where large volumes of data need to be transmitted via unprotected medium. Traditional data storage platforms are centralized in nature. Internet of Things applications fundamentally rely on these platforms. Enhanced encryption algorithms integrated with blockchain technology forms a more suitable and effective way to enhance the security. The encryption scheme should satisfy the three major criteria such as security, access control, and privacy. Attribute Based Encryption technique is identified to provide better access control [4, 5].

Raja et al. [6] had proposed an enhanced Attribute Based Encryption (ABE) scheme for secure wireless communication. The classical key generation phase was optimized using hybrid bio-inspired metaheuristic algorithms. The hybrid combination of Grey Wolf Optimization (GWO) algorithm with Whale Optimization (WOA) algorithm in the global and local searching strategy was incorporated for lightweight dynamic key generation process. The concept of Opposition Based Rule (OBL) was incorporated to enhance the convergence speed of the hybrid algorithm. It prevents the key generation center from gaining knowledge of the user attributes and the corresponding keys. This helped in increasing the user's privacy and data confidentiality.

Christo et al. [7] had proposed a security framework for medical data by integrating ECC and blockchain technology. The user data was collected and stored in Electronic Medical Card (EMC) with microcontroller chip. This data was encrypted using Ellipitic Curve Cryptography (ECC) and stored in blockchain to protect the data from unauthorized users. The authenticated users such as doctors could decrypt the data for further processing.

An improved homomorphic encryption technique was proposed by Tamilarasi et al. [8]. The user data was encrypted using ElGamal encryption technique. Improved grasshopper optimization algorithm was proposed for optimal key generation to enhance the security and privacy of data sharing model. Gaussian mutation and Levy flight ideas were incorporated to enhance the performance of the standard Grasshopper Optimization Algorithm.

Alkhliwi [9] presented a dynamic access control scheme with blockchain technology to prevent data theft and unauthorized access. The transmitted data was encrypted and stored in IPFS. The hash value was again encrypted using CP-ABE scheme. Further, blockchain technology was used to store the data and keys for decryption.

3 Existing System

Devices on the IoT platform rely on wireless communication where the signals are transmitted across mediums that are prone to unauthorized access, and thus, privacy and data security is the need of the hour in this sector. Aiyshwariya Devi et al. [10] had presented an improved ECC scheme to enhance the IoT device security. The typical Elliptic Curve Cryptography (ECC) algorithm was improved by optimizing the underlying key generation step using bio-inspired metaheuristic algorithm. A hybrid combination of Mayfly Optimization Algorithm (MOA) with Black Widow Optimization (BWO) algorithm was introduced for optimal key generation. This includes both private and public keys. The encryption of plaintext was performed using optimal public key and decryption using optimal private key. The ciphertext was further stored in the cloud storage. Thus, the Improved Elliptic Curve Cryptography (IECC) algorithm enhanced the security of data transmitted.

The steps involved in IECC-MA-BWO algorithm for selecting optimal key are as follows:

(1) During initialization, the prime numbers were considered to form a new population size based on male and female mayfly swarms for optimal key selection procedure.

$$\text{Input}_{\text{solution}} = \{A_1, A_2, A_{3,} \ldots A_n\} \tag{1}$$

(2) The fitness function was used throughout the selection of best key with least Mean Squared Error (MSE) to encrypt and decrypt data.

$$\text{Fitness} = \min\{\text{MSE}\} \tag{2}$$

(3) The new set of population were updated to get the best possible solution.

$$\text{Update pop} = \text{pop2} + \text{pop3} \tag{3}$$

(4) After the selection of the optimal key, the process gets terminated.

Figure 1 Illustrates the flowchart of the hybrid MA-BWO algorithm. However, the existing MA-BWO algorithm encounters poor balance between exploration and exploitation phase and leads to slow convergence rate. It suffers from premature convergence and thus affects the overall execution time of the encryption algorithm. And also IECC does not offer one-to-many encryption. Further, in the existing system, the encrypted data is stored in the centralized cloud storage which is prone to malicious attack, single point failure, and data leakage problems

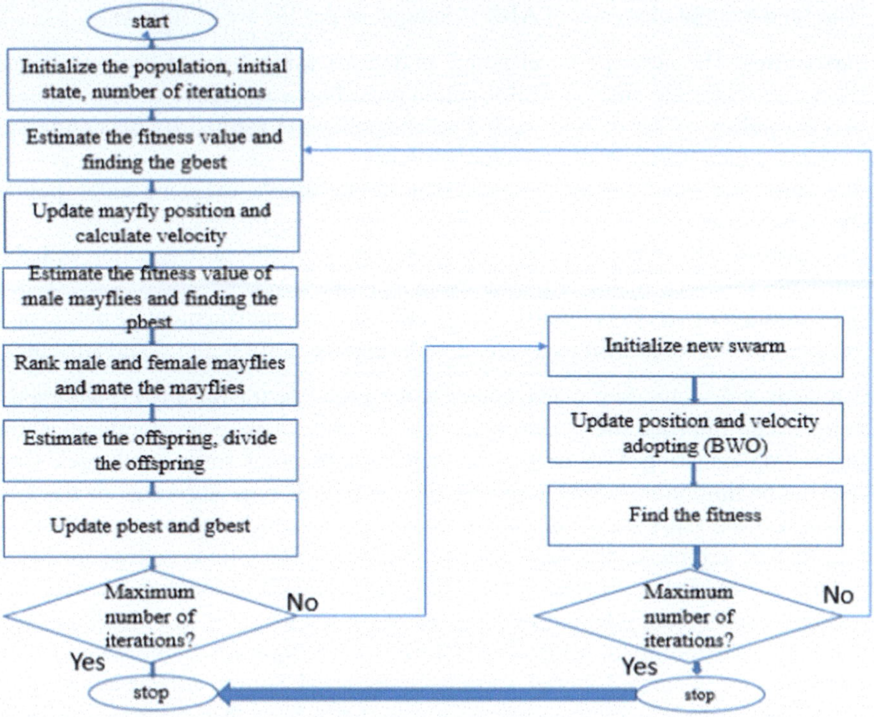

Fig. 1 Flowchart of existing MA-BWO algorithm [10]

4 Proposed System

To overcome the drawbacks of the existing scheme, an Enhanced Attribute Based Encryption (EABE) scheme for blockchain technology is proposed for efficient and secure data sharing using bio-inspired metaheuristic optimization algorithms. The hybrid combination of Green Anaconda Optimization (GAO) and Improved Black Widow Optimization (IBWO) technique is used for lightweight optimal key generation.

4.1 Attribute Based Encryption (ABE) Scheme

ABE technique is considered as the promising encryption models in current crucial sectors such as IoT. It provides one-to-many encryption and fine-grained access control to scrambled data. It ensures the accountability of the user, revocation of the user, scalability, confidentiality of data, and collusion resistance.

The fundamental elements of ABE technique as per [4] are as follows:

- **Attributes**: The access control policy is defined using a set of user attributes. These attributes are any valid information regarding user identity.
- **Access policy**: A set of rules called access policy are framed for accessing the encrypted data based on the selected attributes.
- **Encryption**: Converting the plaintext to the unintelligible text using access policy and a public key.
- **Decryption**: The plaintext is recovered after analyzing the access policy with the attributes provided by the requester using a private key.
- **Key management**: This involves the generation and management of public and private keys to authorized users along with access control policies and attributes.

Efficient and secure key management is the prime factor for any cryptography system. It can improve the performance and security of wireless communication systems. Implementing ABE in real-time applications poses some challenges. One of the major limitation is that ensuring the privacy of user attributes in the key generation phase. Attributes may contain sensitive information about the user. During the key generation phase, the Key Generation Center (KGC) finds out the attributes and the matching keys of the user [6]. This could be used to identify the user and reveal personal and confidential information about the user. So, the traditional key generation procedure involved in Attribute Based Encryption scheme such as how the attributes are generated, stored, and shared needs to be improved to guarantee the user's privacy. Access to the keys by any means of attacks such as brute force attack, side channel attack, physical access of system, weak encryption, and replay attack can collapse the entire wireless communication system [11]. Thus, the traditional key generation procedure is optimized using bio-inspired metaheuristic optimization algorithm.

4.2 Design of ABE-GA-IBWO Algorithm

This proposed model introduced an Enhanced Attribute Based Encryption (EABE) scheme with dynamic key generation using bio-inspired metaheuristic optimization algorithms. Green Anaconda with Improved Black Widow Optimization (GA-IBWO) algorithm is used for optimal key generation.

Green Anaconda Optimization (GAO) algorithm: It is an evolutionary-based metaheuristic algorithm framed based on mimicking the behavioral characteristics of green anacondas. Metaheuristic algorithms usually have two stages of searching process to find the best solution. The global search phase also referred as exploration phase is to find the main optimal region without getting stuck in the local optimum. The next is the local search phase also referred as exploitation phase. It helps to converge toward the best possible solution in the search space [12].

In the global search phase, the green anaconda moves toward the global optimal position, and the position is updated as shown in Eq. (5).

$$a_{i,d}^{L1} = a_{i,d} + t_{i,d}.\left(SF_d^i - I_{i,d}.a_{i,d}\right),$$
$$i = 1, 2, \ldots, N, \text{ and}$$
$$d = 1, 2, \ldots, m \quad (4)$$

$$Y_i = \begin{cases} Y_i^{L1}, \ F_i^{L1} < F_i, \\ Y_i, \quad \text{else} \end{cases} \quad (5)$$

where Y_i^{L1} is the latest suggested position of the ith green anaconda based on the global search process, $a_{i,d}^{L1}$ is its dth dimension, F_i^{L1} is its objective function value.

In the local search phase, the green anaconda randomly moves its position and the position update formula is as follows:

$$Y_i = \begin{cases} Y_i^{L2}, \ F_i^{L2} < F_i, \\ Y_i, \quad \text{else} \end{cases} \quad (6)$$

where Y_i^{L2} is the latest position of the ith green anaconda based on the local search process, $a_{i,d}^{L2}$ is its dth dimension, F_i^{L2} is its objective function value.

Black Widow Optimization (BWO) algorithm: It is an evolutionary-based meta-heuristic algorithm modeled based on the selection, reproduction and mutation strategy of black widow spiders. The algorithm undergoes five stages of mathematical process to find the best solution. This includes initialization of population randomly, procreate, cannibalism, mutation, and population renewal [10].

During initialization process, the population of candidates are randomly selected and the fitness function is calculated as follows.

$$\text{widow} = [z_1, z_2, \ldots, z_S] \quad (7)$$

$$\text{fitness} = f(\text{widow}) = f(z_1, z_2, \ldots, z_S) \quad (8)$$

The population is updated as per the following equation.

$$\text{Update pop} = \text{pop2} + \text{pop3} \quad (9)$$

With the aim to enhance the convergence rate and avoid premature convergence, the hybrid combination of green anaconda and improved black widow optimization algorithms is proposed.

Green Anaconda with Improved Black Widow Optimization algorithm (GA-IBWO): This paper introduces a novel secure data sharing model using Attribute Based Encryption with Green Anaconda and Improved Black Widow Optimization

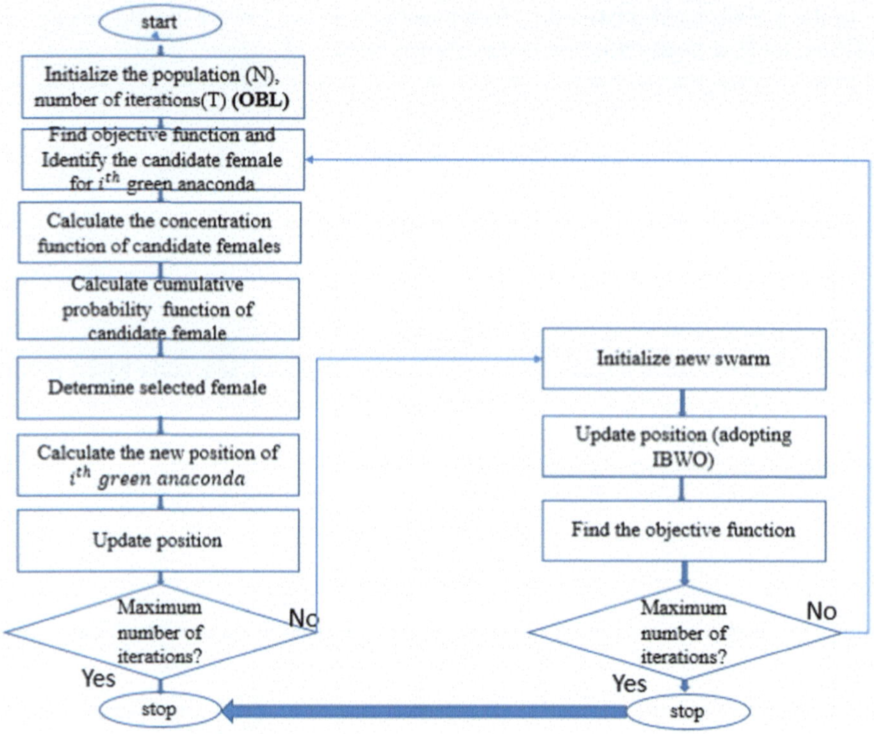

Fig. 2 Flowchart of proposed GA-IBWO algorithm

(ABE-GA-IBWO). During the global search phase, the Green Anaconda Optimization algorithm is used to update the global search position. During the local search phase, Improved Black Widow Optimization algorithm is used to update the local search position to enhance the convergence rate and global optimization approach of the algorithm. The proposed algorithm is used for the optimal key generation in Attribute Based Encryption scheme. Instead of random initialization of population, Opposite Based Learning (OBL) Concept is used to avoid the selection of poor quality of individuals [6, 13] and enhance the searching process during global and local search strategies. Figure 2 Illustrates the flowchart of GA-IBWO algorithm.

4.3 Block Diagram of Enhanced Attribute Based Encryption (EABE) Scheme for Blockchain Technology

The transparent nature of blockchain is prone to data leakage problem. Hence, an additional layer of transaction encryption is required. In the proposed scheme, Enhanced Attribute Based Encryption (EABE) technique is introduced for secure data sharing process. It provides an additional layer of security. Attributes are allocated based on the user identity, location, etc. Access control policy is designed based on the attributes allocated. The Key Generation Center (KGC) plays a vital role in generating and distributing the public and master keys to the authorized users. The traditional key generation phase is optimized using GA-IBWO optimization algorithm in order to enhance the security and privacy of the user attributes.

Step 1 The proposed ABE-GA-IBWO algorithm is used to generate the Public Key PK and a Master Secret Key MSK on the basis of security parameters.

Step 2 The data owner submits the attribute set to the Key Generation Center.

Step 3 The key generation phase uses optimal PK, MSK and user attributes to generate a Secret Key for the data user.

Step 4 The data that is encrypted using optimal PK and Access Policy AP is uploaded to the server.

Step 5 The data user gives a request for data.

Step 6 The server returns the ciphertext with attribute set for decryption.

Step 7 The data user request the Key Generation Center for Secret Key that should match the attribute set.

Step 8 The Key Generation Center returns the Secret Key to the data user for decryption as illustrated in Fig. 3. It provides one-to-many encryption, thus reducing the computational overhead and delay.

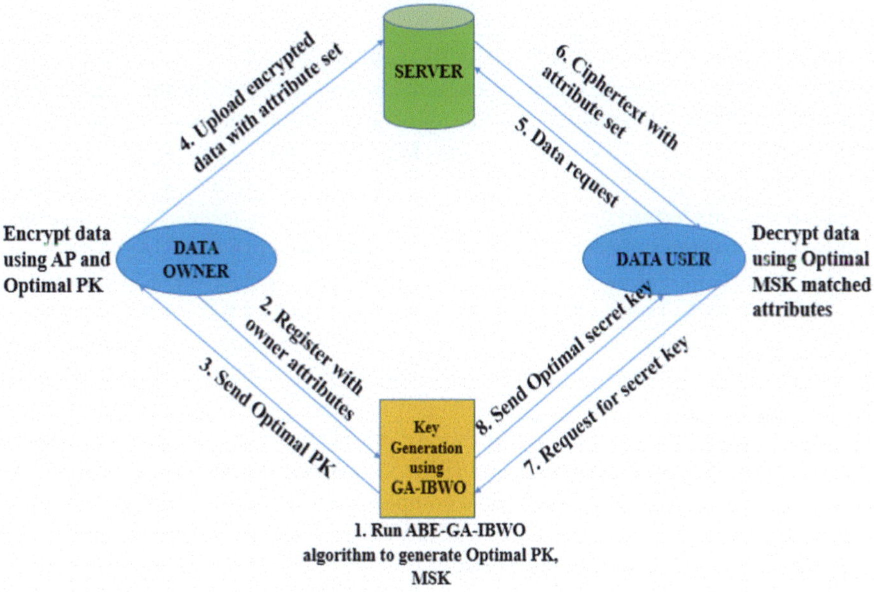

Fig. 3 Block diagram of proposed ABE GA IBWO scheme

5 Results and Analysis

The security and performance analysis of the proposed ABE-GA-IBWO algorithm is presented in this section.

5.1 Security Analysis

The proposed ABE-GA-IBWO model provides fine-grained access control to the data. It ensures authentication and authorization by limiting the access to specific users to view the specific information. The optimized key generation process in the proposed model prevents the Key Generation Center (KGC) from learning the attributes and corresponding keys of the user. It ensures privacy by protecting the sensitive information transmitted over the unprotected internet medium. Therefore, the proposed ABE-GA-IBWO model provides control over data access, data leakage problem, and data loss problem and proves to be more secure when compared to ECC-MA-BWO.

5.2 Performance Analysis

The proposed encryption algorithm is implemented using NS-3.26, and the performance is analyzed. The time taken to generate the keys, encrypt the data, and decrypt the data are analyzed. Upon analyzing the key computation time as shown in Fig. 4, it is evident that ABE-GA-IBWO outperforms ECC-MA-BWO. This significant difference suggests that ABE-GA-IBWO likely employs a more efficient algorithm for key computation. This hybrid optimization technique effectively reduces computational complexity, resulting in faster key computation compared to ECC-MA-BWO.

The comparison is done for Key Generation Time in msvs Packet Size in bytes. The ABE-GA-IBWO has achieved a key generation time of 6.2 ms for packet size of 500 bytes, whereas ECC-MA-BWO has achieved a key generation time of 11 ms. Therefore, the key generation time of ABE-GA-IBWO is 4.8 ms faster than ECC-MA-BWO.

Figure 5 Illustrates the encryption time required for converting a plaintext into a ciphertext. In this context, ABE-GA-IBWO demonstrates superior performance when compared to ECC-MA-BWO. The comparison is done for Encryption Time in msvs Packet Size in bytes. The ABE-GA-IBWO has achieved an encryption time of 26 ms for packet size of 500 bytes, whereas ECC-MA-BWO has achieved 32 ms for encryption. Therefore, the encryption time of ABE-GA-IBWO is 6 ms faster than ECC-MA-BWO

Figure 6 Illustrates the decryption time required for converting ciphertext into a plaintext. In this context, ABE-GA-IBWO demonstrates superior performance when

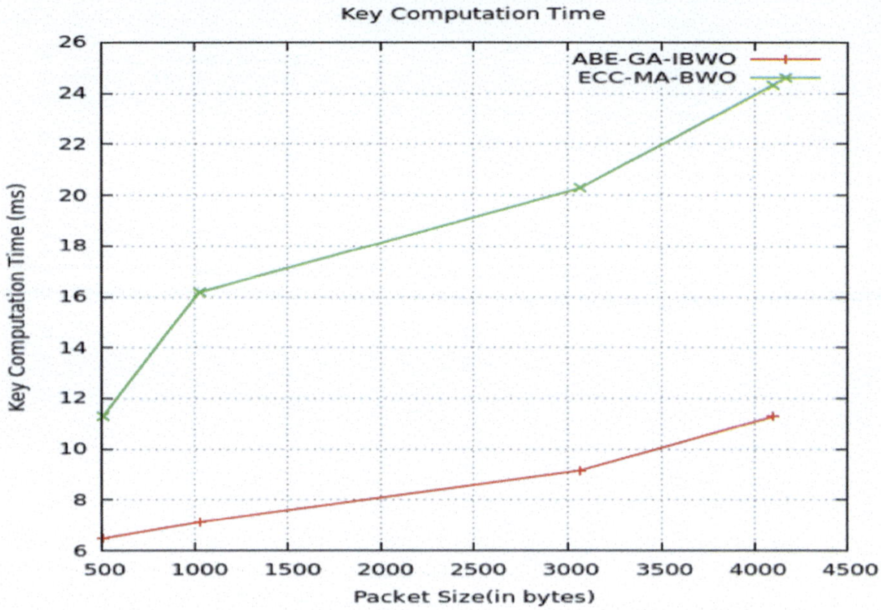

Fig. 4 Key generation time analysis of ABE-GA-IBWO

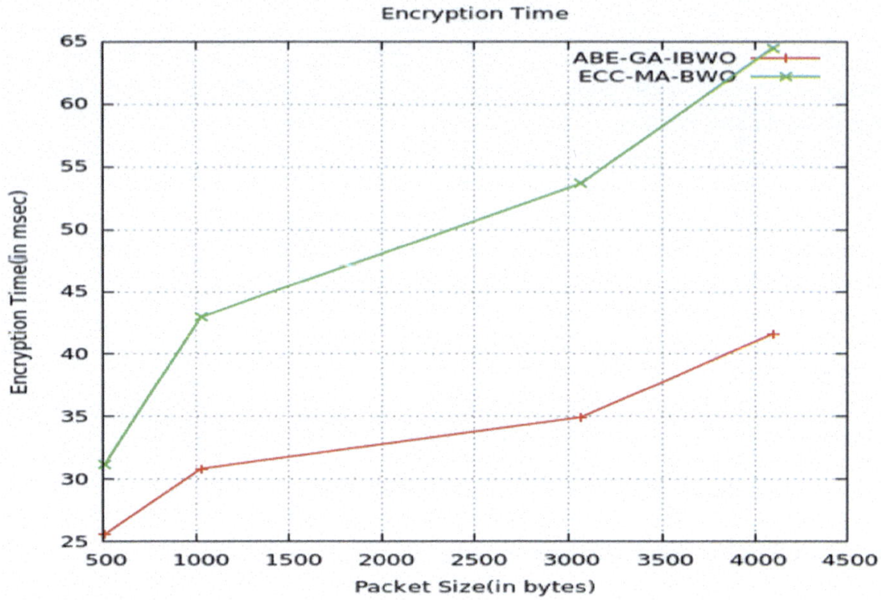

Fig. 5 Encryption time analysis of ABE-GA-IBWO

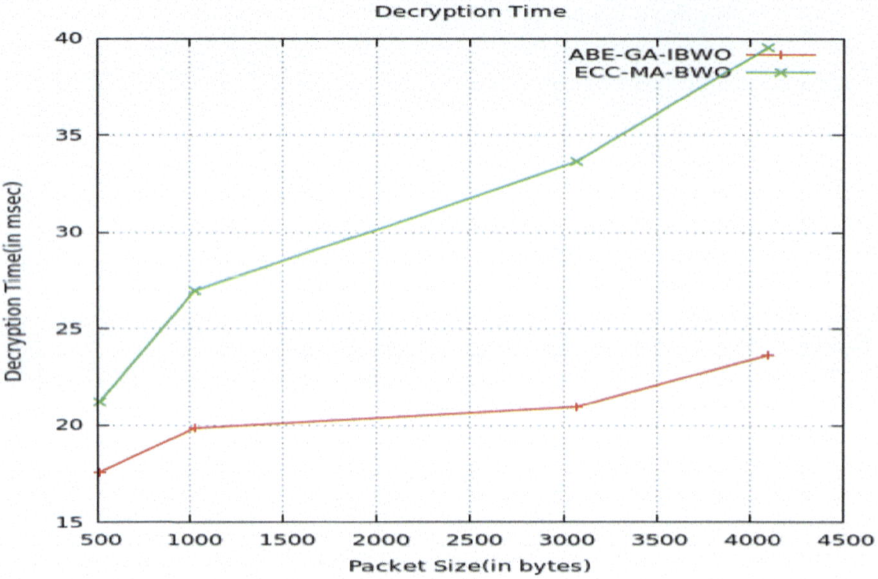

Fig. 6 Decryption time analysis of ABE-GA-IBWO

Table 1 Result analysis of ABE-GA-IBWO model

Parameter	ABE-GA-IBWO (ms)	ECC-MA-BWO (ms)
Key generation time for 2500 bytes	8.6	19
Encryption time for 2500 bytes	34	51
Decryption time for 2500 bytes	21	32

compared to ECC-MA-BWO. The comparison is done for Decryption Time in msvs Packet Size in bytes. The ABE-GA-IBWO has achieved a decryption time of 18 ms for packet size of 500 bytes whereas ECC-MA-BWO has achieved a key generation time of 22 ms. Therefore, the decryption time of ABE-GA-IBWO is 4 ms faster than ECC-MA-BWO

Thus, the proposed ABE-GA-IBWO algorithm has better performance when compared to the existing ECC-MA-BWO algorithm as discussed in Table 1.

6 Conclusion and Future Work

The evolution of blockchain technology is strongly based on cryptography techniques. Enhancement in cryptography techniques is the need of the hour for efficient usage of blockchain technology in real-time industrial applications such as Industrial Internet of Things (IIoT), banking, healthcare application, Unmanned Aerial

Vehicle (UAV), military applications, supply chain management, blue economy, etc. Emerging cryptography and blockchain technologies are reshaping the rapidly growing information system. It enhances the trust and traceability of the flow of large volumes of data over the internet.

In this paper, the key generation phase of classical Attribute Based Encryption Scheme was optimized using Green Anaconda with Improved Black Widow Optimization algorithm (GA-IBWO) for blockchain technology. The proposed ABE-GA-IBWO scheme improved the security and privacy of the user data attributes. The performance of the proposed ABE-GA-IBWO scheme was validated by comparing the key generation, encryption and decryption time of the proposed ABE-GA-IBWO with the existing ECC-MA-BWO algorithm. Further, the proposed ABE-GA-IBWO is to be used as an additional layer of security for blockchain network to prevent the data leakage problem.

References

1. Wang Z, Guan S (2023) A blockchain-based traceable and secure data-sharing scheme. PeerJ Comput Sci 9:e1337. https://doi.org/10.7717/peerj-cs.1337
2. Biswas S, Shaif K, Li F, Nour B, Wang Y (2019) A scalable blockchain framework for secure transaction in IoT. IEEE Internet Things J 6(3):4650–4659
3. Stefanescu D, Montalvillo L, Galan-Garcia P, Unzilla J, Urbieta A (2022) A systematic literature review of lightweight blockchain for IoT. IEEE Access 10:123138–123159
4. Shruti, Rani S, Sah DK, Gianini G (2023) Attribute-based encryption schemes for next generation wireless IoT networks: a comprehensive survey. Sensors 23(13):5921. https://doi.org/10.3390/s23135921. PMID: 37447769; PMCID: PMC10346245
5. Yan L, Ge L, Wang Z, Zhang G, Xu J, Hu Z (2023) Access control scheme based on blockchain and attribute-based searchable encryption in cloud environment. J Cloud Comput 12(61):1–16
6. Raja M, Dhanasekaran S, Vasudevan V (2022) Opposition based joint Grey Wolf-whale optimization algorithm based encryption in secure wireless communication. Wirel Pers Commun 127:635–655
7. Christo MS, Elizabeth Jesi V, Priyadarsini U, Anbarasu V, Venugopal H, Karuppiah M (2021) Ensuring improved security in medical data using ECC and blockchain technology with edge devices. Secur Commun Netw 2021
8. Tamilarasi G, Rajiv Gandhi K, Palanisamy V (2022) Improved homomorphic encryption with optimal key generation technique for VANETs. Intell Autom Soft Comput 33(2):1273–1288
9. Alkhliwi S (2022) An efficient dynamic access control and security sharing scheme using blockchain. Int J Adv Appl Sci 9(8):28–40
10. Aishwarya Devi R, Arunachalam AR (2023) Enhancement of IoT device security using an improved elliptic curve cryptography algorithm and malware detection utilizing deep LSTM. High-Confidence Comput 3(2)
11. Pal O, Alan B, Thakur V, Singh S (2019) Key management for blockchain technology. Korean Inst Commun Inf Sci 7:76–80
12. Dehghani M, Trojovsky P, Malik OP (2023) Green anaconda optimization: a new bio-inspired metaheuristic algorithm for solving optimization problems. Biomimetics 81
13. Semchedine M, Bensoula N (2022) Enhanced black widow algorithm for numerical functions optimization. Int Inf Eng Techn Assoc 36(1):1–11

An Enhanced Text Classification Using Machine Learning

Venkateswarlu Sunkari, Gudikandhula Narasimha Rao, A. Srinagesh, and Elefelious G. Belay

Abstract Content can be grouped using predefined classifications that are applied to plain language content. It's a "pack of words" representation; historical reports include a term that has values and indicate how frequently it appears in the record or not. In any case, because they include abundant or inconsequential data, large reports may face several problems. The impact of several attributes that convey the appropriation of a word in the archive is examined in this research. We refer to these attributes as distributional highlights. Every highlight is chosen based on the tfidf style criteria, and AI techniques are used to combine these highlights. One of the primary factors in distributional highlights—which hold a weighted object set—is term recurrence. Finding unusual information connections is more interesting than mining continuous ones when it comes time to limit a certain score works. The problem of locating rare weighted thing sets, also known as the rare weighted thing set mining problem, is addressed in this study. The classifier selected for order is the one that produces the more precise result. The distributional highlights are useful for text classification, according to experiments.

V. Sunkari (✉) · E. G. Belay
School of Information Technology and Engineering, Addis Ababa Institute of Technology, Addis Ababa University, Addis Ababa, Ethiopia
e-mail: v.sunkari@aait.edu.et

E. G. Belay
e-mail: elefelious.getachew@aau.edu.et

G. N. Rao
Department of CS&SE, Andhra University College of Engineering (A), Andhra University, Visakhapatnam, Andhra Pradesh, India
e-mail: prof.narasimha@andhrauniversity.edu.in

A. Srinagesh
Department of CSE, RVR & JC College of Engineering, Guntur, Andhra Pradesh, India

E. G. Belay
School of Information Technology and Engineering, Addis Ababa Institute of Technology, Addis Ababa, Ethiopia

B. Annappa et al. (eds.), *High Performance Computing, Smart Devices and Networks*, Lecture Notes in Electrical Engineering 1262,
https://doi.org/10.1007/978-981-97-7794-5_2

1 Introduction

One popular exploratory information-digging technique for extracting meaningfully correlated information from massive reports is text mining. Finding regular object sets, or designs whose perception is differentiating event of term recurrence in the material, was the primary goal of the first text mining stage [1]. Applications for Term Frequency were found in a variety of real-world contexts (e.g. market bin analysis, clinical image processing, natural information research, and disease information gathering). However, some standard approaches fail to recognize the value of all that is contained in the examined informational collections. Accordingly, the concept of a scoring thing set has also been introduced, allowing for the treatment of items differently based on their relevance in the content mining measure. Every information item has a score that relates to it and communicates its importance in every interaction [2].

Text categorization isanother type of problem wherein regular language text is assigned predefined classifications based on its relevance attribute sets. Since text order supports characteristic language management, it has attracted more and more attention. This is an administered learning problem for which a variety of classifiers, including Naive Bayes, Fuzzy Classifiers, Decision Trees, Artificial Neural Networks, k Nearest Neighbour (kNN), Support Vector Machine (SVM), and others, have been applied. AdaBoost, SVM and fluffy rationales have recently produced some amazing results [3].

Given that every archive in the extensive informational indexes indicates the quantity of words that appear in the report, every characterization technique consistently presented the "sack of words" representation. In general, it is thought that each word in the report should have a certain set of properties. Finally, it gives the total word count in the report as well as the frequency with which this word appears in the record. When every attribute in the report is identified, that information will be useful for organizing the text [4]. Consider the following example: "How are you" and "How you are" are two distinct words that each represent a comparable vector using recurrence-related attributes, but they have entirely different connotations. This model clearly illustrates the word's presence, repetition, and circulation throughout the record. In order to quantify the presence of a word's dispersion in an archive, this study thus familiarizes itself with plan and builds up some distributional highlights.

In order to measure the characteristics of a word's circulation in an archive, this research aims to plan some distributional highlights. The primary idea is that words tend to appear small. In this case, the minimization calculates if a word appears frequently in a certain section of a record or frequently throughout the archive. The term is regarded as smaller in the last situation than it was in the preceding one, where it was deemed reduced. The realities that follow this idea convince me. There are usually a few sections in an archive. When a term appears more frequently, it is more likely to appear in different contexts and be associated with the report's subject [5].

This work commitment is the following:

(1) Distributional highlights are designed for text order. By employing these highlights, execution effectiveness can be increased with only a small additional cost.
(2) How to make use of distributional highlights in large-scale databases. The display outcomes will increase with the addition of Traditional Term Frequency with Distributional highlights.
(3) Talk to determine which elements are affecting how the distributional highlights are displayed.
(4) The distributional highlights' upward trend is strongly correlated with the number of entries in a corpus information base, and it also supports regular language archives.

The following is how the paper is organized as some related concepts about the extraction of distributional highlights are presented in the following area: The use of distributional highlights is presented in segment 3. Tests in area 4 follow the introduction of the suggested computations' consequences. The information base's support is shown in segment 5, and its conclusions are provided in region 6.

2 Extraction of Distributional Features

"Include" generally has two related meanings, according to the highlights for text classification that are cited. The first one depicts an archive or a report file, whilst the second one focuses on the most effective way to assign a suitable score to a particular component. Take "pack of words" as an example. The component assigned another importance is the term recurrence and converse record recurrence score, while the part with the prior significance is a single word [6]. These highlights are finally used for text configuration based on these two considerations. The score assigned to a particular element originates from two sources: the intra report and the entomb archive, which is the second significance of highlight. Whereas the entomb archive-based score makes use of data from the corpus information base, the intra record-based score uses data from a report.

The record recurrence part for tfidf is a load from an entomb archive source, whereas the term recurrence part can be thought of as a score from an intra report source. It was found that the opposite report recurrence distributed phrase frequencies evenly across the range of 0 to 1. It is applied to the score of every sentence in the whole record to determine its relevance. In particular, there were two ways to measure a sentence's relevance. Ascertaining the relationship between the title and a specific sentence was one task, and adding the significance of every word that appears in this sentence as the final one was another [7]. The initial term recurrence for a word was replaced with a score term recurrence based on the sentence's significance, and each occurrence was positioned according to the sentence's relevance. This section demonstrates how distributional highlights are important for text organization based on word reduction, opposite report recurrence, and term recurrence.

2.1 Technique for Word's Distribution

In terms of word distribution, the report is initially divided into multiple sections. Additionally, it is possible to ascertain the frequency with which this word appears in the document. It is the combination of all the elements in the archive and is often maintained by a cluster. There should be three different types of entries underneath every report. This discussion section discusses logical statements and paragraphs from the archives. As evidenced by content, the semantic entry bargains on the report's importance [8]. Given that each passage compares only one subject or subtopic, this is more accurate. Since everything is equal, the window entry is a grouping that is simple to actualize. This tactic can be applied in this case by the talk section and window entry in addition to semantic entry because it only has a location with the report's significance.

However, the analysis's findings indicate how to look at different document sizes most effectively. Next, we have a model for how "wheat" is expressed in the report. The distribution of "wheat" is depicted in Fig. 1 and comprises ten sentences. At that time, the distributional cluster for "wheat" is [3, 1, 0, 0, 2, 0, 0, 1, 0, 2].

The cycle immediately scans text reports for term recurrences before shipping out the corpus information base. This database separates the archives based on length and adds padding to it. Applying all distributional highlights to those records at that

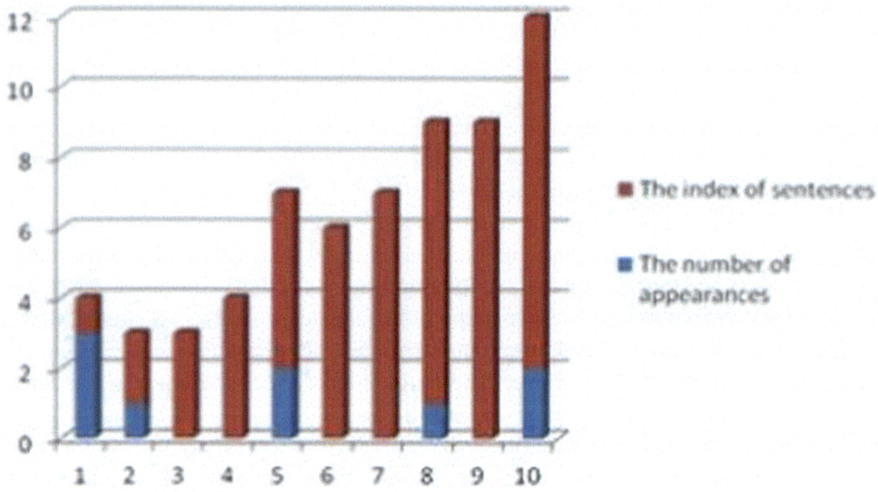

Fig. 1 The distribution of "wheat" in a document

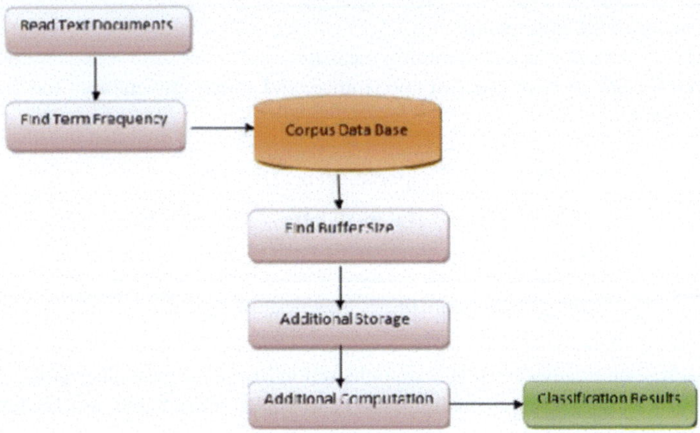

Fig. 2 Extraction of term frequency and distributional features

time determines, in the end, which classification this archive belongs in based on the order in which they are applied. The architecture for extracting word frequency and distributional features for text categorization is shown in Fig. 2 [3].

3 Utilizing Distributional Characteristics

Rather than using the whole number of reports in reverse report recurrence, the word important score is used in the quantity of archives that contain a word to divide the quantity of records without this word. Furthermore, a number of experts and customers acknowledged that the reverse report recurrence obtained directly from text recovery was inappropriate for text arrangement in which the classes of preparation records were available. Numerous controlled scores were displayed before the extraction measure was conducted. Recurrence in tfidf can be thought of as a value that assesses a word's necessity in a particular report [9]. Although a word's significance can be gauged by its term recurrence, it can also be gauged by the context of its first appearance and the smallness of its presence.

Algorithm
Input: Take test prepared set and text archive.

Output: Classification subtleties with most extreme score.

Method:

1. Select all prepared set examples. for example $\{C1, C2, C3, C4, C5, \ldots., Cn\}$
2. Peruse the info text record "*D*."

3. Pre-measure the info record.
4. Applying stop words and stemming measure.
5. Find tfidf, first appearance, last appearance, and conservativeness of the information report.

$$\mathrm{tfidf}(W, D) = \mathrm{importance}(W, D) * \mathrm{idf}(w)$$
$$\mathrm{FA}\ (W, D) = f(\mathrm{First\ App}(W, D),\ \mathrm{length}(d))$$
$$\mathrm{CPFLD}(W, D) = \mathrm{ComPactFLDist}(W, D) + 1/\mathrm{length}(D)$$

6. Analyse input record "*D*" with all classifications of prepared examples.
7. On the off chance that $D \in \{C1, C2, C3, C4, C5, \ldots., Cn\}$ then spot it in to specific classification in any case that is demonstration like as an individual group.
8. Discover score of the information archive which has a place with most noteworthy classification.
9. End the process.

4 Results

In information mining, text classification is arguably the biggest problem. Although it uses a variety of classifiers, the two best classifiers for large informative collections are SVM and kNN. In order to achieve the best results, this paper also illustrates and examines these two classifiers.

4.1 Information Data Sets

Text arrangement fundamentals include informational indexes. This study presents three types of informational indexes: the WebKB corpus, the 20 Newsgroup corpus, and the Reuters-21578 corpus. About 21,578 articles from Reuter's newswire are included in the Reuters-21578 corpus [10]. A subset of the classifications is divided up that both the preparation set and the test set have exactly one record. Assume they don't belong in any class after expulsion reports come to fruition; they will behave as if they belong to their own classification. In this case, 3019 reports were in the test set and 7770 archives were in the preparation set. Some of the orders were completed in Reuters. Following pre-handling, stemming, and stop word removal, the jargon has 12,158 distinct terms that appear in two reports from the Reuters corpus in any case. The contingency for category Ci is displayed in Table 1.

About 19,997 items from Usenet newsgroup assortments are included in the 20 Newsgroup corpus. In this instance, duplicate records are also removed, and reports with different grades are recognized [11]. The 8282 site pages that make up the WebKB corpus were collected from four academic spaces. Following stop word

Table 1 The contingency for category C_i

Category C_i		Expert judgement	
		Yes	No
Classifier judgement	Yes	TP_i	FP_i
	No	FN_i	TN_i

evacuation and stemming, all HTML labels are removed from this section. The jargon now comprises 14,467 distinct words that occur in at least two entries.

4.2 *Measures of Execution*

Their positions about the classified information are the basis for their execution estimations. The top results and strongest correlations are displayed here by the SVM and kNN classifiers. Two order measures and three informational collections came after this cycle. Figure 3 that goes with it displays the typical ranking of every rival in large-scale information indices. Presume that the normal position is lower than what can be considered the best execution proportion for emerging talent. Emerging professionals may be addressed by a combination of TF + CPPN + FAGI or TF + CPPN + FAGLI. When compared to all competitors, the accompanying figure demonstrates that TF + CPPN + FAGI execution is the best [12].

The optimal execution is shown by TF + CPFLD + FALVL in its regular location. The top two displays are TF + CPPN + FLVL and TF + CPPV + FALVL.

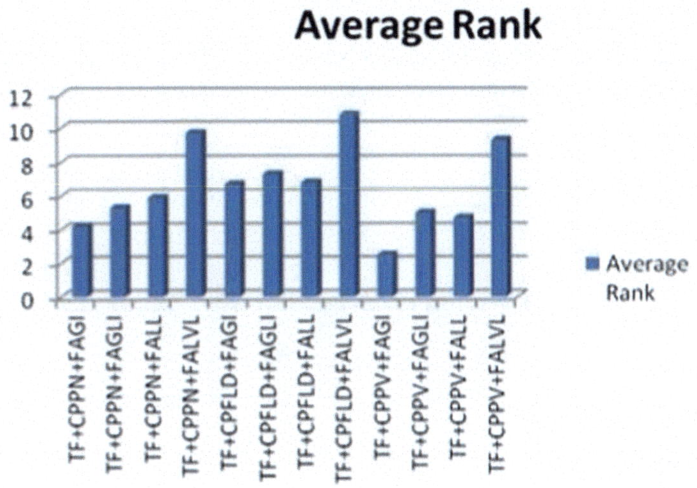

Fig. 3 Average rank of different candidates in a data base

4.3 Investigation for First Appearance Features

The analysis conducted by FA illustrates how we evaluate "*f*" and how we consider a word's primary occurrence. In general, it examined the report sections that FA highlights influenced. Furthermore, each time a word appears in the archive, its weight is assigned using the weighting capacity "*f*" to build a collection of weighted term recurrence (WTR) highlights [13]. The distributional display of the word *t* in the record "*D*" with "*n*" sentences is array $(W, D) = [C1, C2, C3, \ldots., Cn]$. Let us examine this record.

The following is how the weighted term recurrence is calculated:

$$\text{WET}(W, D) = \sum ni = 1\, Ci * f(I, \text{length}(D))/\text{Size}(D)$$

The comparison of three datasets—Reuters, Newsgroup, and WebKB—is depicted in the accompanying Fig. 4. The application of a kNN classifier to three informational collections is shown in Figure.

When it comes to certain informative collections, the Reuters informational collection has flaws at every turn. Furthermore, TF + CPPV + XGI performs more consistently than CPPV + XGI, TF + XGI, and XGI in terms of execution. WebKB's estimation is more accurate in this display than Reuters' and Newsgroup's. Additionally, comparisons between the three datasets—Reuters, Newsgroup, and WebKB—are displayed in Fig. 5 that goes with it. The SVM classifier used it; it produces results similar to kNN when word frequency, compactness, and position variance are combined. In this instance, the execution of Reuters' informational collections was the worst of all. WebKB provides the best execution in comparison with residual informative collections. As a result, the combination performs better than CPPV + XGI, TF + XGI, and XGI.

Figure 6 demonstrates how FA works better than WET, especially on 20 Newsgroup and WebKB, the circumstances under which FA execution is worse than WET

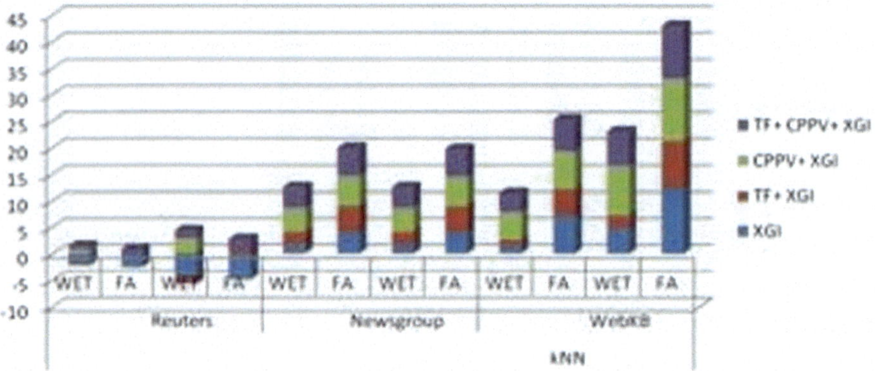

Fig. 4 Comparison between the FA feature and the WET feature by using kNN classifier

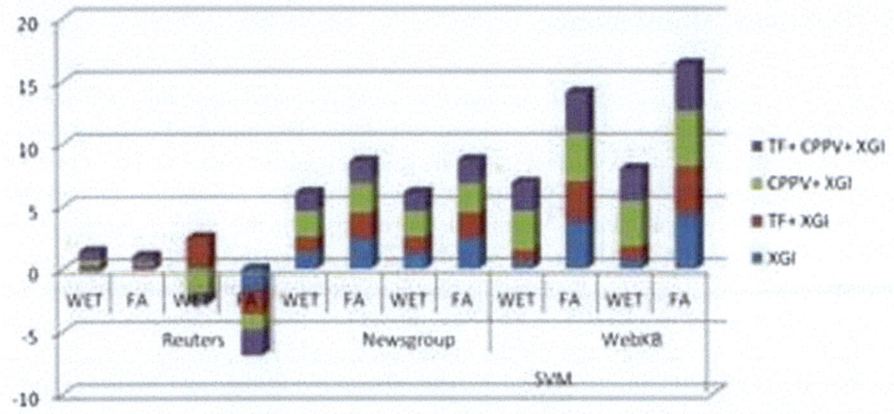

Fig. 5 Comparison between the FA feature and the WET feature by using SVM classifier

execution. Even though WET really makes the gauge better, it is acknowledged that FA's influence on 20 Newsgroup and WebKB comes from its weighting capability as well as its strict process that eliminates any word appearances other than the first. The effects of this harsh process are not immediately apparent to Reuters [14]. It displays the comparison of three informative indexes between kNN and SVM classifiers. Compared to two classifiers, kNN performs better than SVM in terms of execution. Additionally, in these two scenarios, TF + CPPV + WETGI yields better results than WET and TF alone. Based on the results of two classifiers, kNN outperforms SVM as the best classifier in these three sets of archives.

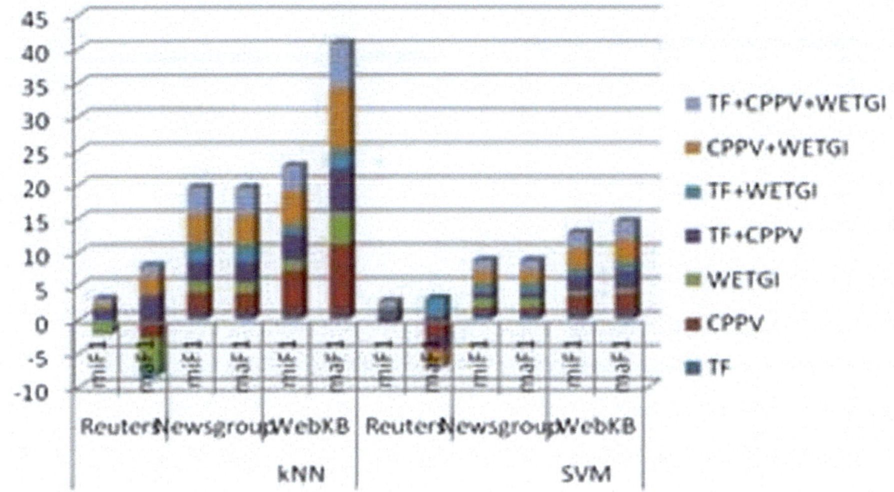

Fig. 6 Comparison between kNN and SVM classifiers

5 Database Maintenance

Select an information record and compare it to every prepared grouping record. That is, every statement in an information document should be compared to every statement in existing records of order. Finally, there may be differences in the weight or score of the information record. It is to consider the most extreme score of the information record; as a result, the document was stored based on the highest characterization score. The information base may simultaneously record the records' joining and departing seasons. Text categorization is displayed in the following figures: Fig. 7a User interface (b) Trained Set; Fig. 8a Test Comparison (b) Classification Score; and Table 2 (the classification result for the input content) [15].

Acquire, Money-tx informational indices also provide the next highest score, although the entire informational collection, in a sense, belongs to Acquire. Here, the wheat informative index talks to zero because it did not receive any related data to incorporate informational index. A comparison of the classification results is shown in Fig. 9.

Assigning predetermined labels or categories to textual input is a key operation in natural language processing known as text classification. Effective text classification techniques are becoming more and more necessary due to the exponential rise of digital content. Compared to data mining algorithms, machine learning techniques have become powerful tools for text categorization, offering a variety of algorithms and methodologies to tackle this issue. This study compares and examines several machine learning approaches used in text classification, noting each one's advantages, disadvantages, and effectiveness over a range of datasets. We compared the performance of various algorithms using common datasets (such as IMDB movie

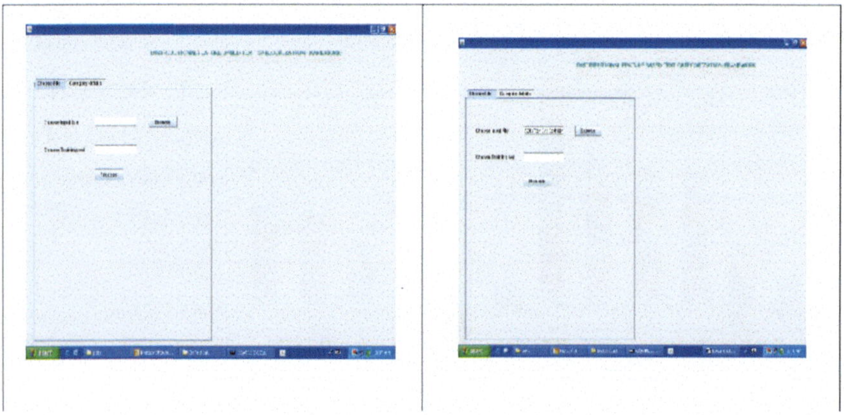

(a) User Interface (b) Trained Set

Fig. 7 Text categorization **a** user interface, **b** trained set

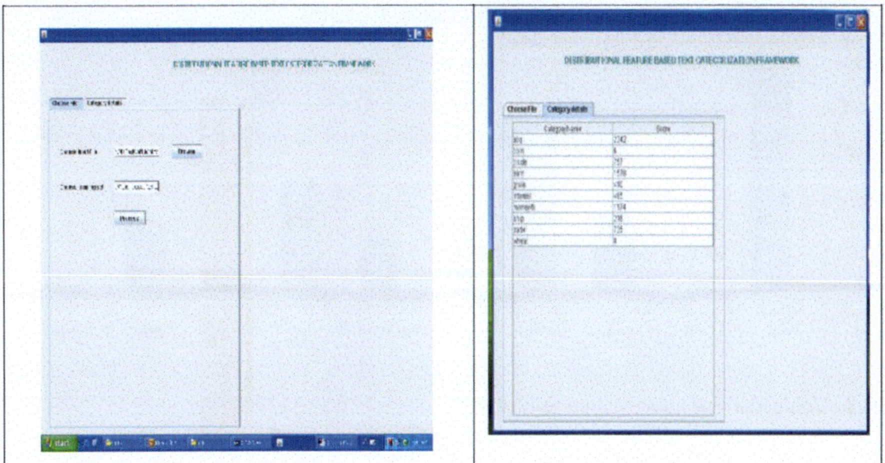

(a) Test Comparision (b) Classification Score

Fig. 8 Text categorization **a** test comparison, **b** classification score

Table 2 Classification result for input document

Trained classifications	Score
Acq	2342
Corn	6
Crude	757
Earn	1578
Grain	410
Interest	495
Money-tx	1174
Ship	216
Trade	735
Wheat	0

reviews, 20 Newsgroups, and Reuters-21578) and examined the interpretability, scalability, and computational complexity of each method. This study attempts to give a thorough overview of the several machine learning approaches that are applied to text classification, along with a comparison of their effectiveness and practicality. Practitioners and researchers can select the best strategy for text categorization problems by being aware of the advantages and disadvantages of various methods.

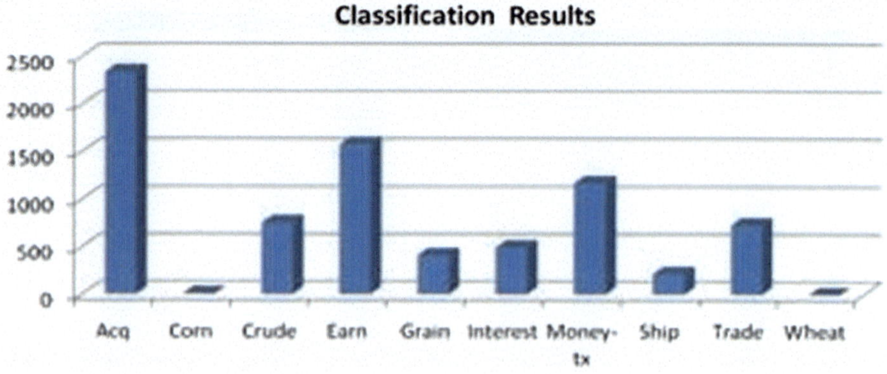

Fig. 9 Comparison of classification results

6 Conclusions

The majority of text categorization relies on word occurrence and recurrence. Such highlights are insufficient to convey all of the information in a particular record. For this reason, the distributional highlights of a word in text layout are shown in this work. These distributional highlights greatly aid in the information's categorization. Distributional highlights include a word's conservativeness in terms of appearances, the principal's position, the word's most recent appearance, and the differentiation between a word's FL appearances in the report. To reflect the different angles, four types of word position appearances and three types of minimization-based highlights are used. Machine learning is used to employ the distributional highlights, and the terms "word recurrence" and "opposite report recurrence style conditions" are used for word recurrence. These highlights are important for all archives in informational collections, regardless of how long the records are and how casually they are written. Finally, these highlights also support regular language archives. The input text (test set) will be stored in a specific database (trained set) based on that score.

References

1. Que HE (2000) Applications of fuzzy correlation on multiple document classification. Unpublished master thesis. Information Engineering Department, Tamkang University, Taipei, Taiwan
2. Rao GN et al (2018) Fire detection in Kambalakonda Reserved Forest, Visakhapatnam, Andhra Pradesh, India: an internet of things approach. J Mater Today Proc 5(1):1162–1168. ISSN: 2214-7853
3. Colas F, Brazdil P (2006) Comparison of SVM and some older classification algorithms in text classification tasks. In: Artificial intelligence in theory and practice, 169–178
4. Rao GN, Ramesh R, Rajesh D, Sekhar DC (2012) An automated advanced clustering algorithm for text classification. Int J Comput Sci Technol 3:2–4

5. Rao GN, Jagdeeswar Rao P (2014) A clustering analysis for heart failure alert system using RFID and GPS. In: ICT and critical infrastructure: proceedings of the 48th annual convention of computer society of India, vol I. Springer, Cham
6. Rao GN et al (2019) Geo spatial study on fire risk assessment in Kambalakonda Reserved Forest, Visakhapatnam, India: a clustering approach. In: Proceedings of international conference on remote sensing for disaster management. Springer, Cham
7. Tai-Yue W, Chiang H-M (2009) One-against-one fuzzy support vector machine classifier: an approach to text categorization. Expert Syst Appl. https://doi.org/10.1016/j.eswa
8. Rao GN, Rao PJ, Duvvuru R (2016) A drone remote sensing for virtual reality simulation system for forest fires: semantic neural network approach. IOP Conf Ser Mater Sci Eng 149(1)
9. Yu B, Xu Z, Li C (2008) Latent semantic analysis for text categorization using neural network. Knowl-Based Syst 2008:900–904
10. Rao GN et al (2016) An enhanced real-time forest fire assessment algorithm based on video by using texture analysis. Perspect Sci 8:618–620
11. Trappey AJC, Hsu FC, Trappey CV, Lin CI (2006) Development of a patent document classification and search platform using a back-propagation network. Expert Syst Appl 2006:755–765; Rao GN, Jagadeeswara Rao P (2014) A clustering analysis for heart failure alert system using RFID and GPS. In: 48th annual convention computer society of India, ICT and critical infrastructure: proceedings of the 48th annual convention of computer society of India, vol II, advances in intelligent systems and computing, vol 249, p 729, CSI 2013. Springer, Visakhapatnam, Andhra Pradesh, India, 13th–15th Dec 2013
12. Beulah K (2019) A fuzzy algorithm for text classification in data science. In: Conference proceedings series on information and communications technology, Taylor & Francis
13. Joachims T (1999) Transductive inference for text classification using support vector machines. In: Proceedings of annual international conference on machine learning, pp 200–209
14. Rao GN, Rao PJ, Duvvuru R, Beulah K, Sunkari V (2019) Internet of things for wildfire disasters. In: Conference proceedings series on information and communications technology, Taylor & Francis
15. Rao GN et al (2023) Neural fuzzy system design in forest fire detection. Microsyst Technol: 1–13

A Powerful GRU-Based Deep Learning Weatherman to Predict Temperature for Visakhapatnam

Deep Karan Singh, G. Lavanya Devi, and P. Samba Siva Raju

Abstract In the contemporary landscape of weather forecasting, a tremendous surge in the utilization of deep learning techniques had been witnessed. This is mainly because deep learning models have the potential to capture complex and non-linear relationships between meteorological variables, making them suitable for predicting weather patterns accurately. One class of such models are the GRUs (Gated Recurrent Units), which have proved high effectiveness in predicting temperature with great precision. This study explores the utilization of the GRU model for predicting the temperature patterns in Visakhapatnam, a coastal city. To train the GRU model, we employed historical temperature data and meteorological variables as input features so that it learns the patterns and dependencies existing in the data. The evaluation encompassed a variety of performance metrics, including the error of the absolute mean, the square of the error of the mean and its root, coefficient of determination, the error of the absolute percentage of the mean, the explained score of variances, as well as training and test loss. The experimental outcomes underscore the high accuracy as well as the efficiency of the GRU-based artificial neural network model in predicting temperature for Visakhapatnam. The GRU model produced a mean absolute error of 0.045 for maximum temperature and 0.049 for minimum temperature, respectively. In addition, the root mean squared error obtained as 0.0653 and 0.667 for the two categories of temperatures in respective manner. These findings highlight the immense promise of employing deep learning methodologies in weather prediction, with a notable emphasis on the effectiveness demonstrated by the GRU model. The research underscores the importance of harnessing sophisticated deep learning models for weather forecasting, offering the potential for enhanced precision and reliability in temperature predictions. Our results indicate that the GRU model stands out as a proficient and reliable tool for forecasting temperatures in Visakhapatnam, and potentially, in other geographical regions too.

D. K. Singh (✉)
India Meteorological Department, Vizag, India
e-mail: deep.karan.singh@imd.gov.in

D. K. Singh · G. Lavanya Devi · P. Samba Siva Raju
Andhra University, Vizag, India

B. Annappa et al. (eds.), *High Performance Computing, Smart Devices and Networks*,
Lecture Notes in Electrical Engineering 1262,
https://doi.org/10.1007/978-981-97-7794-5_3

Keywords Artificial neural networks · Gated recurrent units · Mean absolute errors · Root mean squared errors · Mean squared errors · Mean absolute percentage errors · Explained variance scores · R^2 (coefficient of determination)

1 Introduction

Visakhapatnam, also known as Vizag, is located on the east coast of India and experiences a tropical climate with significant temperature variations throughout the year. The temperature in Visakhapatnam can have a significant impact on the local economy, with several industries, including fishing, agriculture, and tourism, being particularly vulnerable to changes in weather patterns. Local authorities closely monitor the temperature in Visakhapatnam, using advanced weather forecasting techniques to predict and prepare for extreme weather events such as heatwaves and cyclones. Accurately predicting temperature is a crucial element of weather forecasting, with far-reaching implications across industries including agriculture, transportation, tourism, energy, etc.

Traditional statistical models, dynamical models, and numerical weather prediction (NWP) models have long been the standard methods for weather forecasting. These models use mathematical equations and statistical techniques to predict weather patterns based on historical data and meteorological observations. While these models have shown success in predicting weather patterns, they have limitations. Statistical models, for instance, are limited by the assumptions made about the data, and their accuracy diminishes as the complexity of the data increases. Dynamical models rely on physical equations that may not accurately reflect the complex nature of the atmosphere, and they often require significant computational resources. NWP models rely on numerical algorithms to simulate the atmosphere, but the quantity as well as quality of the available observations may limit their utility.

Lately, the emergence of deep learning models has been seen as a promising approach to weather forecasting. These models leverage artificial neural networks, enabling them to learn and forecast, by discerning patterns within the data, without the need for explicit mathematical equations. Deep learning models have shown significant improvement in accuracy and are capable of handling complex and large datasets. Compared to traditional statistical, dynamical, and NWP models, deep learning models excel at grasping intricate data patterns, resulting in heightened prediction accuracy. Moreover, these models have the capacity to integrate diverse data types like satellite imagery, radar data, and social media inputs, enhancing forecast precision. Notably, deep learning models demonstrate superior computational efficiency, demanding fewer computational resources compared to traditional counterparts.

Here, an application of Gated Recurrent Units (GRUs), a class of deep neural network, has been presented, for predicting temperature in Visakhapatnam, wherein we aim to investigate the effectiveness of GRU for temperature prediction. GRU has been shown to perform well in sequence prediction tasks, including weather forecasting. We use the GRU model to analyze historical temperature data from Visakhapatnam and make predictions for future temperature trends. In the subsequent section, previous studies in weather prediction using Gated Recurrent Units (GRU) have been discussed. Section 3 presents the methodology adopted in this work and the dataset utilized. We also explain there, the GRU model architecture and its implementation for temperature prediction. The next Sect. 4 presents the results for the GRU model. The final section concludes the work and discuss potential future directions for this research.

Employing GRU in weather forecasting holds promise for enhancing temperature prediction accuracy, which can lead to better decision-making and planning in various industries.

2 Literature Review

Lately, deep learning has proven to be a versatile asset across various sectors, with a notable impact observed in weather forecasting [1, 2]. The adoption of Recurrent Neural Networks (RNNs) has surged, especially in domains such as statistical machine translation, where the acquisition of phrase representations evolves over time [3]. Diverse adaptations of RNNs, such as Long Short-Term Memory (LSTM) and Gated Recurrent Units (GRUs), have gained prominence for their adeptness in capturing extended dependencies—a pivotal challenge encountered in gradient descent learning [4, 5].

Saha et al. [6] presented a LSTM-based model demonstrating rainfall prediction and applicability of LSTM networks in meteorology. Likewise, Rasp and Lerch [7] have utilized neural networks to refine ensemble weather forecasts through post-processing techniques. Rao et al. [7] successfully applied LSTM-based recurrent neural networks to predict air quality in Visakhapatnam, demonstrating LSTM's utility in environmental science [7]. Shi et al. [8] introduced the convolutional LSTM network for precipitation nowcasting, taking it a step further by incorporating convolutional structures into LSTM networks. This innovation highlights the potential of combining different deep learning strategies to improve forecasting models.

Artificial intelligence has demonstrated its capacity to enhance real-time decision-making concerning high-impact weather events [9]. With continuous advancements in AI, particularly deep learning, new opportunities have emerged to tackle challenges in weather prediction. For instance, using deep learning techniques in Python and R has led to powerful models capable of extracting complex patterns [10, 11]. These languages offer libraries and frameworks specifically designed to facilitate the implementation of deep learning models.

Understanding and visualizing the inner workings of recurrent networks is critical for their successful application and improvement [12]. Thus, there is a need for more interpretability in these models.

Challenges remain in training recurrent neural networks. For instance, Pascanu et al. [13] pointed out the difficulty in training these networks due to issues such as exploding or vanishing gradients. Various strategies, including adaptive subgradient methods, have been developed for the issues. [14].

Deep learning has also been applied beyond weather forecasting. For instance, Vinyals et al. [15] utilized recurrent networks to generate image captions, and Graves et al. [16] applied them to speech recognition.

The convergence of deep learning and earth system science represents a promising frontier. Reichstein et al. [2] discussed how deep learning could enhance our understanding of earth system processes, further emphasizing the relevance and potential of these techniques in weather forecasting and beyond. Zhang et al. [17] conducted an extensive review of the state-of-the-art applications of artificial neural networks in forecasting. The study explored various architectures and learning algorithms used in ANN models and highlighted their superiority in predictive accuracy when compared to traditional statistical methods. Gers et al. [18] introduced the concept of LSTM, which incorporates a memory cell capable of retaining information over extended time intervals. This innovation has proven to be especially valuable in sequential data prediction tasks, as it allows for the effective modeling of both short- and long-term temporal dependencies. Building on the strengths of LSTM networks, Srinivasa Rao et al. [19] applied these models to predict air quality in Visakhapatnam. Their study demonstrated the effectiveness of LSTM-based recurrent neural networks in handling time-series data for environmental monitoring.

While this review demonstrates the ongoing advancements in weather forecasting using deep learning techniques, more research is needed to exploit the full potential of these models, enhance their interpretability, and apply them to other contexts [20].

3 Methodology

This research article presents a methodology for developing and training a GRU-based deep learning model for predicting temperature patterns in Visakhapatnam. The methodology involves several key steps, including collection of the relevant real-world official data, pre-processing of the collected data, development of the model including its training, subsequent evaluation of the developed model with different metrics.

In the first step, historical weather data for Visakhapatnam, including temperature readings and other relevant meteorological observations, are collected and cleaned. The data is then formatted into a suitable input format. Next, the architecture deep learning model is developed, with appropriate hyperparameters and activation functions. Following pre-processing, the model undergoes training on the prepared data utilizing backpropagation algorithms and gradient descent optimization techniques.

Upon the completion of training, the model's efficacy is assessed through a range of metrics. Typically, the model undergoes testing on an independent dataset distinct from the training set to gauge its capacity for generalization and accuracy in temperature prediction.

Overall, the methodology presented in the research article offers a promising approach to improving temperature prediction accuracy and reliability in Visakhapatnam. By leveraging large amounts of data and the ability to capture complex temporal and spatial patterns, GRU-based deep learning models offer a powerful tool for weather forecasting.

3.1 *Architecture of GRU*

The Gated Recurrent Unit (GRU) represents a neural network architecture extensively employed for handling sequential data, including time-series or natural language text. The architecture of the GRU consists of several interconnected layers of cells, with each cell containing a set of neurons that perform computations on the input data.

The architecture of the GRU consists of several interconnected layers of cells, with each cell containing a set of neurons that perform computations on the input data. Every cell within the GRU architecture receives input from the preceding cell and produces an output, which is subsequently transmitted to the subsequent cell in the sequence.

Essential elements of a GRU comprise the reset gate, update gate, and candidate output, strategically employed to regulate information flow and enhance neural network performance. The reset gate decides the extent of retention from the previous state and incorporation of new input. Conversely, the update gate governs the proportion of the previous state to preserve and the relevance of the candidate output. The candidate output, derived from the current input and previous hidden state, contributes to computing the current cell's output.

Mathematical equations for the GRU architecture include:

$$\text{The reset gate: } r_t = \text{sigmoid}(W_r * [h_{t-1} \cdot x_t])$$
$$\text{The update gate: } z_t = \text{sigmoid}(W_z * [h_{t-1} \cdot x_t])$$
$$\text{The candidate output: } h_t\prime = \tanh(W_h * [r_t * h_{t-1} \cdot x_t])$$
$$\text{The output of the cell: } h_t = (1 - z_t) * h_{t-1} + z_t * h_t\prime$$

Here, h_{t-1} represents the previous hidden state, x_t denotes the current input, r_t signifies the reset gate, z_t indicates the update gate, $h_t\prime$ stands for the candidate output, h_t is the current hidden state, and W_r, W_z, and W_h are weight matrices.

The reset gate modulates the extent to which the previous state is forgotten and the influence of the new input, computed through the dot product of the previous hidden state h_{t-1} and the current input x_t, followed by a sigmoid function. Similarly, the update gate determines the proportion of the previous state to retain and the relevance of the candidate output, calculated through the dot product of h_{t-1} and the current input x_t and processed via a sigmoid function. The candidate output emerges

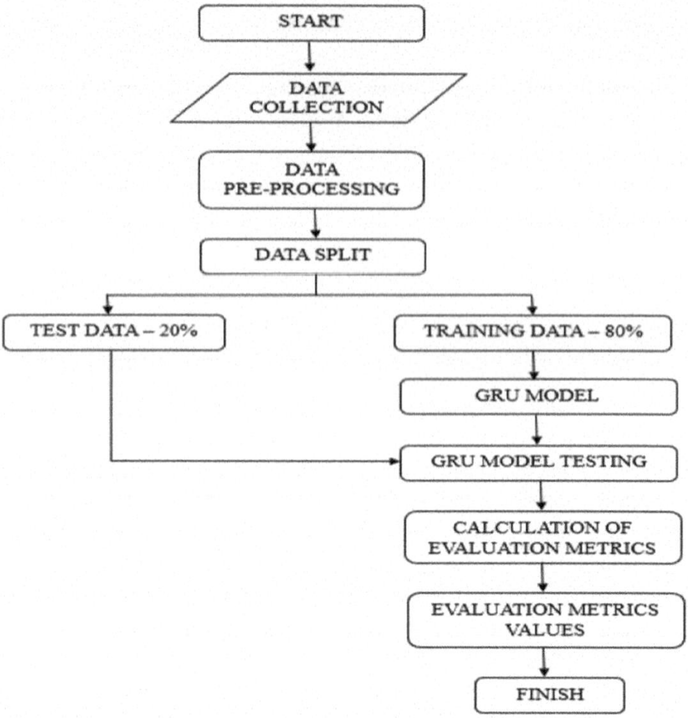

Fig. 1 GRU model's workflow

from combining the current input and the previous hidden state, passed through a hyperbolic tangent (tanh) function. Lastly, the cell's output is obtained as a weighted summation of h_{t-1} and $h_t\prime$, with the weights determined by the update gate z_t (Fig. 1).

3.2 Data Set

The employed dataset consists of a vast collection of real-world weather data spanning over four decades, from January 1969 to January 2015, at Visakhapatnam Airport. The dataset comprises more than 16,000 observations, with each observation representing a single day of recorded weather data. The data contains vital information such as the date, maximum temperature, minimum temperature, and other meteorological parameters recorded at Visakhapatnam Airport, making it a rich source of information for understanding weather patterns in the region.

Given the comprehensive nature of the dataset, it serves as a valuable resource for developing and testing weather forecasting models. These models can be instrumental in improving safety and planning in various industries, including agriculture, transportation, and tourism. The authors of this research article leverage the maximum and minimum temperature data present in the dataset to create an intelligent weather forecasting model (Fig. 2; Table 1).

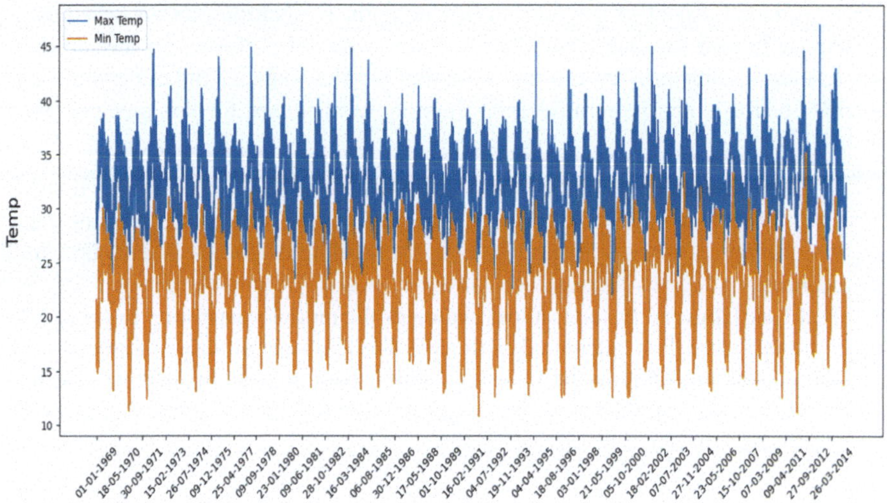

Fig. 2 Temperature dataset—maximum and minimum

Table 1 Visakhapatnam airport temperature dataset (°C)

Date (DD-MM-YYYY)	Maximum temp. (°C)	Minimum temp. (°C)
01-01-1969	28.2	19.9
02-01-1969	28.8	21.6
03-01-1969	29.8	18.7
04-01-1969	30.6	19.3
05-01-1969	30.5	20.8
…	…	…
…	…	…
31-01-2015	30.4	19.4
01-02-2015	29.8	21.8
02-02-2015	28.8	22.2
03-02-2015	30.2	18.4
04-02-2015	32.4	18.6

3.3 Evaluation Metrics

Evaluating the performance of DL models in temperature prediction using historical data is a pivotal aspect of research in this domain. Evaluation metrics serve as an indispensable tool for understanding how well a model performs on unseen data. This research article utilizes several evaluation metrics to gauge the effectiveness of the GRU-based temperature model. Evaluation metrics used for this piece of work are as follows:

(a) **Training Loss**: One of the primary evaluation metrics used is the training loss, which quantifies the model's capacity to conform to the training data. It's computed as the average loss over all training examples at each epoch of training.
(b) **Test loss**: Conversely, the test loss evaluates the model's capability to generalize to unseen data. It is calculated as the average loss over all test examples.
(c) **Mean Absolute Error (MAE):** It quantifies the average magnitude of errors in predictions, calculated as the mean of the absolute variances between the predicted values and the actual values.

$$\text{MAE} = \frac{1}{n}\sum_{1}^{n}|y - \hat{y}|$$

where

n denotes number of samples in the dataset.
y is the actual value of the target variable.
$\hat{y}$ represents the predicted value of the target variable.

(d) **Mean Squared Error (MSE)**: It assesses the average squared disparity between predicted values and actual values, assigning greater significance to larger errors compared to MAE.

$$\text{MSE} = \frac{1}{n}\sum_{1}^{n}(y - \hat{y})^2$$

(e) **Root Mean Squared Error (RMSE)**: Root Mean Squared Error (RMSE), conversely, is the square root of the Mean Squared Error (MSE), offering enhanced interpretability as it aligns with the units of the target variable.

$$\text{RMSE} = \sqrt{\frac{1}{n}\sum_{1}^{n}(y - \hat{y})^2}$$

(f) **Coefficient of Determination ($\mathbf{R^2}$)**: The R-squared (R^2) score evaluates the model's goodness of fit relative to a baseline model that consistently predicts the mean of the target variable. It ranges between 0 and 1, with 1 signifying a perfect fit.

$$R^2 = 1 - \frac{\mathrm{SS}_{\mathrm{res}}}{\mathrm{SS}_{\mathrm{tot}}},$$

where $\mathrm{SS}_{\mathrm{res}}$ is the sum of squared errors and $\mathrm{SS}_{\mathrm{tot}}$ is the total of sum of squares, given by:

$$\mathrm{SS}_{\mathrm{res}} = \sum \left(y_{\mathrm{true}} - y_{\mathrm{pred}}\right)^2$$

$$\mathrm{SS}_{\mathrm{tot}} = \sum \left(y_{\mathrm{true}} - y_{\mathrm{mean}}\right)^2$$

y_{true} represents the true values of the target variable,
y_{pred} represents the predicted values, and
y_{mean} represents the mean of the true values.

(g) **Mean Average Percentage Error (MAPE)**: It, conversely, quantifies the average percentage deviation between predicted values and actual values, calculated as the mean of the absolute percentage differences.

$$\mathrm{MAPE} = \frac{100}{n} * \sum_{1}^{n} \left| \frac{y - \hat{y}}{y} \right|$$

where

n represents the total number of observations.
Σ denotes sum of the absolute percentage error for all observations.
y is the actual value.
$\hat{y}$ is the predicted value.

(h) **Explained Variance Score**: It quantifies the fraction of variance in the target variable elucidated by the model. It ranges between 0 and 1, where 1 denotes that the model accounts for all the variance in the target variable.

$$\mathrm{EVS} = \frac{\mathrm{Var}(y_{\mathrm{true}} - y_{\mathrm{pred}})}{\mathrm{Var}(y_{\mathrm{true}})}$$

where

y_{true} the true target values.
y_{pred} the predicted target values.
Var() the variance of the inputs (in this case, the difference between true and predicted values).

Evaluation metrics also enable researchers to detect any potential issues with the model, such as overfitting or underfitting. Moreover, the metrics help in selecting appropriate hyperparameters for the model and assessing the uncertainty and variability of the predictions. The use of these evaluation metrics is crucial in developing reliable and accurate deep learning time-series forecasting models, making them invaluable in decision-making applications. This research article underscores the significance of employing suitable evaluation metrics to evaluate the efficacy of the GRU-based temperature prediction model, thereby enhancing the accuracy and reliability of such models.

4 Results

This section offers valuable insights into the precision of the GRU-based weather forecasting model in predicting both the maximum and minimum temperatures within the Visakhapatnam region. Illustrated graphically in Figs. 3 and 4, respectively, the results underscore the model's proficiency in capturing inherent data patterns and trends. Importantly, predictions were conducted on test data that remained separate from the model's training phase, ensuring an assessment of the model's performance on unseen data.

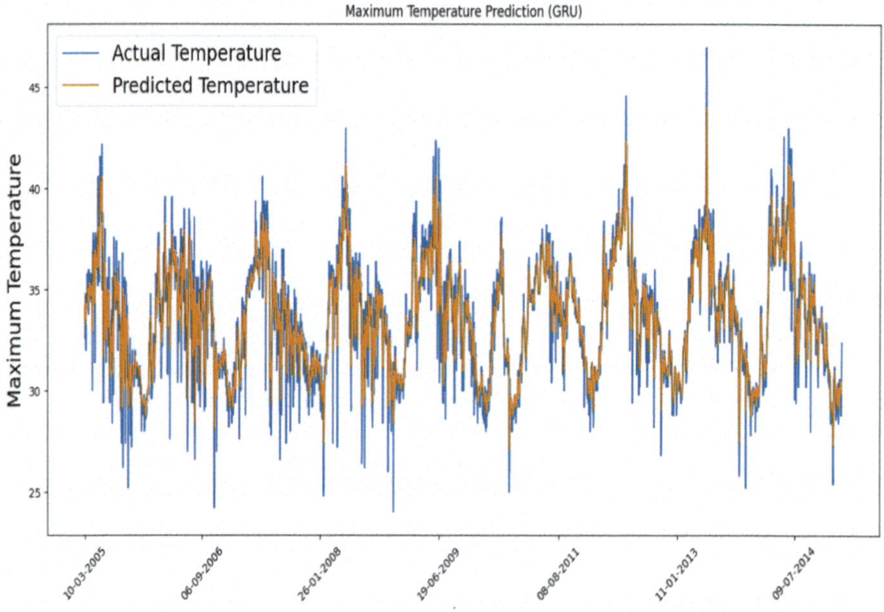

Fig. 3 Maximum temperature prediction

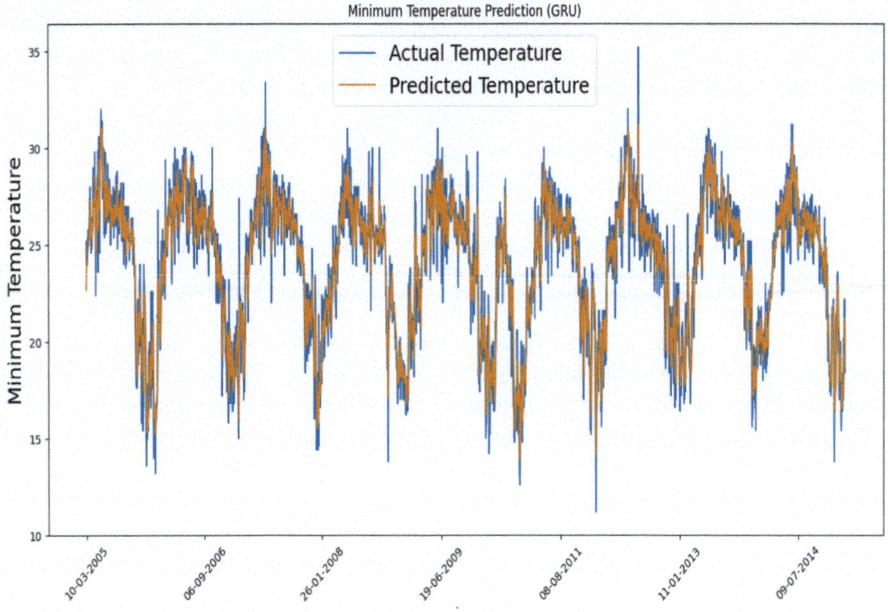

Fig. 4 Minimum temperature prediction

Visually, the graphical representations of the predicted temperature trends against the actual temperature values give evidence that the performance of the model is significantly better than traditional models used in weather forecasting. The deviations between the predicted and actual values appear relatively small, indicating the model's precision in capturing the underlying trends within the data. Visual inspection further supports this, demonstrating the model's accuracy in capturing the inherent patterns and trends, thus suggesting its efficacy in forecasting temperature trends within the Visakhapatnam region.

4.1 Performance Measure

The measure of performance is a crucial aspect of any predictive model, and it is no different in the case of the GRU-based weather forecasting model developed in this study. The model is evaluated with multiple statistical evaluation metrics, which provide quantitative measures of its accuracy and reliability in prediction results.

In the development of the GRU-based weather forecasting model within this study, both training loss and test loss serve as vital metrics for evaluating its performance across training and testing phases.

Training loss assesses how effectively the model fits the training data throughout the training phase. It is computed as the average loss over all training examples at each epoch of training. The objective during the training phase is to minimize the training loss, indicating that the model is progressively improving in its ability to predict maximum and minimum temperatures based on historical data.

Figures 5 and 6 depict the fluctuation of training loss with batch for training with maximum and minimum temperatures, respectively.

Conversely, test loss evaluates the model's capacity to generalize to unseen data during the testing phase. It is determined as the average loss over all test examples. The primary purpose of the testing phase is to assess the model's accuracy in predicting maximum and minimum temperatures using data not encountered during the training phase. The significance of test loss lies in its provision of a metric for the model's ability to make precise predictions on unseen data, which holds paramount importance in real-world applications.

Figures 5 and 6 also show the overall test loss incurred during the testing phase of the model for maximum and minimum temperatures, respectively.

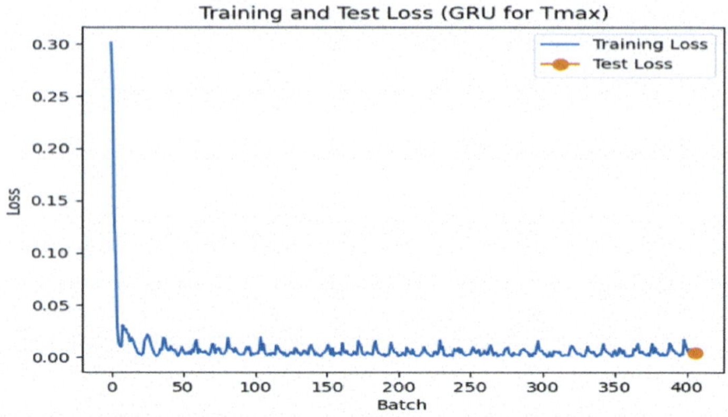

Fig. 5 Training and test loss for maximum temperature

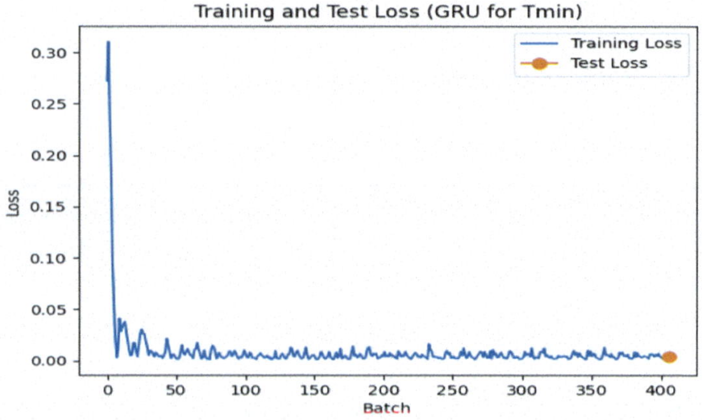

Fig. 6 Training and test loss for minimum temperature

The performance metrics for a machine learning model that predicts the maximum and minimum temperatures are presented in Table 2.

The evaluation metrics for the GRU-based weather forecasting model, as presented in the table, suggest commendable performance in the designated prediction task for which it was developed. Both training and test loss values are notably low, signifying the model's adeptness in fitting the training data well and effectively generalizing to new data. This indicates the model's capacity to learn and capture patterns and relationships inherent in the training data, subsequently applying them to accurately predict maximum and minimum temperatures for new, unseen data.

Furthermore, the Mean Absolute Error (MAE) and Mean Squared Error (MSE) values for both maximum and minimum temperatures are relatively minimal, underscoring the model's low errors in its predictions. The Root Mean Squared Error (RMSE) values, likewise, are comparatively low for both maximum and minimum temperatures, indicating a favorable overall magnitude of error aligning with the target variable's units.

Moreover, the coefficient of determination (R^2 score) values for both maximum and minimum temperatures are moderately high, indicating the model's capability to explain a significant portion of the variance in the target variable. With values close to 1 for both temperatures, the model demonstrates a strong fit for the data.

The Mean Absolute Percentage Error (MAPE) values for both maximum and minimum temperatures are relatively low, suggesting that the model's predictions closely align with the actual values in terms of percentage. Similarly, the Explained Variance Score (EVS) values for both temperatures are high, indicating the model's ability to account for a substantial amount of the variance in the target variable.

For a comparative visualization of the various evaluation metrics for maximum and minimum temperature, refer to Figs. 7 and 8, respectively. These visualizations further support the model's efficacy in weather forecasting.

The GRU-based weather forecasting model developed for predicting maximum and minimum temperatures in the Visakhapatnam region demonstrates notable performance in terms of accuracy and reliability. Trained on a dataset spanning over

Table 2 Performance metrics

Metric	Maximum temperature	Minimum temperature
Training loss	0.004401893354952335	0.004368509631603956
Test loss	0.0042710076086223125	0.0044553508050739765
MAE	0.04504628614643458	0.049031242286034525
MSE	0.004271007576818973	0.004455351632703862
RMSE	0.06535294619846127	0.0667484204510029
R^2 score	0.7502841636444406	0.8307862083220827
MAPE	30.84660868057036	38.60907298725832
EVS	0.7504807925021552	0.8308010474576788

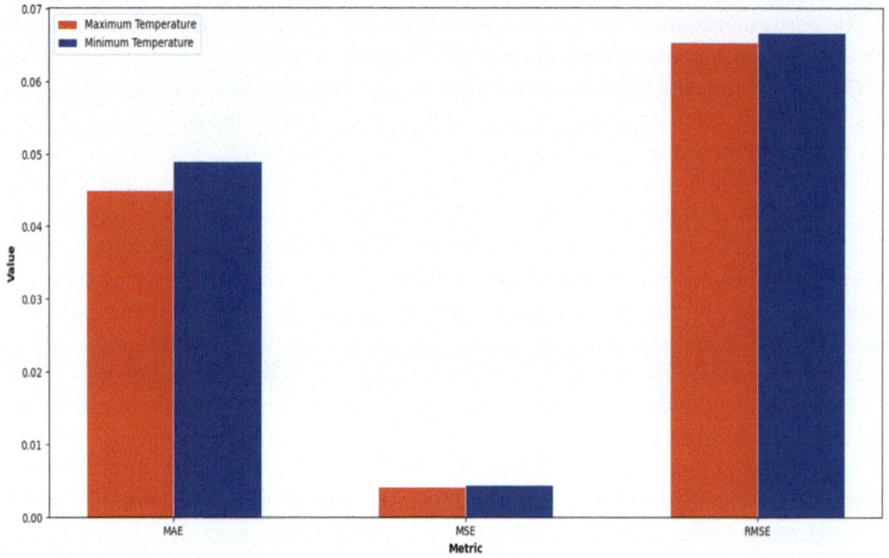

Fig. 7 MAE, MSE, and RMSE comparative

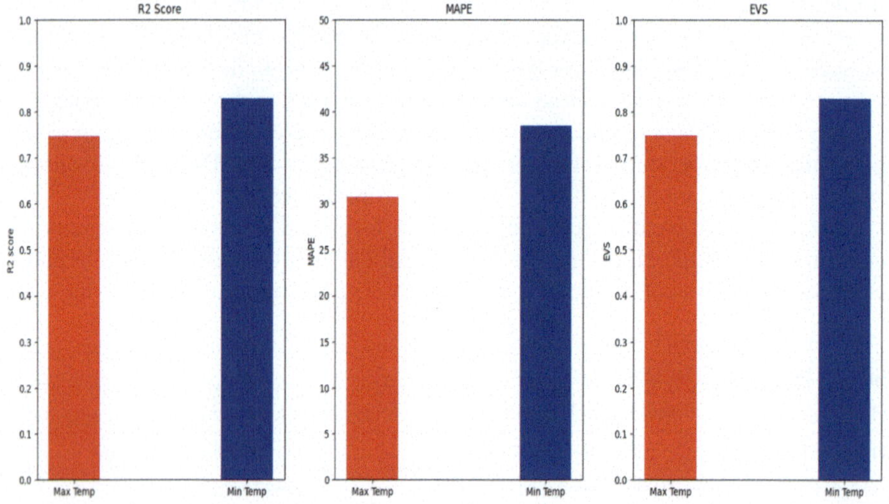

Fig. 8 R^2 score, MAPE, and EVS comparative

45 years and comprising more than 16,000 observations, the model undergoes evaluation using various metrics including training loss, test loss, MAE, MSE, RMSE, R^2 score, MAPE, and EVS.

The results reveal a close alignment between the model's predictions and actual observations, affirming its ability to generalize effectively to unseen data. Evaluation metrics further indicate the model's accuracy and efficacy in predicting temperature trends, showcasing its potential utility in decision-making across industries such as agriculture, transportation, and tourism.

In summary, the GRU-based weather forecasting model presents a promising approach for time-series forecasting in the Visakhapatnam region, with implications for enhancing predictive capabilities and informed decision-making in diverse sectors.

4.2 Contribution of the Research

The contribution of the above research lies in several key aspects:

(a) Demonstrates the applicability of GRU in weather forecasting, particularly for predicting temperature trends.
(b) Offers insights into the effectiveness and reliability of GRU in handling historical temperature data and making accurate predictions for future trends.
(c) Positions the current research within the broader context of existing knowledge, highlighting the novelty or advancements brought by the study.
(d) Provides transparency, allowing other researchers to understand and potentially replicate the study, fostering reproducibility in the field.
(e) Indicates that the use of GRU in weather forecasting can have real-world applications, various industries rely on accurate temperature predictions.
(f) Identifies areas where further investigation or improvements can be made, guiding future research efforts in the domain of weather forecasting using neural networks.

5 Conclusion

To sum up, a GRU-based DL model has been presented for predicting the temperatures (maximum and minimum) in the Visakhapatnam region. It is trained on a comprehensive dataset spanning over 45 years and has been evaluated using various performance metrics. One significant advantage of our proposed model is its ability to outperform traditional statistical, dynamical, and NWP models for temperature prediction. This highlights the superiority of deep learning models over traditional approaches, primarily due to their capability to discern intricate patterns and dependencies within the data. The architecture of the model is crafted to capture extended dependencies within the data and to adeptly adjust to pattern variations, rendering it particularly efficacious in time-series forecasting. The results demonstrate that the model performs well in predicting temperature trends and is useful in many fields, like agriculture, transportation, energy, tourism, etc. Evaluation metrics suggest that the model exhibits accuracy and reliability in forecasting temperature trends, and can be used to improve safety and planning in a variety of industries. The GRU-based deep learning model offers a promising approach to time-series forecasting in the Visakhapatnam region, and it can be extended to other weather forecasting applications as well. The model's proficiency in capturing intricate temporal patterns renders it a potent tool for accurately predicting future weather trends. Further refinement of the model's architecture is in the pipeline, and exploring the use of additional meteorological parameters to enhance its predictive capabilities is planned. Overall, the GRU-based deep learning model shows great potential in the field of weather forecasting and can lead to significant advancements in the prediction of weather patterns.

References

1. Goodfellow I, Bengio Y, Courville A (2016) Deep learning. MIT Press
2. Reichstein M, Camps-Valls G, Stevens B, Jung M, Denzler J, Carvalhais N, Prabhat (2019) Deep learning and process understanding for data-driven earth system science. Nature 566(7743):195–204
3. Cho K, Van Merriënboer B, Gulcehre C, Bahdanau D, Bougares F, Schwenk H, Bengio Y (2014) Learning phrase representations using RNN encoder-decoder for statistical machine translation. arXiv preprint arXiv:1406.1078
4. Bengio Y, Simard P, Frasconi P (1994) Learning long-term dependencies with gradient descent is difficult. IEEE Trans Neural Netw 5(2):157–166
5. Hochreiter S, Schmidhuber J (1997) Long short-term memory. Neural Comput 9(8):1735–1780
6. Saha SK, Pokhrel S, Chatterjee C (2018) Long short-term memory (LSTM) based state-of-the-art model for rainfall prediction. Procedia Comput Sci 132:1239–1249
7. Rasp S, Lerch S (2018) Neural networks for postprocessing ensemble weather forecasts. Mon Weather Rev 146(11):3885–3900
8. Shi X, Chen Z, Wang H, Yeung DY, Wong WK, Woo WC (2015) Convolutional LSTM network: a machine learning approach for precipitation nowcasting. In: Advances in neural information processing systems, pp 802–810
9. McGovern A, Elmore KL, Gagne DJ, Haupt SE, Karstens CD, Lagerquist R, Smith T, Williams JK (2017) Using artificial intelligence to improve real-time decision-making for high-impact weather. Bull Am Meteor Soc 98(10):2073–2090
10. Chollet F (2018) Deep learning with python. Manning Publications Co.
11. Chollet F, Allaire J (2018) Deep learning with R. Manning Publications Co.
12. Karpathy A, Johnson J, Fei-Fei L (2015) Visualizing and understanding recurrent networks. arXiv preprint arXiv:1506.02078
13. Pascanu R, Mikolov T, Bengio Y (2013) On the difficulty of training recurrent neural networks. In: International conference on machine learning. PMLR, pp 1310–1318
14. Duchi J, Hazan E, Singer Y (2011) Adaptive subgradient methods for online learning and stochastic optimization. J Mach Learn Res 12:2121–2159
15. Vinyals O, Toshev A, Bengio S, Erhan D (2015) Show and tell: a neural image caption generator. In: Proceedings of the IEEE conference on computer vision and pattern recognition, pp 3156–3164
16. Graves A, Mohamed AR, Hinton G (2013) Speech recognition with deep recurrent neural networks. In: 2013 IEEE international conference on acoustics, speech and signal processing. IEEE, pp 6645–6649
17. Zhang G, Patuwo BE, Hu MY (1998) Forecasting with artificial neural networks: the state of the art. Int J Forecast 14(1):35–62
18. Gers FA, Schmidhuber J, Cummins F (1999) Learning to forget: continual prediction with LSTM. Neural Comput 12(10):2451–2471

19. Srinivasa Rao K, Lavanya Devi G, Ramesh N (2019) Air quality prediction in Visakhapatnam with LSTM based recurrent neural networks. Int J Intell Syst Appl (IJISA) 11(2):18–24
20. Wilks DS (2019) Statistical methods in the atmospheric sciences, vol 100. Academic Press

A Novel Approach for Facemask Detection and Alert System Using Machine Learning

Samrat Krishna Gaddam, T. S. RaviKiran, and A. Srisaila

Abstract The Covid-19 pandemic has changed the whole world. It has effected economically and socially and has changed the human lives unexpectedly. It spreads from the infected persons and is contagious. We can reduce the exposure to the virus using facemask and social distancing. Although government has given outrules that wearing facemasks and keeping social distance is obligatory because of the social inattention, numerous are still being affected by the virus especially the transmission of virus in closed rooms or indoors in more when compared to outdoors. Wearing of facemasks incorrectly is one of the major problems nowadays, and monitoring each and every person all the while is a bit tidy task. As the technology is advancing we have latest advancements in the field of artificial intelligence and face recognition techniques that can be used to identify the persons who are wearing mask incorrectly and reduce the virus transmission. In this paper, we recommend an approach to automatically identify the persons who are wearing the masks incorrectly. We us OpenCV and Keras to identify the persons wearing facemask incorrectly and notify the authorities about the person not wearing the facemask and stop their entry at the entrance itself. We can implement this approach in the surveillance cameras, and thus, this can be implemented without requirement of any additional hardware. We have achieved high training accuracy of 96.26% and validation accuracy of 92.59%, respectively, on two different datasets than our earlier models and implemented this model successfully with alert system.

Keywords Facemask · Covid · Covid detection · Facemask alert

S. K. Gaddam (✉) · T. S. RaviKiran · A. Srisaila
P. B. Siddhartha College of Arts and Science, Vijayawada, AP, India
e-mail: gsamratkrishna@pbsiddhartha.ac.in

T. S. RaviKiran
e-mail: tsravikiran@pbsiddhartha.ac.in

A. Srisaila
e-mail: a.srisaila@vrsiddhartha.ac.in

A. Srisaila
Velagapudi Ramakrishna Siddhartha Engineering College, Vijayawada, India

B. Annappa et al. (eds.), *High Performance Computing, Smart Devices and Networks*, Lecture Notes in Electrical Engineering 1262,
https://doi.org/10.1007/978-981-97-7794-5_4

1 Introduction

The World Health Organization (WHO) has declared that 30 million people are infected globally in 2019 which caused deaths of 0.7million all over the world [1]. The people who are infected with the virus have had symptoms of serious illness causing respiratory problems like difficulty in breathing or shortness of breathing is one of them. Higher risks are for the elderly people who are having lung diseases like pneumonia [2]. 229-E, HKU-1, OC4-3, and NL-63 are some of the common coronaviruses that infect public around the world. Before in capacitating folks, viruses like nCoV-2019, CoV-SARS, and CoV-MERS pass on a disease to animals and grew to human coronaviruses [3].

Individuals who have respirational difficulties can make anyone vulnerable to the virus. As close contact with the infected person or touching the surfaces which the infected person have touched or even the droplets of the infected persons can cause severe exposure to the virus [4]. Wearing a clinical facemask can control such respiratory problems. The public should have the awareness to wear masks when they go to public places or when they are in indoors. WHO has recommended that wearing of the surgical masks and PPE kits is mandatory for serving the virus-infected persons [4]. So, the face mask recognition became a vital task in current world. Using facemask detection, we can check whether a person has masked or not.

This task is similar to detection of general objects. Using face detection a lot more applications can be developed, and they include self-driven vehicles, edification, surveillance, etc. [5]. In this paper, we propose a cut down approach to use it for facemask detection and alerting using packages like TensorFlow, Keras, OpenCV, and FastSMS. Section 2 of this paper gives the brief of the related work which was carried out. Section 3 shows the sample of datasets used for developing the model. Various packages used to develop the model are explained in Sect. 4. Algorithm used and the proposed work for developing the model are briefed in Sect. 5. Section 6 shows results and analysis. Future scope and conclusion of the model proposed are explained in Sect. 7.

2 Related Work

In face detection using an image, a face has to be identified from it. As per the earlier research, face detection requires expression recognition and face tracking [6]. From a given image, identification of face is a challenge. Detection of face from an image is a tough job as the faces vary in dimension, nature, color, etc., and they are indefinable. When an image is unclear and has some other objects in the image, it becomes a tidy process to identify the face. Earlier researchers found it difficult to find the datasets with people images with and without masks and secondly the elimination of expressions from the face in the covered part. By using locally linear embedding algorithm and models trained using vast collection of faces with masks and faces without masks. As per the authors [7], convolutional neural network (CNNs) used in computer vision has a stringent restriction concerning the input image size. To fit them into the network, the images are resized. In this process, the main tidy task is to recognize a face from the image properly and then to find if the person in the image is wearing a mask or not. To monitor all this method, we are proposing is capable of detecting the faces with mask in motion.

3 Dataset

We have used two datasets in current process for experimenting. Dataset1 consists of 396 images of people with face masks. Figure 1 consists of images of people with facemasks.

Dataset2 consists of 384 images of people without facemask. Figure 2 shows the sample dataset of people without masks. Both the datasets have the images with and without masked people with different environment, backgrounds, and colors.

Fig. 1 Dataset of persons with facemasks

Fig. 2 Dataset of persons without facemasks

4 Packages Used

4.1 *TensorFlow*

It is an interface used in machine learning and is used to implement machine learning systems in fabrication in a cluster of areas like sentiment analysis, voice recognition, geographic information extraction, computer vision, text summarization, information retrieval, computational drug discovery, and flaw detection to carry out thc research [8]. Sequential CNN architecture makes use of TensorFlow.

4.2 *Keras*

The basic reflections and structured parts Keras offers fundamental replications and constructing units for designing and hauling of ML provisions with towering iteration velocity. The capabilities of scalability and cross platform of TensorFlow are used by Keras, and the foundations of Keras are models and layers [9]. Keras is used to implement model in all layers of CNN. To compile the model conversion of class vector to binary matrix is used.

4.3 *OpenCV*

To distinguish and identify faces, identify objects, crowd actions in recordings, sketch progressive module, track eye motion, track camera events, drive out red eyes from pictures taken using flash, find relative pictures from an image database, distinguish landscape, and set up markers to overlay it with increased reality, open source computer vision library (OpenCV), an open-source computer vision and ML software library, is utilized [10]. In the proposed model, we use the features of OpenCV to resize images and conversion of colors of images.

5 Proposed Work

Model Proposed

The proposed model has a cascade classifier and a pre-trained CNN which consists of two 2D convolution layers connected to layers of dense neurons. Face mask detection algorithm is as follows:

Input: Datasets including faces with and without masks.
Output: Identifying a person without mask and alerting.

Algorithm

1. For each image in datasets
    (a) Visualize the image into two categories and label them with mask and without mask.
    (b) Convert RGB image into grayscale image.
    (c) Resize the image in 110 × 110.
    (d) Standardize the image and change it into four-dimensional array.
2. end
3. For constructing the CNN model do
    (a) Add a convolutional layer of 200 filters.
    (b) Add the 2nd convolutional layer of 100 layers.
    (c) A flatten layer is added to the network classifier.
    (d) A dense layer of 64 neurons is added
    (e) Final dense layer with two outputs for two categories is added.
4. end
5. Split the data and train the model.

5.1 Preprocessing of Data

Data preprocessing is the conversion of data from a specified format to user friendly and significant format. The converted data can be of any format like tables, videos, images, and graphs. This prepared information fits in with composition or information model and identifies the relationship among various entities [11]. The proposed model deals with video and image data with NumPy and OpenCV.

(a) Data Visualization

It is the method used to transform conceptual data to meaningful data using knowledge communication using encodings. It helps a particular part to be read in the dataset [12]. The images present in the dataset are visualized in two categories with mask and without mask. The statement Data_Dir = os.listdir(Datasetpath) specifies the directories in the particular path. The categories retrieved are 'with mask' and 'without mask'. The variable that are mapped can be represented as {with mask:0, without mask:1}.

(b) Conversion of RGB image to gray image

Latest technologies based on image recognition works mostly on grayscale images by using the method of converting color images to grayscale. This method is used as the after effect in very minimal after conversion of color image to grayscale. By using additional information, the volume of training data will increase which results in achieving good performance. Grayscale is used to extract descriptors in place of working on color images [13] (Fig. 3).

Fig. 3 Converting RGB image to grayscale to size of 150

To change the color, we use the function grayscale_img = cv2.cvtColor (img, flag). Flag defines the conversion type [14]. In this scenario, cv2.COLOR_ BGR2GRAY is the flag used for conversion of the color image into grayscale.

As a fixed size of input image is needed for DeepCNN, we need to have a preset image size for all the images in the dataset. We are resizing the images to 110 × 110.

A fixed input image size is required in DeepCNN. So, a fixed communal dimension of all the images in the dataset has to be preserved. Bycv2.resize() the grayscale image is converted into 100 × 100 dimensional image.

(c) Image Reshaping

When using a 3D feature tensor, all the images should have same equal size. Here images and their subsequent feature tensors are habitually coextensive [15]. The serious problem arises when the CNN can accept only images which are fine-tuned. Even this problem can be eliminated by again configuring the input images before inserting in the network [7]. Pixel range is constrained between 0 and 1 by normalizing the images. Later they are converted to 4D arrays using images = np.reshape(images,(images.shape[0],img_rows, img_cols,1)) in this grayscale is denoted by 1. The categorical.

5.2 Training of Model

(a) **Model construction using CNN**

CNN is now a growing in the field of computer vision [16]. Our model uses sequential CNN. Rectified linear unit (ReLU) and maxpooling layers follow the first convolutional layer. We use 200 filters to make the convolutional layer learn. Then 2D convolutional window height and width is specified when the kernel size is defines as 3 × 3. Input shape information is provided to the first layer in the model as the input expected should be known by the model. Automatic shape computation is done by the layer beneath. The dimension of data index1 is defined as data, shape [1]. Conv2D is set to ReLU which is a part of activation parameter. The gradient descent methods optimize the linear models which possess the linear function. The activation functions are compared with metrics and generalization in deep learning

[8]. The spatial dimensions for the output can be reduced using maxpooling. Kernal size is defined as 3 × 3, and the convolutional layer consists of 100 filters followed by maxpooling and ReLU. The input is accepted and passed through flatten layer for transforming features matrix to a vector which later can be sent a fully linked neural network classifier so that the data can be inserted in to CNN. The minimize the over fitting a layer with 50% probability to set zero's as input is incorporated in the model. ReLU function is activated with 64 neurons at the dense layer. The last layer is activated with two categories and two outputs (Fig. 4).

(b) **Splitting the data and training the CNN model**

The model we proposed should be trained using a dataset and tested using live streaming of video. A good model has to produce precise results alongside making the prediction. Here 50% of the training data was used as validation data. 50 epochs are used for training this model to maintain an exchange among the chance of overfitting and accuracy (Fig. 5).

(c) **Alerting System**

In this paper, we used a free SMS provider 'FastSMS' where we used the API specified by the SMS provider. Whenever a person without mask is identified by the model, it automatically sends an SMS to the number provided by the authorities so that they can act accordingly.

6 Results and Analysis

Using the dataset, the model is trained, validated, and tested. Figures 6 and 7 shows the training and validation and accuracy graphs. After inputting our data, our model nearly attains a higher training accuracy of 96.26% and validation accuracy of 92.59%.

The main reason for attaining high accuracy is because of maxpooling. It is made easy maxpooling to reduce the no. of parameters which the model should learn (Figs. 8 and 9).

This model can proficiently identify faces without facemask or faces covered partially with facemask. It differentiates among mask, and face covered using hand can be differentiated using four regions like mouth, nose, chin, and eyes. So, a mask which covers both nose and jaw only will be measured as 'with mask' by the model. Apart from identifying the persons without mask, it also sends an alert to the concerned authorities that a person has been identified without wearing a mask or partially wearing a mask.

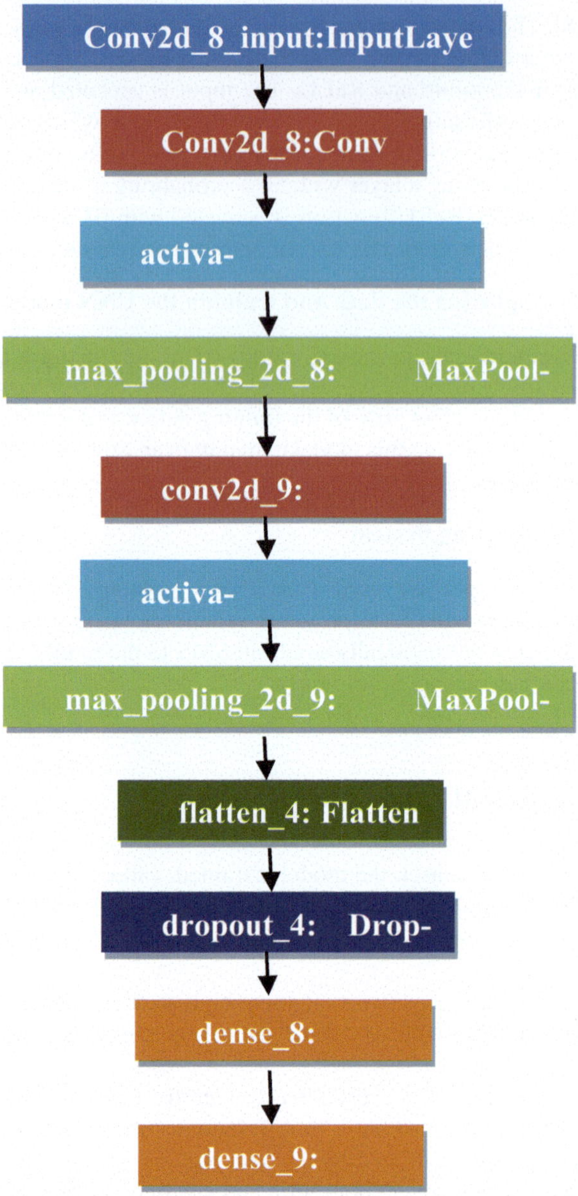

Fig. 4 Algorithm used for developing the model

7 Future Scope and Conclusion

In this paper, we proposed a model which performed with high accuracy using the machine learning tools and cut down techniques. We can use the developed model in various applications. Wearing a facemask may be enforced in the coming times

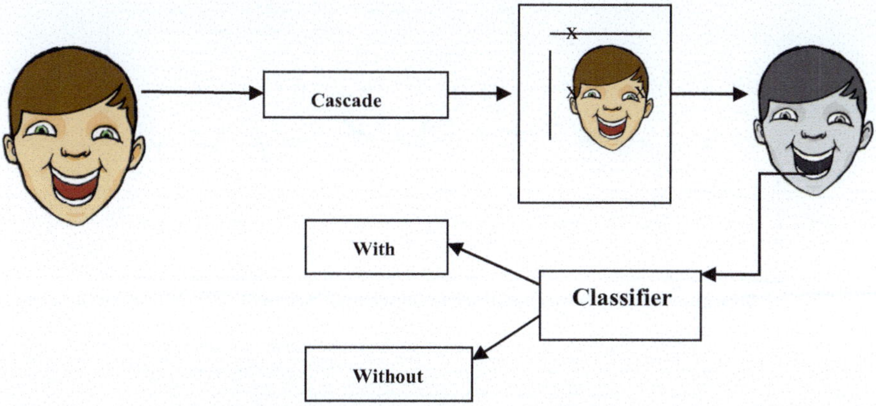

Fig. 5 Overview of model used in proposed work

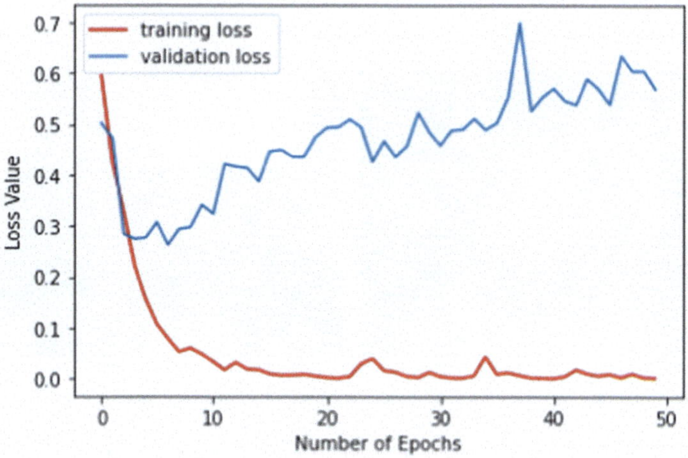

Fig. 6 Training loss versus validation loss

as it was an effect of Covid-19. Many places where the crowds are huge and public places where gatherings will be more wearing a mask will protect us from the Covid virus, and to use the services provided by service providers, visiting malls, shopping, schools, and colleges, wearing a mask is mandatory. The model proposed will immeasurably help the public sector. As the model can be implemented with the existing infrastructure and without any further investment, it makes easy to be deployed. In future, we propose to develop a model which identifies the level to which the mask protects from the Covid-19 virus like detection of mask type like surgical, N-95, etc., or not.

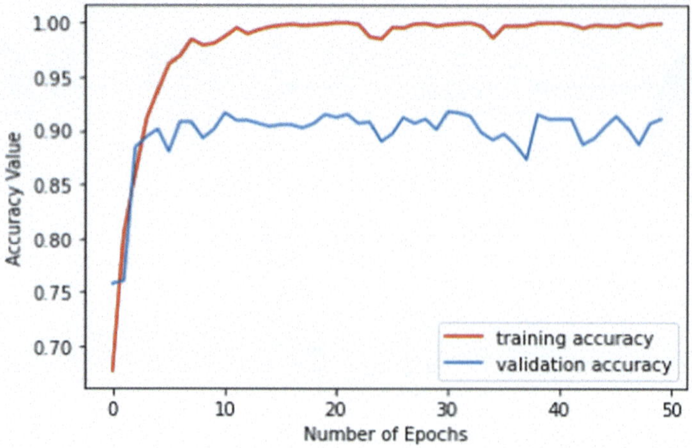

Fig. 7 Training accuracy versus validation accuracy

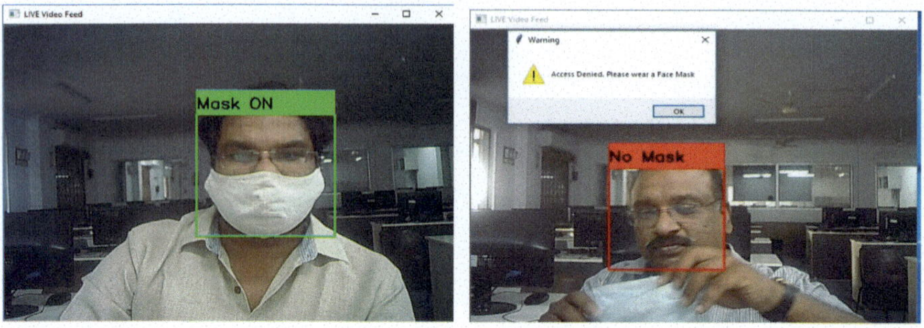

Fig. 8 Model detecting the persons with facemask and without facemask

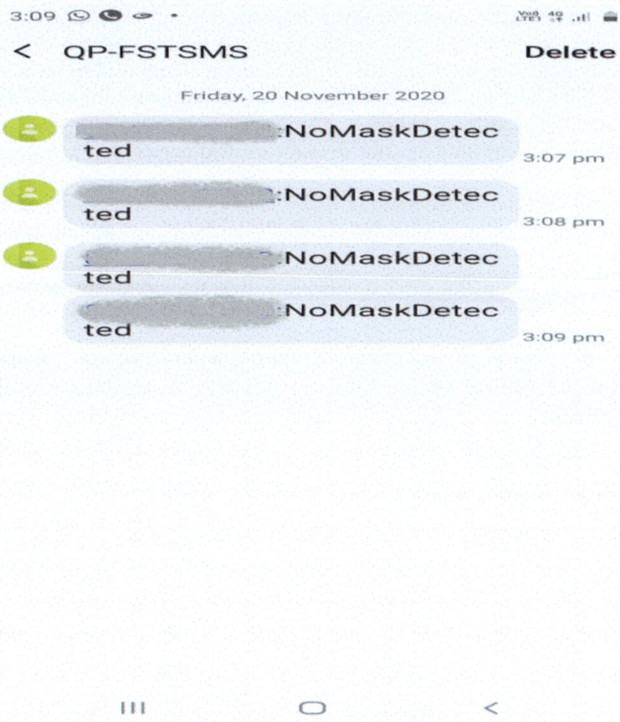

Fig. 9 Model alerting after detecting the person without facemask

References

1. W.H.O. (2020) Coronavirus disease 2019 (covid-19): situation report, 205
2. Coronavirus Disease 2019 (COVID-19)—Symptoms. Centers for Disease Control and Prevention [online]. Available https://www.cdc.gov/coronavirus/2019-ncov/symptomstesting/symptoms.html
3. Coronavirus—human coronavirus types—CDC (2020). Cdc.gov [online]. Available https://www.cdc.gov/coronavirus/types.html
4. W.H.O. (2020) Advice on the use of masks in the context of COVID-19: interim guidance
5. Jiang M, Fan X, Yan H (2020) "RetinaMask: a face mask detector [online]. Available https://arxiv.org/abs/2005.03950
6. Ge S, Li J, Ye Q, Luo Z (2017) Detecting masked faces in the wild with LLE-CNNs. In: 2017 IEEE conference on computer vision and pattern recognition (CVPR), Honolulu, HI, pp 426–434. https://doi.org/10.1109/CVPR.2017.53
7. Ghosh S, Das N, Nasipuri M (2019) Reshaping inputs for convolutional neural network: some common and uncommon methods. Pattern Recogn 93:79–94. Available https://doi.org/10.1016/j.patcog.2019.04.009
8. TensorFlow white papers. TensorFlow (2020) [online]. Available https://www.tensorflow.org/about/bib
9. K. Team (2020) Keras documentation: about Keras. Keras.io (2020) [online]. Available https://keras.io/about
10. OpenCV. Opencv.org (2020) [online]. Available https://opencv.org/

11. Suvarnamukhi B, Seshashayee M (2018) Big data concepts and techniques in data processing. Int J Comput Sci Eng 6(10):712–714. https://doi.org/10.26438/ijcse/v6i10.712714
12. Hohman F, Kahng M, Pienta R, Chau DH (2019) Visual analytics in deep learning: an interrogative survey for the next frontiers. IEEE Trans Vis Comput Graph 25(8):2674–2693. https://doi.org/10.1109/TVCG.2018.2843369
13. Kanan C, Cottrell G. Color-to-grayscale: does the method matter in image recognition? PLoS One
14. Opencv-python-tutroals.readthedocs.io (2020) Changing Colorspaces—Opencv-Python tutorials 1 documentation [online]. Available at https://opencv-python-tutroals.readthedocs.io/en/latest/pytutorials/pyimgproc/pycolorspaces/pycolorspaces.html
15. Hashemi M (2020) Enlarging smaller images before inputting into convolutional neural network: zero-padding vs. interpolation. J Big Data 6(1). Available https://doi.org/10.1186/s40537-019-0263-7
16. Yamashita R, Nishio M, Do R, Togashi K (2018) Convolutional neural networks: an overview and application in radiology. Insights Imag 9(4):611–629. Available https://doi.org/10.1007/s13244018-0639-9

Protecting the IoT Devices from DDoS Attack Using Blockchain-Based Multi-code-Driven Trust Mechanism

D. Jayanarayana Reddy, Addula Jahnavi, Pothula Indhu, and Mekala Hema Latha

Abstract The Internet of things (IoT) offers a promising platform for developing mission-critical programs. However, these programs often depend on a centralised storage architecture, which poses significant challenges such as security, privacy, and a single point of failure. Recently, blockchain technology has emerged as a key component in IoT application development. Blockchain can address issues related to data privacy, security, and third-party dependence in IoT applications. Both individuals and communities can benefit from integrating blockchain with IoT. Nonetheless, a DDoS attack on a mining pool in 2017 exposed significant vulnerabilities in blockchain-enabled IoT systems. Additionally, the data generated by these programs is vast. Machine learning (ML) is an effective analytical tool that facilitates fully autonomous analysis and decision-making with large datasets. This research proposes a novel distributed intrusion detection system (IDS) utilising fog computing to detect DDoS attacks on mining pools in blockchain-enabled IoT networks, addressing the aforementioned challenges. It introduces an energy-efficient, decentralised, blockchain-based multi-mobile code-driven method for detecting internal threats in sensor node-powered IoT. The results indicate that the proposed approach outperforms existing alternatives.

Keywords Internet of Things · Distributed Denial of Service · Blockchain-based multi-mobile code driven solution · Machine learning

D. Jayanarayana Reddy (✉) · P. Indhu · M. H. Latha
Department of CSE, G. Pullaiah College of Engineering and Technology, Kurnool, India
e-mail: djnreddy@gmail.com

A. Jahnavi · P. Indhu · M. H. Latha
Kurnool, India

B. Annappa et al. (eds.), *High Performance Computing, Smart Devices and Networks*, Lecture Notes in Electrical Engineering 1262,
https://doi.org/10.1007/978-981-97-7794-5_5

1 Introduction

As the Internet has evolved, a new technology known as the Internet of things (IoT) has become deeply embedded in our daily lives. Examples of IoT-based applications include supply identity management systems, which are directly empowering individuals and communities [1]. The security of IoT devices is notoriously inadequate at the present time, and cybercriminals are taking advantage of this widespread vulnerability to launch devastating cyberattacks [2, 3]—especially distributed denial-of-service (DDoS) [4]. By fusing cloud computing and machine learning, the underlying technology is showing promise for data analysis and modelling [4]. Innovations based on the Internet of things are fuelling expansion in many different industries. The majority of centralised storage and compute architecture [5] is used by applications built using IoT systems. There are not any potential vulnerabilities or privacy issues with the centralised storage strategy. Future IoT-based system growth is constrained by the underlying working model [6]. Thus, a decentralised or distributed storage approach is required to solve these problems. Blockchain technology is one of the new decentralised systems that is beginning to emerge [7].

Initiated by a peer node in the network, blockchain is a decentralised and immutable ledger that records all transaction data. A distributed ledger is a digital ledger that is not stored in a single location. Each transaction processed by the ledger must be validated by the consensus of the network to be considered legitimate. Currently, Bitcoin is the most well-known application of this technology [8]. Blockchain's decentralised storage mechanism can synchronise IoT devices and provide real-time data to each IoT node, among other advantages of integrating the two technologies. This integrated model eliminates reliance on third parties, thereby removing a potential vulnerability [9]. In addition, a decentralised client–server paradigm is unnecessary for IoT addition with blockchain to allow for peer-to-peer messaging, real-time data and file exchange, and nodes [10]. Although its immutability and transparency, blockchain technology may be compromised in a number of ways. Standalone IoT applications have been greatly impacted by the merging of IoT with blockchain. Unfortunately, this trend has been accompanied with a rise in the frequency of assaults. As DDoS attacks on blockchain networks sometimes result from memory pool flooding, legitimate users suffer serious effects [11, 12].

This article exemplifies the interactions that occur from using blockchain-based multi-mobile code-driven trust stored on IoT devices (BMCTM). For example, in a smart healthcare system, consensus is achieved across fog nodes when a new node joins the network, or it can be used to identify and isolate nodes within the network. As centralised techniques may struggle to handle the scalability and distributed nature of fog-based IoT systems, blockchain offers decentralised security solutions to support fog computing. We discuss the benefits of blockchain for the Internet of things, such as reducing energy consumption, securing connected devices, decreasing latency, and maintaining scalability. Blockchain's foundational principle enables collaboration between anonymous, potentially untrustworthy parties. Similar to modern cloud computing designs, it enhances the decentralised nature of the Internet of things by eliminating the need for a centralised security and authentication authority.

This is how the rest of the paper is organised: In chapter "An Enhanced Text Classification Using Machine Learning", the relevant research is presented. The suggested model is obtainable in Sect. 3. In Sect. 4, we label our methodology and the outcomes of our experiments. Section 5 provides a summary and discussion of findings and future directions.

2 Related Works

Safe direct-to-device (D2D) interactions in a smart home are proposed by Yakubu et al. [13]. A decentralised prototype and distributed ledger system can be developed on the Ethereum blockchain. DDoS attacks are mitigated using a combination of a single-server queuing system concept and an authentication method. Twenty simulation runs were conducted, each with a different set of devices (ranging from 1 to 30). Each requesting device simultaneously sends a random request with specific resource needs. This approach allows the system's reliability to be evaluated across various hardware configurations. Additionally, experimental results show that the proposed protocol excels in response times, effectively resisting colluding attacks.

To transmit signals between controllers, Sandhia et al. [14] offer a novel "Simplified Byzantine Fault Tolerance (SBFT)" algorithm based on the Simplified Byzantine Multiple Network (SDMN) and conduct an analytical study to evaluate SBFT's security and performance. Despite some limitations, SDMN helps mitigate issues such as DDoS/DoS attacks, unauthorised access, and single points of failure, while offering several other benefits. Blockchain, an emerging disruptive technology, promises innovative solutions for ensuring network security. This research introduces a Simple Byzantine Fault Tolerance (SBFT) model leveraging SDMN to enhance mobile privacy.

IoT devices used with public blockchain networks may have their battery life extended, transaction costs lowered, and data integrity protected thanks to a protocol proposed by Khor et al. [15]. To test and ensure the efficacy of the proposed data storage mechanism, a resistant of concept is shown utilising an ESP32S2 device. A formal smart contract analysis tool is used in the creation and evaluation of a smart contract. The collected blockchain will be displayed and verified via a decentralised web application. The suggested solution's impact on power use, memory requirements, and safety is assessed. Data integrity can be maintained even for low-power sensor nodes that link to public blockchains through a Wi-Fi network, as shown by the assessment findings.

Kaur and Singh [16] investigate how susceptible smart devices are to a wide variety of assaults, more. Security for IoT data is provided through a wide variety of protocols, including DTLS, IPv6, and many more. Yet, sniffing or modifying header information is a common vector for attack in both encrypted and unencrypted protocols. Intrusion detection systems (IDSs) based on machine learning are one approach of protecting against attacks that use header data. Trust in the ML/DL algorithms and the reliability of the training data (IoT data) are essential to the safety of these IDSs. The emergence of blockchain as a new sophisticated technology has several potential applications in the (IoT) space, particularly in the area of security. We merge blockchain and ML/DL approaches to IoT data security to provide a robust blockchain-based framework for securing data in a diverse IoT ecosystem.

In order to justify their proposal of an integrated approach for creating security mechanisms for smart city IoT networks, Rangelov et al. [17] provided a short summary of current assaults and defensive measures. It provided a blueprint for the design and deployment of cybersecurity solutions for urban IoT networks, including a pipeline of actions that would be taken in the development and implementation processes. Following the recommended method would result in the implementation and ongoing enhancement of suitable IoT security created in conjunction with industry partners, and a local government was used to demonstrate the technique in action.

Ahmad et al. [18] emphasise the importance of a multi-factor authentication (MFA) environment. Initially, they define "smart cities" and discuss the associated privacy and safety concerns. The article then delves into how multi-factor authentication can protect various smart city components and services. In this context, we introduce BAuth-ZKP, an innovative approach to multi-factor authentication based on blockchain technology, designed to secure transactions within a smart city. The core idea is to establish smart contracts between entities in a smart city and use authentication methods that safeguard users' privacy during transactions. The final section explores the potential, future directions, and scope of MFA's application in the smart city ecosystem.

3 Proposed System

3.1 Pre-processing of Data

The size of the data flowing over an IoT network might vary widely. Pre-processing data aids in converting raw data into a form more suited for modelling. This method helps cut down on the amount of time spent training and testing the detection system. By recognising attack observations in an IoT setting, it increases the overall effectiveness of the detection system.

3.2 Data Normalisation

IoT networks that use blockchain technologies get varying levels of traffic. The feature values are normalised with the help of the StandardScaler method in this model. Incoming traffic distribution is one example of a characteristic that may be transformed using this method, which assigns a mean value of 0 and a standard deviation of 1 to observed data [19]. Since this technique removes incoming traffic bias without altering the detection system's mathematical features, it contributes to the system's practicality. The transformation is shown by Eq. (1).

$$s_k = \frac{\text{val}_k - \mu_k}{\sigma_k} \tag{1}$$

where s_k represents the detection system's standard score for features, and $k = (k_1, k_2, k_3, \ldots, k_n)$. IoT traffic characteristics are represented by the val_k. The average and standard deviation of characteristics under consideration are denoted by _k and _k, respectively, in the formula below:

$$\mu_k = \frac{\sum_{k=1}^{K} \text{val}_k}{N} \quad \text{and}\, \sigma_k = \sqrt{\frac{1}{N}\sum_{k=1}^{N}(\text{val}_k - \mu_k)^2} \tag{2}$$

3.3 Building Blocks

BMCTM's Structure Blocks Are Full in the Next

For detecting, acting upon, and exchanging information, SNs are often placed in various contexts (later referred to as systems), homes. In order to accomplish a goal, it is frequently necessary for SNs from various systems to share information with one another via communication. It may be necessary to cut off all communication if one of the SNs suddenly becomes malignant and launches a blackhole or greyhole assault. Network performance and longevity are severely degraded in both assault scenarios. By forcing the victim and intermediary nodes to resend the lost packets, they might cause message and energy overheads. Hence, a trustworthy and dependable environment is necessary to enable safe communications. Thus, the trustworthiness of an SN is a prerequisite for exchanging data. The information will be used appropriately by the receiver, who will not, for example, alter it or misuse it. The projected model eliminates the possibility of a single point of failure (which is common in centralised trust-based mechanisms), facilitates consensus regarding can scale to keep up with the distributed and rapidly expanding IoT-based infrastructures. We establish three distinct cutoffs based on an SN's reliability to provide this assurance (Table 1).

To create and preserve a trustworthy setting, many modules make up the fog layer, which includes blockchain-enabled fog nodes that do trust assessment. They are sent by genuine service providers to carry out their duties. They provide services such distributed device communication, storage, and processing in addition to security (i.e. encryption) [20–22]. During each transaction, devices develop various trip plans

Table 1 Classes of device trust

Trust value variety	Trust kind
75 ≤ trust value ≤ 100 && remaining energy ≥ 50	Highly trusted
50 ≤ Trust value ≤ 74 && remaining energy > 50	Moderately trusted
Trust value ≤ 49 && remaining energy < 50	Malicious

and launch numerous MCs in accordance with the itineraries designed, blockchain. When used to the storage of trust value information, blockchain technology creates an immutable ledger that protects both historical and real-time data for all involved nodes. It protects information against tampering and forgeries in contexts where confidence cannot be assumed. By preserving the honesty of MAs for assessing fault-free trust values, it prevents the manipulation of data that might have fatal consequences and guarantees the reliability of data aggregation (i.e. collecting trust parameters). The subsequent sub-modules and repositories are used by each fog node.

One such device is the "device detector," which can determine the overall number of SNs active in a given cluster. The gateway does this by disclosing info about each connected device, including its address and kind (sensor or manager).

The genuine SNs and their corresponding gateways are sent from the device detection module to the device list manager. It stores information about gateways and other networked gadgets. Information such as physical coordinates, service areas, and system affiliations will be provided. To further facilitate the system's ability to device, it additionally registers these SNs and blockchain. The user is then sent to a dynamic itinerary plan generator for action.

The dynamic itinerary plan generator module creates a network cluster of SNs, and the mobile code produces several MCs to accommodate the many itineraries that may be created from those SNs. Each MC's trip to the selected cluster is generated separately by module.

Lastly, we have a dynamic itinerary plan generator that can make itineraries for several MCs to go over the network. When additional nodes join or depart the network, it will adjust accordingly and continue to produce fresh strategies. The MC disseminator module receives these plans and uses them to guide MCs as they tour SNs in search of information about the SNs' forwarding behaviours and the corresponding parameters. When an MC is created, a security token is also chosen and appended to the transaction. At the beginning of each MC, the security tokens are assigned in a completely random manner. The travel itinerary for each MC is then exchanged in a blockchain transaction. As they arrive at the doorway, they undergo a series of checks to verify their authenticity.

Module for spreading mobile code (or MCs) from the fog layer to their corresponding gateways. After obtaining information on trust, the MCs relay their findings to it. It works in tandem with the initiator module to keep track of which gateways/ SNs have been entered and exited.

The sixth module is the analyser, which gets information from the MC distributor and checks out MCs as they come back, as well as updating the device repository. It is also in charge of determining how credible information is under investigation. Then, the trust value is recorded on the distributed ledger.

Several MCs are started for each plan based on the sum of SNs.

The disseminator module is the next stop on the road to the gateway for the guilty MC(s).

The MCs are sent to the specified gateway through the disseminator module, number 8. Each MC's legitimacy is checked by having the gateway check its security token.

If an MC is legitimate, it will make a round-trip to the specified SN cluster to get the information it needs. It returns to the portal after gathering information.

After number 10, it is disseminated back into the fog layer.

Eleven, the details are sent from the disseminator module to the analyser module, where they are processed and the level of confidence is determined. The next section elaborates on the method used for computing the trust's value.

The determined trust value for each SN is then added to the blockchain through appropriate transactions.

- Phase of Trust Calculation: The collected data is then sent to the trust analyser module, which is at the MCs' disposal. Each SN's forwarding behaviour and energy usage data is included here. The mathematical formula for determining how much energy is still available is shown in (1). If an SN is compromised, its energy ingesting will spike, forcing it to evacuate sooner than usual.

The next stage, after the collection of trust values, is to include those values in the blockchain. The rules for adding or changing trust values on the blockchain through smart contract, as they apply to a given SN. This method takes as inputs the trust value, node ID, system ID, and fog address.

After a fog node has successfully calculated the trust value in accordance with the smart contract regulations, it will perform a blockchain transaction to store the trust worth. The SN ID and associated trust value are included in the exchange. The contract first verifies the mapping between the requesting fog and the system. The existence of the fog node may be confirmed if this file is found. If the device is a legitimate fog node, its registration may be verified in the distributed ledger. Next, we will make sure the gadgets are talking to one other and the rest of the system. A mapping between trust and SN is constructed if it does not already exist, and the trust value of the associated SN is incremented in the blockchain if it does not already do so. The primary goal of the limitation is to prevent an attack on the whole system. For instance, if the fog node of a smart home system is hacked, it should not have any effect on the trust levels of any other IoT devices in use elsewhere. Another limitation is that no other IoT device may add to or change the trust value. A hacked node cannot change the trust value of an established IoT gadget. This is accomplished by validating the fog node's signature during a system transaction.

The suggested approach concludes with a phase dedicated to identifying and isolating malicious nodes. Before exchanging data with another SN from a different system, an SN from the same network will make sure it can be trusted. To facilitate communication between nodes A and B, for example, node A will conduct a transaction with the fog layer blockchain on its own. In response to this exchange, node A will get the most recent trust value of node B from the fog layer. In addition, as was previously noted, blockchain is also implemented on gateways. Yet SNs can only make the first move in a conversation. In contrast, the gateway blockchain stores just the thin data (the headers of each blocks). The mining-related consensus mechanism is set up on the fog layer.

The presence of a node in the blockchain is verified after an initial check for the node and system mapping. The blockchain is then queried to see whether the requesting gateway exists. The trust value is returned if all the conditions are met.

The smart agreement will determine whether a node is reliable after calculating its trustworthiness. If node B's trust value is more than 75 in relation to the preceding example, then it is a highly trusted node. A SN is regarded not trustworthy and remote or deleted from the network if its trust value is less than 50. This algorithm determines whether or not fog is linked to the system. The existence of the SN and the linked system are then verified. After verification, the given SN ID is deleted from the system.

4 Results and Discussion

4.1 Experimental Results and Discussion

This section presents experimental research on the use of fog computing and distributed intrusion detection systems to mitigate networks. In order to conduct the experiment, Python is used. We evaluate our suggested model's efficacy using the BoTdataset. Using Intel(R) Xeon(R) Silver 4114 processors at 2.20 GHz, 128 GB Memory, and 2 TB hard drive, the Tyrone PC does the experiments..

4.2 Description of BoT-IoT Dataset

The suggested distributed discovery approach is tested on a select dataset for accuracy. In the UNSW Canberra Center, a situation was created and used in the creation of the BoT dataset [23–25]. This dataset is a replacement for blockchain-based IoT solutions, and it is built using the protocol to establish connections between devices talking to one another. Table 2 displays the many attributes included in the BoT-IoT dataset. This is a short summary of the numerous forms of assault:

1. Denial-of-Service (DoS) Attacks: These attacks occur when hostile cyber danger devices block legitimate IoT devices' access to network resources like servers and databases. The goal of a denial-of-service (DoS) assault is to prevent legitimate

Table 2 BoT-IoT dataset features report

Feature number	Description	Features
F1	Rows identification	pkSeqID
F2	Protocols of Transaction shows network flow	Proto
F3	Source IP address	saddr
F4	Port number of source	Sport
F5	Destination IP address	daddr
F6	Port sum of destination	dport
F7	Record entire duration	seq
F8	Aggregated records with its standard deviation	stddev
F9	Source ip inbound connection	N_IN_Conn_P_SrcIP
F10	Minimum time duration for aggregate transactions	Min
F11	Numerical state features selection	State_number
F12	Aggregate record average time duration	Mean
F13	Destination ip inbound connection	N_IN_Conn_P_DstIP
F14	Per second period for terminus to source packet	drate
F15	Per second time for basis to destination packets	srate
F16	Aggregated record extreme duration time	Max
F17	Level 1 is for attack and level 0 for usual	Attack
F18	It signify traffic kinds	Category
F19	It demonstrations traffic sub kinds	Subcategory

(IoT) devices from connecting to the target network or host by overwhelming them with so much data that they crash or become unresponsive.

When several (IoT) sensor devices work composed to overwhelm a single target, this is known as a (DDoS) assault. Botnets allow attackers to deploy distributed denial-of-service (DDoS) attack scenarios on a massive scale, compromising a whole network of Internet-connected IoT devices. Attackers of security holes or flaws in the system by taking control of the devices via a series of instructions. As more and more gadgets join the Internet of things, DDoS assaults become increasingly dangerous. There is a common security flaw with IoT devices: they utilise default keys encrypted using tone. When an attacker compromises an Internet of things (IoT) system, they may be able to do it without the network owners even realising it has happened.

2. Reconnaissance (Recon) is the act of acquiring or probing information to determine the weak points of a target network and launch a successful attack.
3. Reconnaissance attacks include things like monitoring network traffic, examining network packets, probing network ports, and querying IP addresses.

Fourthly, theft attacks are an effort by a hacker to get access to private information stored in an IoT system by bypassing its security measures. Theft attacks include data leaking and keylogging. Attackers try to break into a distant Internet of things

Table 3 BoT-IoT dataset delivery for usual and attack cases in training usual

Class-wise division	Class frequency (in %) (%)	No. of occurrences
Theft	0.0022	65
Total	100	2,934,817
DDoS	52.5183	1,541,315
DoS	44.9823	1,320,148
Recon	2.4846	72,919
Normal	0.0126	370

Table 4 BoT-IoT dataset delivery for usual and attack examples in testing set

Class-wise division	Class frequency (in %) (%)	No. of occurrences
DDoS	52.5155	385,309
DoS	44.9925	330,112
Recon	2.4755	18,163
Normal	0.0146	107
Theft	0.0019	14
Total	100	733,705

device in order to grab data and send it to their own computer. The attacker uses the remote host as a keylogger to steal sensitive information from unsuspecting users.

Tables 3 and 4 provide the dataset, including their distribution, cumulative facts of attack, and characteristic properties. The total sum of examples in the training dataset is 2,934,817. There have been 1,541,315 distributed denial-of-service (DDoS) attacks, 1,320,148 distributed denial-of-service (DoS) attacks, 72,919 reconnaissance (Recon) attacks, 370 normal assaults, and 65 theft attacks. Following then, the testing dataset has 733,705 samples. There were 385,309 distributed denial-of-service attacks, 330,112 denial-of-service attacks, 18,163 Recon attacks, and 107 denial of theft bouts. In this subsection, we specify the metrics utilised for measuring the proposed model.

4.3 Evaluation Metrics

Our approach is ranked and its classification performance is measured using established measures. Specificity, sensitivity, and accuracy are three measures used to assess a categorisation system (AC). The benchmarks for success are specified as:

$$\mathrm{SE} = \frac{\mathrm{tp}}{\mathrm{tp} + \mathrm{fn}} \tag{3}$$

$$SP = \frac{tn}{tn + fp} \tag{4}$$

$$AC = \frac{tp + tn}{tp + fp + tn + fn} \tag{5}$$

where tp, tn, fp, and fn signify the sum of a true positive, true negative, and false negative.

4.4 *Performance Validation of Projected Model*

Table 5 presents the analysis of projected perfect with existing techniques for binary classification and for multiclass attacks are presented in Table 6.

For binary classification, the projected technique achieved 89.96% of accuracy, 80.46% of SE, and 90.23% of SP. But the existing models such as DBN, RNN, CNN, and LSTM achieved nearly 76–86% of ACC, 66–76% of SE, and 83–89% of SP.

For multiclass attacks, the projected perfect achieved better performance than existing models on all metrics. For instance, proposed model achieved 93.65% of accuracy, DBN achieved 81% of AC, RNN achieved 83.54%, CNN achieved 87.02%, and LSTM achieved 90.25%. When the models are tested with SP, the proposed model achieved 96.24%, DBN has 92%, RNN has 93.64%, CNN has 94.88%, and LSTM achieved 95.12%. Figures 1, 2 and 3 depict the graphical analysis of various techniques.

Table 5 Analysis for binary class

Method	ACC (%)	SE (%)	SP (%)
DBN	76.67	66.67	86.67
RNN	79.35	69.48	89.58
CNN	83.65	73.98	83.81
LSTM	86.44	76.34	86.48
Proposed	89.96	80.46	90.23

Table 6 Analysis for multiclass attacks

Method	ACC (%)	SE (%)	SP (%)
DBN	81.00	70.00	92.00
RNN	83.54	73.65	93.64
CNN	87.02	76.95	94.88
LSTM	90.25	80.56	95.12
Proposed	93.65	83.78	96.24

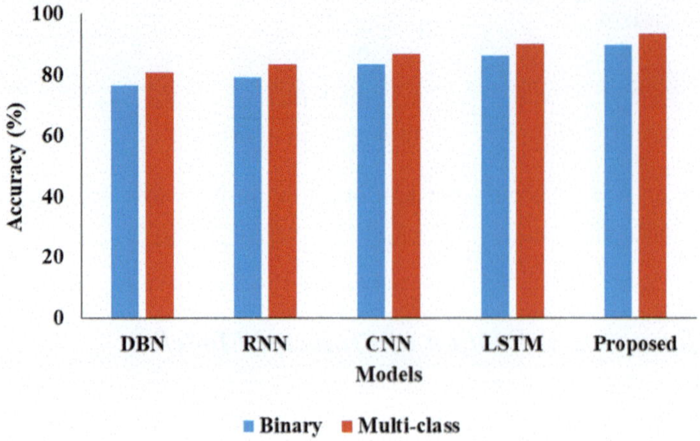

Fig. 1 Accuracy comparison for various classifiers

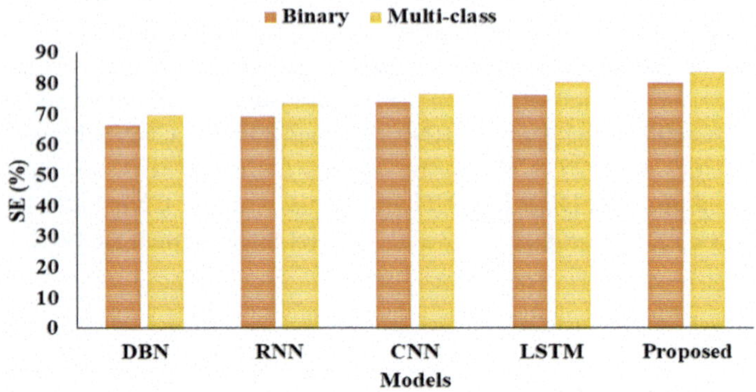

Fig. 2 SE analysis for different classifiers

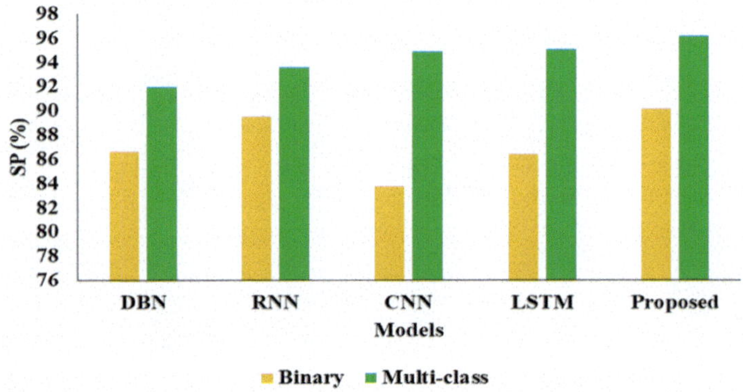

Fig. 3 SP analysis

5 Conclusion

Information gathering and dissemination are two common uses of networks. As a result, only reliable information sources and data should be used throughout the information-gathering process. As the SNs are the data sources, it is possible that an attacker may take a sensor or that an SN could become unreliable. These costs drain the energy of a single SN, which in turn endangers the whole network. Thus, this study introduces a trust device for identifying insider threats in IoT systems powered by sensor nodes. The number of SNs is used to construct dynamic itineraries for different mobile codes, with the goal of minimising the traffic congestion issue and the associated message overhead. The information is sent into the fog nodes. To accommodate the decentralised SN placement, enable single-point-of-failure issue, trust is computed and transacted on the blockchain. The results showed progress in areas like message overhead node finding speed, scalability, and network lifespan and performance optimisation. The use of blockchain technology to verify the identity of fog nodes and their administrators and encrypt inter-node communications would be an additional feature. To that aim, we want to put the suggested blockchain-based mechanism through its paces in a practical setting, like smart health care, to gauge its efficacy.

References

1. Shah Z, Ullah I, Li H, Levula A, Khurshid K (2022) Blockchain based solutions to mitigate distributed denial of service (DDoS) attacks in the internet of things (IoT): a survey. Sensors 22(3):1094
2. Gaur V, Kumar R (2022) Analysis of machine learning classifiers for early detection of DDoS attacks on IoT devices. Arab J Sci Eng 47(2):1353–1374
3. Almaraz-Rivera JG, Perez-Diaz JA, Cantoral-Ceballos JA (2022) Transport and application layer DDoS attacks detection to IoT devices by using machine learning and deep learning models. Sensors 22(9):3367
4. Chaganti R, Bhushan B, Ravi V (2022) The role of blockchain in DDoS attacks mitigation: techniques, open challenges and future directions. arXiv:2202.03617
5. Ali MH, Jaber MM, Abd SK, Rehman A, Awan MJ, Damaševičius R, Bahaj SA (2022) Threat analysis and distributed denial of service (DDoS) attack recognition in the internet of things (IoT). Electronics 11(3):494
6. Papalkar RR, Alvi AS (2022) Analysis of defense techniques for DDos attacks in IoT—a review. ECS Trans 107(1):3061
7. Ibrahim RF, Abu Al-Haija Q, Ahmad A (2022) DDoS attack prevention for internet of thing devices using ethereum blockchain technology. Sensors 22(18):6806
8. Huang L (2022) Design of an IoT DDoS attack prediction system based on data mining technology. J Supercomput 78(4):4601–4623
9. Kumar R, Kumar P, Tripathi R, Gupta GP, Garg S, Hassan MM (2022) A distributed intrusion detection system to detect DDoS attacks in blockchain-enabled IoT network. J Parallel Distrib Comput 164:55–68
10. Jing H, Wang J (2022) Detection of DDoS attack within industrial IoT devices based on clustering and graph structure features. Secur Commun Netw

11. Hayat RF, Aurangzeb S, Aleem M, Srivastava G, Lin JCW (2022) ML-DDoS: a blockchain-based multilevel DDoS mitigation mechanism for IoT environments. IEEE Trans Eng Manage
12. Zhou L, Zhu Y, Zong T, Xiang Y (2022) A feature selection-based method for DDoS attack flow classification. Futur Gener Comput Syst 132:67–79
13. Yakubu BM, Khan MI, Khan A, Jabeen F, Jeon G (2023) Blockchain-based DDoS attack mitigation protocol for device-to-device interaction in smart home. Dig Commun Netw
14. Sandhia GK, Nithyaselvakumari S, Saidulu V, Sulaiman NHB, Salameh AA (2023) Enhancing the security of software defined mobile networks (SDMN) based on blockchain technology. Int J Interactive Mob Technol 17(4)
15. Khor JH, Sidorov M, Ong MT, Chua SY (2023) Public blockchain-based data integrity verification for low-power IoT devices. IEEE Internet Things J
16. Kaur J, Singh G (2023) A blockchain-based machine learning intrusion detection system for internet of things. In: Principles and practice of blockchains. Springer, Cham, pp 119–134
17. Rangelov D, Lämmel P, Brunzel L, Borgert S, Darius P, Tcholtchev N, Boerger M (2023) Towards an integrated methodology and toolchain for machine learning-based intrusion detection in urban IoT networks and platforms. Future Internet 15(3):98
18. Ahmad MO, Tripathi G, Siddiqui F, Alam MA, Ahad MA, Akhtar MM, Casalino G (2023) BAuth-ZKP—a blockchain-based multi-factor authentication mechanism for securing smart cities. Sensors 23(5):2757
19. Kumar P, Kumar R, Gupta GP, Tripathi R (2021) A distributed framework for detecting DDoS attacks in smart contract-based blockchain-IoT systems by leveraging fog computing. Trans Emerg Telecommun Technol 32(6):e4112
20. Reddy KUK, Shabbiha S, Rudra Kumar M (2020) Design of high security smart health care monitoring system using IoT. Int J 8
21. Ramana K et al (2022) Leaf disease classification in smart agriculture using deep neural network architecture and IoT. J Circ Syst Comput 31(15):2240004. https://doi.org/10.1142/S0218126622400047
22. Zhang Y, Kasahara S, Shen Y, Jiang X, Wan J (2018) Smart contract-based access control for the internet of things. IEEE Internet Things J 6:1594–1605
23. Dwaram JR, Madapuri RK (2022) Crop yield forecasting by long short-term memory network with Adam optimizer and Huber loss function in Andhra Pradesh, India. Concurr Comput Pract Exp 34(27):e7310
24. Kishen Ajay Kumar V, Rudra Kumar M, Shribala N, Singh N, Gunjan VK, Siddiquee KN, Arif M (2022) Dynamic wavelength scheduling by multiobjectives in OBS networks. J Math 2022:10. https://doi.org/10.1155/2022/3806018
25. Koroniotis N, Moustafa N, Sitnikova E, Turnbull B (2019) Towards the development of realistic botnet dataset in the internet of things for network forensic analytics: BoT-IoT dataset. Future Gener Comput Syst 100:779–796

Cosine Similarity-Based Android Application for Intelligent Driver Recommendations

Naga Deepika Ginjupalli, Sandeep Vemuri, Kiran Babu Bandi, and Varshini Murukutla

Abstract In several fields, machine learning techniques are employed extensively to address everyday issues. There are numerous issues that need to be re-solved in the essential industry of transportation. Finding the nearest and most suitable driver is one of the problems that the majority of people deal with. In general, everyone needs a professional and courteous driver, but some people run into issues when they hire a driver without knowing about him. The pro-posed solution is to develop a mobile application that connects customers with drivers effectively. By utilizing cosine similarity, the app suggests a list of drivers based on the similarity between profile vectors of new drivers and the customer's previously hired driver. The app also dynamically updates the ranking of the driver in the list of suggested drivers, based on the cumulative rating of the driver which is calculated by considering feedback from other customers. This innovative approach simplifies the process for customers by eliminating the need for them to actively search for drivers who best match their preferences.

Keywords Cosine similarity · Dynamic ranking · Android mobile application · User feedback

1 Introduction

Many people require a driver for their car. Even though some of them have their own cars, they may not be interested in driving due to their hectic schedules, driving phobia, or a variety of other factors. They prefer to travel in their own vehicle because they possess one and do not want to call a cab or taxi. Instead, they require a chauffeur to operate their vehicle on a temporary or permanent basis. It is preferable for those with their own vehicles to hire a driver rather than call a cab because it is less

N. D. Ginjupalli (✉) · S. Vemuri · K. B. Bandi · V. Murukutla
Velagapudi Ramakrishna Siddhartha Engineering College, Vijayawada, India
e-mail: nagadeepikaginjupalli@gmail.com

B. Annappa et al. (eds.), *High Performance Computing, Smart Devices and Networks*, Lecture Notes in Electrical Engineering 1262,
https://doi.org/10.1007/978-981-97-7794-5_6

expensive. Everyone needs a skilled and polite driver. Problems arise when a driver is merely hired without any background checks.

In order to address this issue and minimize hazards, we came up with an idea of developing an interface between drivers and those who need a driver. So that each individual can choose the driver of their choice. For this purpose, we developed a mobile application in which customers and drivers initially register themselves by entering the required details in the app. Our app will recommend the nearest and most suitable list of drivers to customers based on the feedback given to the driver by his previous customers where he has worked in the past. In order to speed up the process of recommendation of drivers to customers, machine learning algorithm cosine similarity is used; this will recommend customers with the list of new drivers whose profile is nearer to the profile of the customer's previously hired driver. The ranking of the driver in the list of recommendations will also change dynamically based on user feedback.

The K-nearest neighbours (KNN) algorithm for movie recommendation also uses cosine similarity [1]. Based on movie popularity and/or genre, generalized recommendations are given to every user using the KNN algorithm and the cosine similarity principle. CNN-based music recommendation algorithm [2]. Cosine similarity and Euclidean distance are both used in this recommendation system to measure how similar pieces of music are, and cosine similarity metric performs better compared to Euclidean distance. The system for recommending products uses content-based filtering techniques [3]. In this case, cosine similarity is employed to suggest goods that are comparable to those that buyers have already purchased. This product recommendation system will suggest a product that is comparable to the item the customer has requested if the product they are looking for is currently not available.

Figure 1 describes the cosine similarity distance between the profile vector of drivers. The Android application developed will recommend the most suitable list of drivers to customers based on the cosine similarity distance between the profile of the customer's previously hired driver and new drivers.

2 Related Works

There are numerous applications related to our project that provide a driver on demand to the customer whenever he needs it, for example, DriveU, Drive4U, and Hire4drive: Car Drivers and Cabs. All these applications help customers hire a driver in online mode without explicitly searching for a driver. But these applications will not allow customers to select the driver of their choice, and customer has to hire the driver recommended by the system. Our approach is slightly different from previously developed applications. Our application uses cosine similarity distance for recommending drivers to customers based on user feedback. The recommendation system in our proposed application will recommend a list of drivers to the customer based on the profile of his previously hired driver.

Fig. 1 Cosine similarity between driver profiles

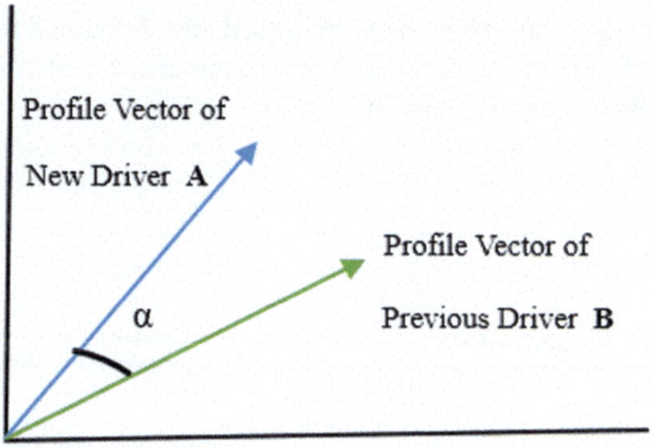

$$\text{Cosine Similarity of A and B} = \frac{|A||B| * \cos(\alpha)}{|A||B|}$$

Vinay N. Kadam developed an Android application "DriveMyCar" [4] which is a driver booking system. The main objective of this application is to provide drivers on rental to customers who have their own cars but are unable to drive. In this study, their approach is to develop a driver booking system using a server-based approach and also to ensure the safety of passengers. Through this app, customers can easily search for drivers and hire them on duty. But they cannot choose a driver of their choice they can only hire the driver recommended by the system.

Mullika Phanhong developed a spot recommendation system using Monte Carlo optimization [5]. This recommendation system improves the transportation system in Bangkok by reducing traffic problems. In this study, their aim is to develop a taxi recommendation system in order to improve the job of taxi drivers by suggesting where to stop and wait for passengers. Their main objective is to find the optimal waiting locations that will minimize the waiting time and maximize the loading distance. The results of this recommendation system are taxi drivers could drive an average time of 17.85% less than before without any passengers in cars. This recommendation system will reduce the waiting time of both drivers and passengers at taxi stops. So it is beneficial to both taxi drivers and passengers.

Md Rayhan Hasnat developed a Website for hiring drivers by conducting interviews. Online driver recruitment system [6] is aimed at developing a Website for recruiting drivers based on their performance in interviews. Some of the features of this Website include showing vacancies, storing applicants data, scheduling interviews, storing interview results of applicants, and finally hiring the driver. Drivers had to check for vacancies on the Website and apply for the job by uploading all the required documents. Based on the profile of the driver, he will be informed to attend the interview. After the completion of interview, the recruiting team will hire the drivers based on the results of the interview. But this is a lengthy process because every application has to be processed and scheduled for the interview. This is mostly suitable for the companies who maintain a lot of cars, taxis, etc., and provide the vehicle on rental to the customers.

Hairong, Yan had implemented the driver recommendation algorithm in enterprise transportation [7]. Cargo transportation plays an important role in the manufacturing industry by providing efficient drivers to deliver goods quickly and safely, thereby enhancing the company's reputation. If the goods do not reach the right customer, the customer will get a loss and the company also lose its reputation. The main objective of this driver recommendation algorithm is to address the problems that occur in the manual allocation of distributed tasks to drivers. The algorithm collects data on the drivers during the delivery of goods to customers and constructs a driver behaviour characteristics model. This model is used to automatically allocate new orders with suitable drivers and improves delivery efficiency and customer satisfaction by ensuring appropriate driver allocation for each order.

Apostolos and Skotis proposed a method for evaluating bias in user online reviews by using past ratings. The paper identifying drivers of evaluation bias in online re-views of city destinations [8] proposed an opinion of mining approach to identify content-specific, intrinsic terms with rating levels. Using the opinion mining approach, they evaluate the positive impact and negative impact of user online reviews. Based on the impact of the rating levels, regression and classification methods were used to classify the reviews into five clusters. The proposed study is a simplified segmentation approach that is useful for user profile rating based on previous online ratings.

Jayant, Vyas developed a driver recommendation system based on driver behaviour and fuel efficiency to ensure safe ride for customers. In the paper DriveBFR: Driver Behaviour and Fuel-Efficiency-Based Recommendation System [9], recommendation system is developed by using hidden Markov model machine learning algorithm, predicted the driver behaviour using MOME model by using UAH-DriveSet dataset, and obtained fuel efficiency and behavioural score using HMM model. They achieved overall 95%accuracy.

There are different applications that provide driver on demand to customers. And these applications will not allow customer to choose the driver of their choice instead customer has to choose the driver recommended by the system. Also these applications are not useful for people who are having their own cars. So an Android application is developed to help the people who have their own vehicles but are unable to drive. This application uses cosine similarity distance to recommend customers with a list of drivers who are nearer to the customer location and whose profile is best matched with customer's previously hired driver.

3 Methodology

The architecture of the proposed system is discussed in this section. In Fig. 2, the user interface of the mobile application is described clearly. Firstly, drivers and customers register in the app by providing the required details. Drivers have to upload their driving license certificates during registration, and the details are stored in their respective databases. Later the customer can log into the app and search for drivers and hire a suitable driver; the next time when the user logs into the app, he/she will be recommended with the nearest and most suitable list of drivers whose profile is similar to the previously hired driver of the customer. The customer can select a driver from the list of drivers and hire him if required, and also customer will give feedback to the driver before un-hiring him from duty. Based on the rating of the driver, next time he will be recommended to other customers.

Every time the details of the hired driver are stored in the customer profile, based on the profile vector of the previous driver, customer will be recommended with the list of new drivers whose profile vector is similar to the previous driver and who

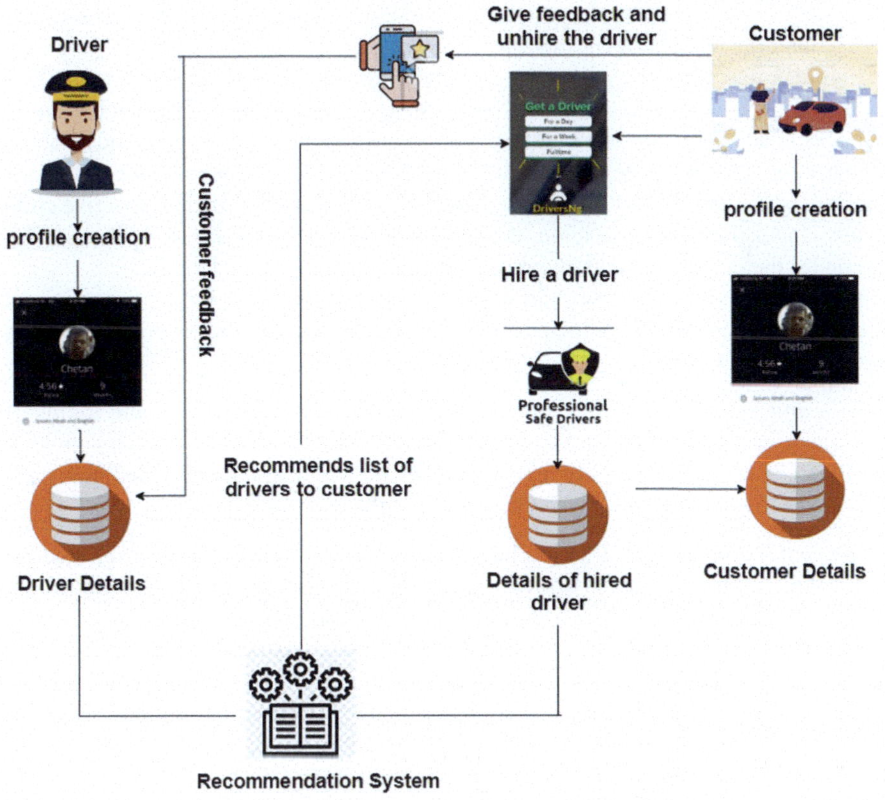

Fig. 2 Architecture of cosine similarity-based driver recommendation system

Fig. 3 Real-time database used in the application

are nearer to the customer location. And feedback is taken from the customer to under-stand the behaviour of the driver.

Figure 3 describes the hierarchy of data stored in the real-time database. This application is developed in Android Studio using XML and Java. The database used in the backend is Firebase to store the details of customers and drivers. In the Firebase database, we created three child nodes, namely customers, drivers, and my drivers. The data of customers and drivers are stored as a child node in their respective child nodes. Under the my driver node, the details of the customer are stored as a child, and under the customer node, the details of the hired driver are stored.

Figure 4 describes the flow chart of the proposed application; it mainly consists of three modules, namely Sign-In and Sign-Up module, the Recommendation Module, and the Feedback Module.

Sign-In and Sign-Up Module: This module will include registrations and logins. If at all the user is new to the application, then he can register into the app by clicking on signup. If the user is already registered, then he will be directed to the home page if he is not logged out, otherwise, the details of the user are asked for login.

Recommendation Module: This module describes how the customer is recommended with a suitable list of drivers who are nearer to his location. The recommendation system implemented in this Android application uses cosine similarity distance. Cosine similarity distance is calculated between the customer's previously hired driver and the new driver. The final score is obtained by multiplying the cosine similarity score with the cumulative rating of the driver to recommend the most efficient and suitable driver. Also, the rank of driver in the recommended list of drivers is updated dynamically based on the final score.

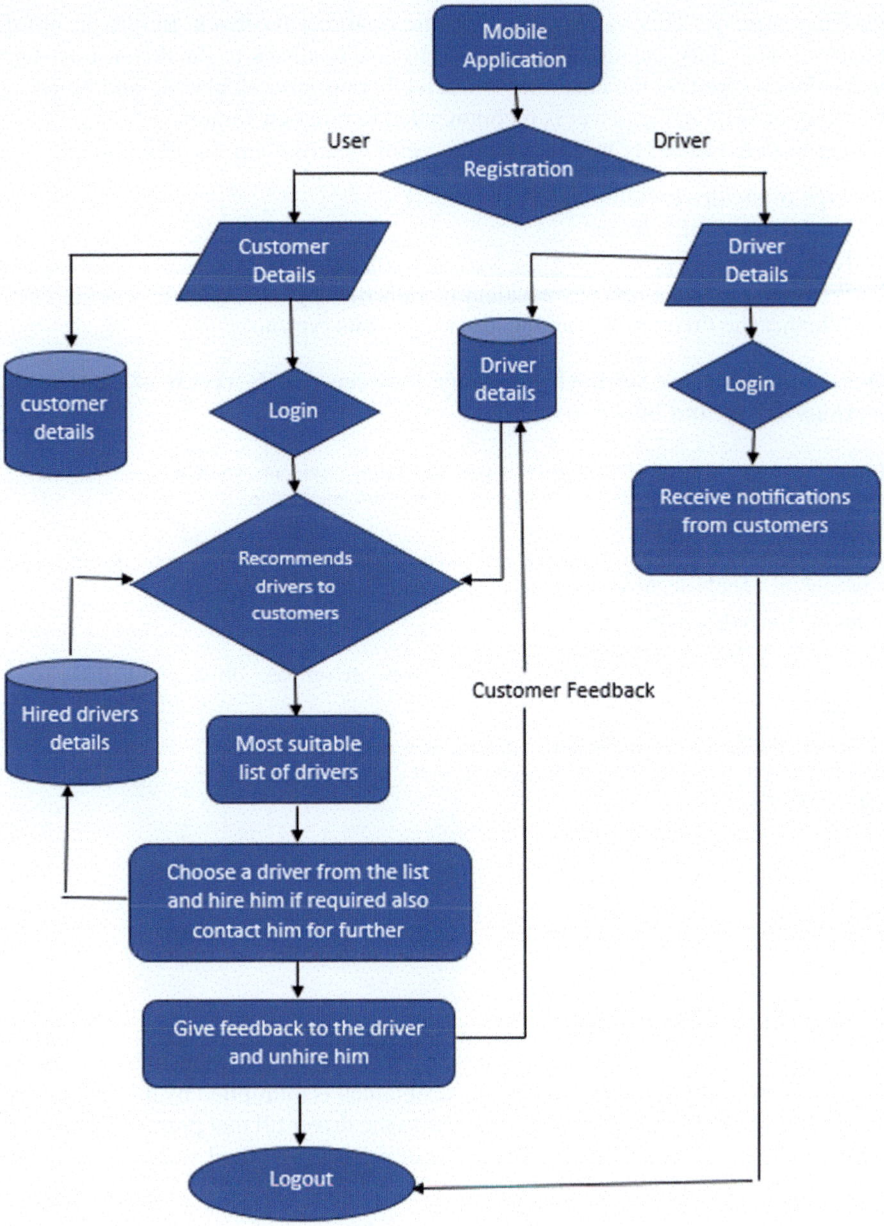

Fig. 4 Flow chart of proposed application

Feedback Module: This module describes the customer feedback. Before un-hiring the driver from duty, customer is supposed to give feedback to the driver. Based on the feedback given by the customer, the driver's rating is calculated. And based on the rating next time, the driver is recommended to other customers.

Factors considered for building profile vector of drivers are listed below:

1. Type of the driver (temporary/permanent).
2. Experience of the driver.
3. Rating of the driver.
4. Whether the driver had any accident records in the past (yes/no)?
5. Whether the driver had drinking/smoking habits (yes/no)?

Algorithm 1: Cosine similarity distance between profile vector of customer's previous driver and new drivers

Input: Previous_driver = {typeofdriver, experience, rating, accidents, smoking/drinking} New_driver = {typeofdriver, experience, rating, accidents, smoking/drinking}
Output: similarity_scoreC

1. Initialization: New_Array = list(Previous_driver.union(New_driver)) previous_driver_profile = [0] * len(New_Array), new_driver_profile = [0] * len(New_Array), *a*, *b*

2. for i in range(len(New_Array)):
If New_Array[i] in Previous_driver:
previous_driver_profile[i] = 1 **if New_Array[i] in New_driver**: new_driver_profile[i] = 1

3. End for

4. Cosine_similarity(previous_driver_profile, new_driver_profile)
previous_driver_profile = Anew_driver_profile = B
$dot\_product = \vec{A} \cdot \vec{B}$
$a = |A|$
$b = |B|$
dot_product
$C = a \times b$
return C

Algorithm 1 is used to determine how close the profile of the new driver is to the profile of the customer's previous driver. This algorithm demonstrates the use of the cosine similarity principle. The value C obtained is multiplied by the cumulative rating (average rating given by the customer to the driver till that point) of the driver to obtain the most suitable and efficient list of drivers. The final score is calculated using the below formula.

$$\text{final\_ score} = \frac{R}{5} \times C \tag{1}$$

Here in Eq. (1), R is the cumulative rating of the new driver, and C is the cosine similarity distance between the new driver and the customer's previous hired driver.

4 Results and Analysis

This section displays the outcome of the proposed system. The outcome of this application is used to provide the nearest and most suitable driver to the customer based on user feedback. This is obtained by using a cosine similarity-based recommendation algorithm to suggest the best list of drivers to customers based on their profile vectors. The execution screenshots of the proposed application are shown below from Figs. 5, 6, 7, 8 and 9.

Figure 5 shows the account page of the user, if the user is already logged in, then he will be directed to this page. He can edit his profile, by clicking on view drivers he can view the list of recommended drivers, and by clicking on my drivers, he will move to the profile page of the current driver.

Figure 6 displays the profile page of current hired driver. User can give feed-back to the driver, and also, he can un-hire the driver if not required.

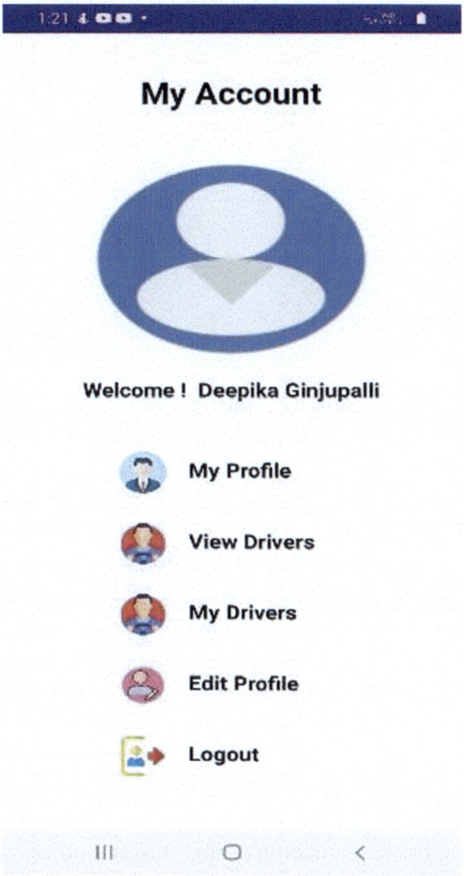

Fig. 5 User account page

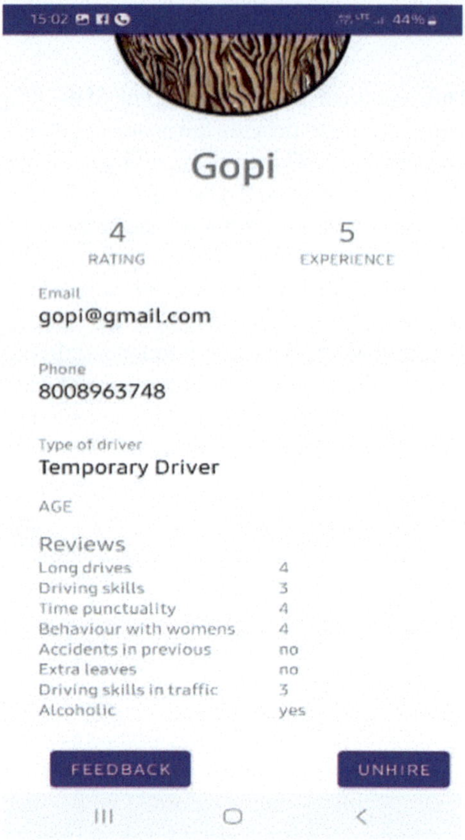

Fig. 6 Profile page of a current hired driver

Figure 7 displays the nearest and most suitable list of drivers who are best matched with the profile of the customer's previously hired driver shown in Fig. 6. From the list of card drivers, customers can click on any one of the cards to see the profile of the new driver which is shown in Fig. 8. Also, the customer can use the filter option at the top of the page to search for a driver.

Figure 8 displays the profile page of the new driver recommended by the sys-tem to the user. The user can check all the details of the driver including his license card; he can also contact the driver for further discussions by clicking on the call button. If the user is willing to hire the driver, then he can click on the hire button, now the driver is hired by the customer, and he no longer appears in the list of recommended drivers for any other customer until the driver is un-hired by his previous owner.

Figure 9 displays the feedback page. Before un-hiring the driver from duty, customer is supposed to give feedback on the driver. Feedback given by the customer is used to calculate the cumulative rating of the driver. Based on the rating, next time the driver is recommended to some other customers.

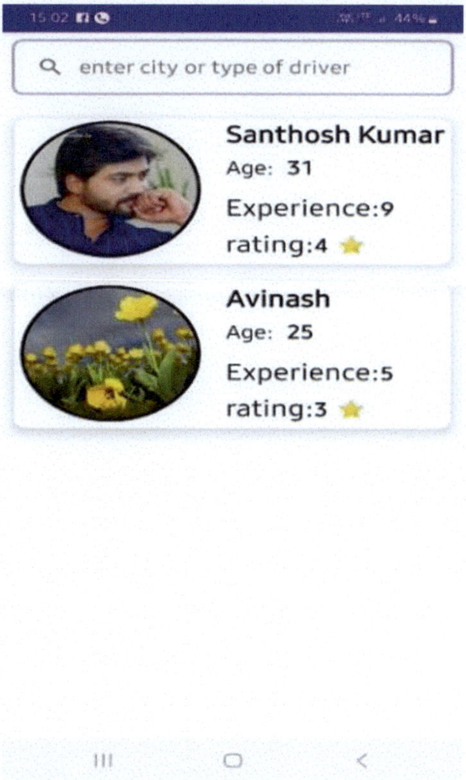

Fig. 7 Page of suitable list of drivers

5 Conclusion and Future Work

The proposed methodology resulted in good and accurate results for driver recommendations by using cosine similarity. The profile vector of the driver is created by considering some basic requirements of the customer, and then, the similarity of the profile vector of the new driver is measured with the customer's previously hired driver using the cosine similarity principle, and the final score is obtained by multiplying cosine similarity distance with the cumulative rating of the driver; this score is used to recommend the most suitable and efficient list of drivers to customers. And whenever the new customer gives feedback to the driver, the cumulative rating of the driver may change which leads to a change in the final score of the driver. This final score is used to dynamically update the list of recommended drivers. There are certain limitations that drivers are recommended on the cosine similarity score and rating of the driver, so sometimes the driver who has high cosine similarity and low rating maybe recommended; in this case the customer is recommended by an inefficient driver where the system fails.

In the future, this system can be expanded to different modes of transportation, and the algorithms used in the recommendation system can be modified such

Fig. 8 Profile page of the new driver

that it can overcome the limitations of this proposed study and also for efficient recommendations.

Fig. 9 Feedback page

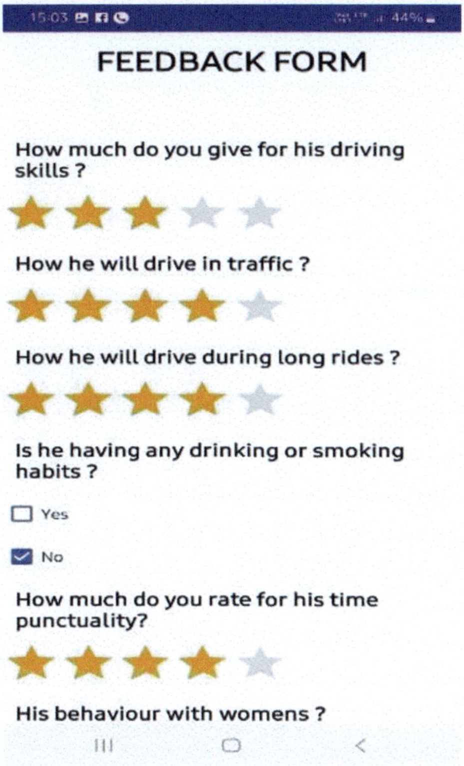

References

1. Singh RH et al (2020) Movie recommendation system using cosine similarity and KNN. Int J Eng Adv Technol 9(5):556–559
2. Sheikh Fathollahi M, Razzazi F (2021) Music similarity measurement and recommendation system using convolutional neural networks. Int J Multimedia Inf Retrieval 10:43–53
3. Fiarni C, Maharani H (2019) Product recommendation system design using cosine similarity and content-based filtering methods. IJITEE (Int J Inf Technol Electr Eng) 3(2):42–48 (2019)
4. Kadam V, Rao Y, Pawar S (2021) Drive my car android application
5. Phanhong M et al (2020) A spot-recommendation system for taxi drivers using Monte Carlo optimization. In: 2020 1st international conference on big data analytics and practices (IBDAP). IEEE
6. Hasnat MR, Ali MN, Martuza MA, Banik SC (2017) Development of an online based driver recruitment system
7. Yan H, Wang H (2021) Design and implementation of driver recommendation algorithm in enterprise transportation. In: 2021 IEEE 5th advanced information technology, electronic and automation control confer-ence (IAEAC), vol 5. IEEE
8. Skotis A, Morfaki C, Livas C (2023) Identifying drivers of evaluation bias in online reviews of city destinations. Int J Inf Manage Data Insights 3(2):100184
9. Vyas J, Das D, Chaudhury S (2021) DriveBFR: driver behavior and fuel-efficiency-based recommendation system. IEEE Trans Comput Social Syst 9(5):1446–1455

Evolving Safety Protocols: Deep Learning-Enabled Detection of Personal Protective Equipment

Mustafa Alahmid, Kishankumar Bhimani, Khushbu Saradva, Sushil Ghildiyal, Sayed Aryan Saeedi, Saraa Ali, Han Yan, and Evgeny Filippov

Abstract To give shift in safety protocols, we have employed advanced deep learning algorithms and frameworks (Shrestha and Mahmood in IEEE Access 7:53,040–53,065, 2019 [25]) to construct an innovative AI model. The designed model detects the usage of personal protective equipment (PPE) (Personal protective equipment. Geneva: World Health Organization, 2020 [18]) by workers in high-risk industries such as construction and manufacturing. We have used Google's TensorFlow object detection API (Sai and Sasikala in Object detection and count of objects in image using tensor flow object detection API, pp 542–546, 2019 [22]) to modify and train a model for dual purposes: PPE detection and face recognition. The state-of-the-art of this research is to substantially enhance safety compliance by addressing the prevalent issue of PPE non-compliance. To emphasis this, we have developed a pioneering software prototype that synergizes PPE detection with a face recognition-based clock-in system. This prototype demonstrates impressive object detection metrics with a mean average precision (mAP) of 0.9 for vests and 0.85 for helmets. Moreover, it exhibited efficient face recognition with a successful

M. Alahmid · S. A. Saeedi · H. Yan · E. Filippov
Faculty of Computer Science and Technology, Saint Petersburg Electrotechnical University "LETI", Saint Petersburg, Russian Federation

K. Bhimani (✉) · K. Saradva
Laboratory on AI for Computational Biology, Faculty of Computer Science, HSE University, Moscow, Russian Federation
e-mail: info.bhimani@gmail.com; kbhimani@hse.ru

S. Ghildiyal
Department of Computer Science and Engineering, Indian Institute of Technology Ropar, Punjab, India

S. Ali
Laboratory of Methods for Big Data Analysis, Faculty of Computer Science, HSE University, Moscow, Russian Federation

S. A. Saeedi
Deggendorf Institute of Technology, Bavaria, Germany

B. Annappa et al. (eds.), *High Performance Computing, Smart Devices and Networks*, Lecture Notes in Electrical Engineering 1262,
https://doi.org/10.1007/978-981-97-7794-5_7

threshold range of 17–20%. The implementation of AI in our system promises significant enhancements to worker safety, while concurrently reducing the financial burden associated with big hazards and accidents. Beyond the development and performance of the system, this paper provides a thorough exploration of the encountered challenges, potential real-world applications (particularly in employee monitoring and clock-in systems), and the future implications of this study on research and practical applications in the field of AI-integrated safety compliance.

Keywords Personal protective equipment (PPE) · AI · Object detection · Open CV · TensorFlow object detection API · Faster R-CNN

1 Introduction

1.1 Safety Protocols and History

Safety is a paramount concern for both day laborers and skilled workers. Every year, the construction industry is plagued by numerous accidents, many of which can be attributed to a lack of personal protective equipment (PPE) or by intensely avoiding to utilize the provided PPE. PPE is an equipment designed to shield employees from health and safety hazards at work. It is deployed to mitigate employee exposure to risks when administrative and engineering interventions are not feasible or effective in reducing these hazards to acceptable levels. Such risks may include falling debris, slippery surfaces, and various industrial hazards. PPE typically includes protective vests, helmets, and safety glasses.

Construction workers have a high rate of injury; there are roughly 340 million accident injuries per year, according to the International Labour Organization Statistics [11]. It is crucial to ensure PPE compliance. Therefore, we propose the design of deep learning model capable of detecting whether workers are equipped with the necessary safety gears and PPE.

1.2 Personal Protective Equipment (PPE) Challenges

Significant compliance issues with the use of personal protective equipment (PPE) have been revealed by a survey [34] of one hundred professionals in high-risk industries such as construction, which is the problem of interest in our research, the employees non-compliance issue was one of the main issues that 38% of the professionals have mentioned the problem according to the survey, and it is on the top of the PPE concerning challenges.

The report by the artificial intelligence company Cortexica [30]said that the vast majority (84%) of businesses that have work in a high-risk environment incurred losses from injuries due to the PPE non-compliance in the past year. The report also claims that 84% of them still rely on manually checking for employees for PPE non-compliance.

1.3 History and Evolution of Object Detection

Object detection, a crucial component of machine learning and deep learning, has transformed drastically over the years. Presently, deep learning-based approaches lead the field, thanks to the robustness and accuracy they provide. The rise of deep learning in object detection [32] coincides with advancements in neural network architectures [23], especially convolutional neural networks (CNNs) [15]. Deep learning algorithms, typically defined as layered models of inputs, aim to emulate the human understanding of the world. This allows them to perform tasks like image and video recognition, image classification, natural language processing, and many more.

Nowadays, the approaches of deep learning are the state-of-arts [17] in this field. CNN is the cornerstone of most deep learning models today. Its fundamental components include convolution layers and pooling operations, both contributing to its high efficacy in image and video processing.

Today's object detection systems boast accuracy, speed, and efficiently classifying objects. This precision is achieved through extensions of image classification models and the introduction of new APIs, like Google's TensorFlow. API includes pre-built architectures and weights for models like SSD with Inception V2 [1], Single Shot Multibox Detector [13], faster RCNN with Inception ResNet V2 [3], faster RCNN with ResNet 101, and region-based fully convolutional networks (R-FCN) [2].

1.4 Face Detection and Recognition

Face recognition, a computer vision problem consisting of face detection and face identification [33], is dominant in modern user authentication mechanisms. As an example, tech giant Baidu leverages face recognition instead of traditional ID cards for employee office access [14].

The 1990s saw a significant leap with Eigenfaces, implemented by Sirovich and Kirby [26], which developed a low-dimensional representation of facial images. This approach established that less than a hundred values were needed to accurately code a normalized image of a face.

More recently, AlexNet [12], using a deep convolutional neural network (CNN), achieved an impressive error rate of 15.3%, surpassing other algorithms [9] of its time by over 10% in the ImageNet Large Scale Visual Recognition Challenge (ILSVRC) [21]. The technology further improved, reaching an error rate of just 3.6% by 2015 with ResNet [6].

FaceNet [24], a system for face recognition, verification, and clustering, utilizes a Euclidean embedding for image representation through a deep convolution network. The architecture facilitates the matching of faces based on minimized distances for the same person and maximized distances for different individuals. Its structure consists of a batch input layer, a convolutional neural network, L2 normalization, and a triplet loss function. After creating the embeddings represented by a 128-feature vector, the recognition task becomes a KNN classification problem, which could be solved using a SVM classifier.

2 Holistic Methodology for PPE Identification and Face Recognition

2.1 Role and Importance of PPE Monitoring in Workplace Safety

In an effort to mitigate workplace accidents, particularly in the construction industry, constant monitoring of PPE usage among employees is vital. However, manual monitoring presents issues such as cost, time, and human error. This paper proposes an AI-based model to detect PPE usage among workers, thus ensuring continuous safety compliance.

The use of PPE such as helmets, vests, gloves, and safety footwear (see Fig. 1) plays a critical role in protecting employees from various on-site hazards. Employers have an obligation to ensure the provision and correct usage of PPE. Monitoring is a primary duty within many industries, where workers are mandated to wear PPE.

2.2 Detection Model: Faster R-CNN, SSD, and Inception

Transfer Learning: Transfer learning [31] is a technique where a pre-trained model for a specific task is leveraged as an initiation for another related task. In the context of deep learning and computer vision, transfer learning allows reuse of learned features from one model to benefit a new model trained on a different task. Two common approaches to transfer learning are feature extraction and weight initialization.

TensorFlow Object Detection API: The TensorFlow object detection API [29] offers a comprehensive framework for building, training, and deploying object detection models. This research utilizes the TensorFlow object detection API to develop a PPE detection model.

Faster R-CNN: The faster R-CNN model [20] comprises two interconnected modules: the region proposal network (RPN) and the fast R-CNN detector. The

Fig. 1 Personal protective equipment spectrum: helmets to accessories (essential gloves, cones, safety glasses, protective boots, and vests)

RPN predicts object presence and generates bounding boxes, while the fast R-CNN uses these characteristics to classify objects and refine their bounding boxes. This unified architecture enables efficient and accurate object detection.

Inception Model: The Inception model [28] can tackle the challenge of handling objects with varying sizes in an image. By employing filters of different sizes at the same network level, the model can capture both global and local information. Inception models, including the widely-known GoogleNet [27], demonstrate improved computational efficiency through smart factorization methods.

Single Shot Detector (SSD): The SSD model represents a single deep neural network capable of performing object detection and classification in a single feed forward pass. By producing fixed-size bounding boxes for objects and associated scores, the model streamlines the detection process. Suppression techniques are then applied to select high-scoring detections above a threshold.

These methodologies form the basis for practical implementation of PPE detection.

2.3 Dataset

The dataset consisted of images collected from the Internet source using three different methods including Google Search PPE Images, Google Open Dataset Images [16], and video frames extracted from construction documentaries using Python script. In total the dataset contains 8351 objects distributed as shown in the (see Fig. 2).

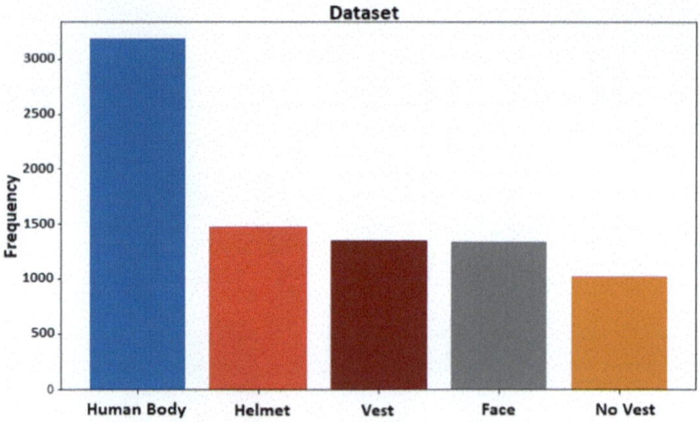

Fig. 2 Dataset statistics—with 8351 objects—helmets, vests, and construction workers faces from Google Images, Google Open Dataset, and video frame extraction

Labeling the Dataset: In order to feed the model with data, the data should be labeled, and we use LabelImg (from Label Studio) [7] tool for labeling images. The final dataset was annotated with around 1300 images. The figure (see Fig. 3a) displays random examples of images from the dataset and their corresponding annotations achieved through the LabelImg tool in (see Fig. 3b).

Computational Environment: Python 3.10 or above version with the following libraries latest version: TensorFlow, Protosun, Pillow, NumPy, lamp, OpenCV, and Pandas.

(a) Dataset Images (b) Dataset Labeled Images

Fig. 3 **a** Shows random images examples from the dataset, **b** labeled using the LabelImg tool [7], resulting in around 1,300 annotated images

3 Experiment

3.1 TensorFlow Object Detection API

Experiment 1: Worker Classification in Overlapping Body Images

- The dataset contained images of workers with and without helmets and vests.
- The objective was to detect human bodies and classify them as workers.
- Challenged by significant overlap between worker and non-worker bodies, leading to difficulty in accurate classification.

Experiment 2: Model Performance on Synthetic Dataset

- The model trained on a dataset of 500 images from Google (vest and helmets).
- Achieved accurate detection of helmets, but low accuracy in predicting vests.
- The dataset lacked real-life images, resulting in reduced accuracy when applied to real-life videos or images.

Experiment 3: Model Training and Optimization

- Trained faster R-CNN Inception v2 and SSD Inception v2 models on an image dataset with five classes: vests, no vests, helmets, persons, and faces.
- Both models required approximately 50,000 training steps to achieve the lowest total loss.

3.2 Evaluation

Mean Average Precision (mAP): [8] is a widely used evaluation metric for object detection and segmentation models. It assesses the accuracy of predicting both the class and location of objects in images. The ground truth includes class labels and precise bounding box coordinates for each object. During training and validation, images are annotated similarly. The model generates multiple predictions, but only those above a confidence threshold are considered. The resulting detections are then evaluated using Intersection over Union (IoU) to measure the correctness of bounding boxes. mAP provides a performance measure for accurately predicting object classes and locations in images (see Fig. 4).

Interpolated AP: PASCAL VOC [4] is a popular object detection dataset. In PASCAL VOC, a prediction is considered positive if IoU $\geq$ 0.5. Multiple detections of the same object are treated as negatives, except for the first one. The average precision (AP) in Pascal VOC2008 [10] is calculated using 11-point interpolation, where precision is computed for 11 equally spaced recall values from 0 to 1.

Table 1 gives a summary of models performance with mAP score.

Fig. 4 mAP: evaluating object detection and segmentation with boundary box coordinates

Table 1 mAP score with various models

Model	Vest	No vest	Face	Human body	Helmet	mAP
No. of images	1347	1019	1338	3179	1468	–
SSD model	0.9056	0.43	0.66	0.59	0.721	0.661
Faster R-CNN	0.96	0.69	0.78	0.72	0.85	0.804

3.3 Face Detection and Recognition

Here, we focus on developing a face detection and recognition system, comprising these two main components. Faster R-CNN [20], a highly effective face detection model, is utilized to identify faces as objects of interest. Additionally, FaceNet [24] is employed to extract feature embeddings for each face image, enabling accurate face recognition.

Face Detection: Faster R-CNN [20], originally designed for PPE detection, demonstrates exceptional performance in detecting faces. By incorporating face detection into our system, we leverage the capabilities of this model.

Face Recognition: The face recognition process consists of two steps: creating embeddings (feature vectors) and predicting the face ID using SVM or KNN algorithms. FaceNet, a face recognition system, generates embeddings that map face images to a Euclidean space where distances reflect face similarity. These embeddings enable accurate face identification and can be further classified using an SVM algorithm.

Data Preprocessing: The face dataset utilized for training includes images of 19 different celebrity classes collected from Google Images. The faces are cropped using the faster R-CNN model, and the resulting cropped images are stored for subsequent processing. FaceNet is then employed to create 128-dimensional feature embeddings for each face image. The dataset is divided into training and testing sets.

Model Selection: This paper presents two powerful PPE detection models. Figure (see Fig. 5) shows the SSD model's accuracy in face detection (a) and PPE detection on data images without and with vests (b), respectively. Similarly, (see Fig. 6) demonstrates the faster R-CNN model's proficiency in face detection (c) and PPE detection without and with vests (d). Notably, the faster R-CNN model excels in both PPE (vest) detection and face recognition.

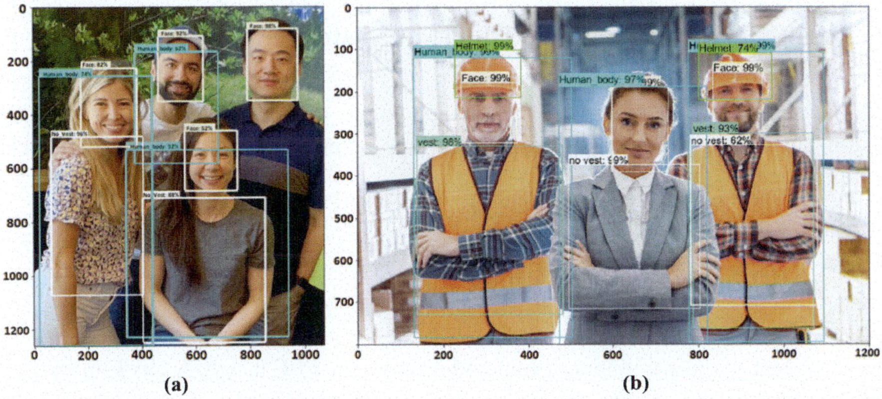

Fig. 5 PPE detection using SSD model

Fig. 6 PPE detection with the faster R-CNN model

Face Recognition system Evaluation: The evaluation process involves calculating the false acceptance rate (FAR) and the false rejection rate (FRR) using a test set of 200 face images, including 100 imposter pictures (50%). The evaluation is performed by varying the threshold (predicting probability) from 1 to 100%. The obtained results are used to generate a FAR-FRR curve, enabling the determination of the equal error rate (EER). In our experiments, the EER is nearly zero, indicating optimal performance at a threshold of 16%. The confusion matrix [19] illustrates the correct and misclassified face instances.

The faster R-CNN model demonstrates high accuracy and speed compared to other models, making it suitable for our requirements. The face recognition module achieves good accuracy within the threshold range of 17–20%.

4 Result: Monitoring and Clock-in System

PPE is necessary at construction sites, and many countries mandate its use. AI vision can assist in monitoring workers and implementing efficient systems for PPE compliance.

4.1 Monitoring and Clock-in System

A PPE detector can aid in identifying and monitoring PPE usage. A clock-in system can be implemented using the PPE detector and biometric technologies such as fingerprint or face recognition (see Fig. 7a).This system ensures that workers can only sign in (see Fig. 7b) when wearing the required PPE, enforcing compliance.

4.2 Application Objective

The objective is to create an accurate attendance record and enable precise payroll calculations. The clock-in system combines PPE detection with face recognition to verify worker identity and enforce PPE compliance effectively. By integrating AI and biometrics, the application enhances safety measures and enforces the PPE mandate.

1. **PPE Detection**: The system utilizes a camera to capture images and employs a detector to identify safety equipment such as vests, helmets, and faces. By analyzing the image, the system ensures that workers are wearing the required PPE.

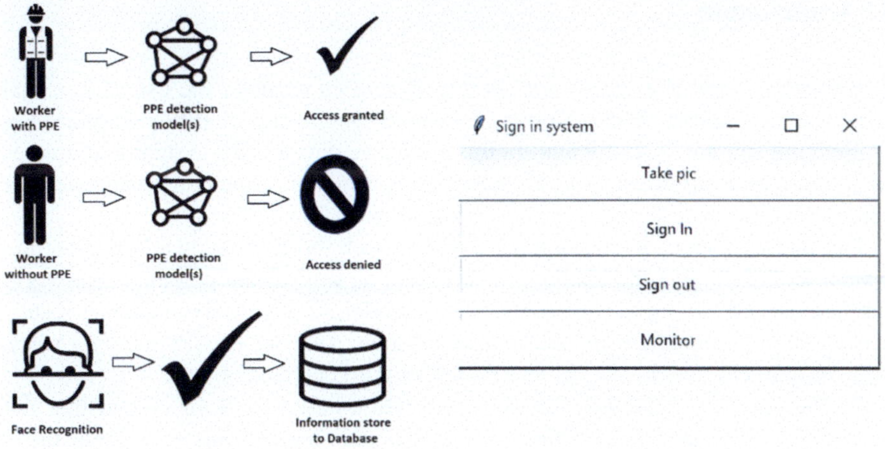

(a) (b)

Fig. 7 Application pipeline and functioning (**a**) access protocol, (**b**) clock-in system

2. **Face Recognition and ID**: The application extracts the face from the image and utilizes facial recognition technology to predict the worker's ID [5]. It records the ID along with the time of clock-in and clock-out, creating an accurate attendance record.
3. **Field Monitoring**: The system uses a camera and a PPE detection model to monitor the field (see Fig. 8). By detecting PPE, it assists in monitoring worker compliance and ensuring safety protocols are followed.

Fig. 8 Real-time field monitoring and PPE detection on workers

5 Conclusion

Construction work is a hazardous activity that often requires the use of personal protective equipment (PPE). However, manual monitoring of PPE compliance is challenging and not cost-efficient. In this study, we developed a system that utilizes AI to recognize and identify PPE, determining whether workers are wearing it or not.

Our research demonstrates a promising approach for developing a PPE detection system and highlights the challenges involved, as well as the potential real-world applications. By employing the faster R-CNN model, we achieved satisfactory results in detecting vests and helmets, with a mean average precision (mAP) of 0.9 for vests, 0.85 for helmets, and an average mAP of 0.8 for all detected objects.

Furthermore, we developed a face recognition-based clock-in system that not only detects PPE but also verifies worker identities and records their login times in a database. The proposed system consists of two parts: PPE detection, where we utilized the faster R-CNN to train the model to recognize both faces and PPE, achieving a respectable mAP of 0.78; and face recognition, which identifies worker IDs and stores them along with login timestamps.

6 The Future of AI in PPE

The future of AI in PPE holds great promise, particularly in addressing challenges related to worker density and camera or surveillance distance from workers. By leveraging advancements in AI technology, especially deploying liquid neural network, Detectron2, and InsightFace in detection applications significantly improves result.

References

1. Bose SR, Kumar VS (2020) Efficient inception V2 based deep convolutional neural network for real-time hand action recognition. IET Image Proc 14(4):688–696
2. Dai J et al (2016) R-FCN: object detection via region-based fully convolutional networks. In: NIPS'16. Curran Associates Inc., Red Hook, NY, USA, pp 379–387
3. Devi SK, Subalalitha CN (2022) Assisting the visually challenged people using faster RCNN with inception ResNet V2 based object detection model. In: Ranganathan G et al (eds) Pervasive computing and social networking, pp 171–181. https://doi.org/10.1007/978-981-16-5640-8_14
4. Everingham M, VanGool L, Williams CKI, Winn J, Zisserman A (2008) The PASCAL visual object classes challenge 2008 (VOC2008) results. http://www.pascal-network.org/challenges/VOC/voc2008/workshop/index.html
5. Hassan N, Ramli DA, Suandi SA (2014) Fusion of face and fingerprint for robust personal verification system. Int J Mach Learn Comput 4:371–375. https://doi.org/10.7763/ijmlc.2014.v4.439

6. He K, Zhang X, Ren S, Sun J (2016) Deep residual learning for image recognition. In: 2016 IEEE conference on computer vision and pattern recognition (CVPR). IEEE, Las Vegas, NV, USA, pp 770–778. https://doi.org/10.1109/CVPR.2016.90
7. Heartexlabs: Tzutalin: LabelImg. https://github.com/heartexlabs/labelImg
8. Henderson P, Ferrari V (2017) End-to-end training of object class detectors for mean average precision. In: Lai S-H et al (eds) Computer vision—ACCV, pp 198–213. https://doi.org/10.1007/978-3-319-54193-8_13
9. Hochreutiner C (2020) The history of facial recognition technologies: how image recognition got so advanced. https://anyconnect.com/blog/the-history-of-facial-recognition-technologies
10. Hoiem D, Divvala SK, Hays JH (2009) Pascal VOC 2008 challenge. World Literature Today Citeseer 24:1
11. International Labour Organization (ILO). World statistic. occupational safety and health (OSH). https://www.ilo.org/moscow/areas-of-work/occupational-safety-and-health/WCMS_249278/lang--en/index.htm
12. Krizhevsky A, Sutskever I, Hinton GE (2017) ImageNet classification with deep convolutional neural networks. Commun ACM 60(6):84–90
13. Liu W, Anguelov D, Erhan D, Szegedy C, Reed SM, Fu C-Y, Berg AC (2016) SSD: single shot multibox detector. In: Leibe B et al (eds) Computer vision—ECCV, pp 21–37. https://doi.org/10.1007/978-3-319-46448-0_2
14. Marr B (2018) How Chinese internet giant Baidu uses artificial intelligence and machine learning. https://www.forbes.com/sites/bernardmarr/2018/07/06/how-chinese-internet-giant-baidu-uses-artificial-intelligence-and-machine-learning/?sh=6b66fdc2d557
15. O'Shea K, Nash R (2015) An introduction to convolutional neural networks
16. Open Images Dataset V7 and Extensions. https://storage.googleapis.com/openimages/web/index.html
17. Pal SK, Pramanik A, Maiti J, Mitra P (2021) Deep learning in multi-object detection and tracking: state of the art. ApplIntell. 51(9):6400–6429
18. Personal protective equipment. Geneva: World Health Organization; Licence: CC BY-NC-SA 3.0 IGO (2020)
19. Ren J, Wang Y (2022) Overview of object detection algorithms using convolutional neural networks. J Comput Commun 10(1):115–132
20. Ren S, He K, Girshick R, Sun J (2015) Faster R-CNN: towards real-time object detection with region proposal networks. In: NIPS'15, vol 1. MIT Press, pp 91–99. https://doi.org/10.1109/TPAMI.2016.2577031
21. Russakovsky O, Deng J, Su H, Krause J, Satheesh S, Ma S, Huang Z, Karpathy A, Khosla A, Bernstein MS, Berg AC, Fei-Fei L (2015) ImageNet large scale visual recognition challenge. Int J Comput Vision 115(3):211–252
22. Sai BNK, Sasikala T (2019) Object detection and count of objects in image using tensor flow object detection API. In: 2019 ICSSIT, Tirunelveli, India, pp 542–546. https://doi.org/10.1109/ICSSIT46314.2019.8987942
23. Schmidhuber J (2015) Deep learning in neural networks: an overview. Neural Netw 61:85–117. https://doi.org/10.1016/j.neunet.2014.09.003
24. Schroff F, Kalenichenko D, Philbin J (2015) FaceNet: a unified embedding for face recognition and clustering. In: 2015 IEEE conference on computer vision and pattern recognition (CVPR). IEEE, Boston, MA, USA, pp 815–823
25. Shrestha A, Mahmood A (2019) Review of deep learning algorithms and architectures. IEEE Access 7:53040–53065. https://doi.org/10.1109/ACCESS.2019.2912200
26. Sirovich L, Kirby M (1987) Low-dimensional procedure for the characterization of human faces. J Opt Soc Am A 4(3):519. https://doi.org/10.1364/JOSAA.4.000519
27. Szegedy C, Liu W, Jia Y, Sermanet P, Reed SM, Anguelov D, Erhan D, Vanhoucke V, Rabinovich A (2015) Going deeper with convolutions. In: 2015 IEEE conference on computer vision and pattern recognition (CVPR), Boston, USA, pp 1–9
28. Szegedy C, Vanhoucke V, Ioffe S, Shlens J, Wojna Z (2016) Rethinking the inception architecture for computer vision. In: 2016 IEEE conference on computer vision and pattern recognition (CVPR). IEEE, Las Vegas, NV, USA, pp 2818–2826

29. TensorFlow Object Detection API. https://github.com/tensorflow/models
30. Tomorrow's Health and Safety Team, Cortexica (2018) Overcoming the barriers of PPE compliance with the use of artificial intelligence. https://www.tomorrowshs.com/improper-ppe-cost-businesses-seventy-nine-billion--1038.html
31. Torrey L, Shavlik J (2010) Transfer learning. IGI Global, pp 242–264
32. Xiao Y, Tian Z, Yu J, Zhang Y, Liu S, Du S, Lan X (2020) A review of object detection based on deep learning. Multimed Tools Appl 79(33–34):23729–23791
33. Zhafran F, Ningrum ES, Tamara MN, Kusumawati E (2019) Computer vision system based for personal protective equipment detection, by using convolutional neural network. In: 2019 international electronics symposium (IES), Indonesia, pp 516–521. https://doi.org/10.1109/ELECSYM.2019.8901664
34. Zou Z, Chen K, Shi Z, Guo Y, Guo Y (2023) Object detection in 20 years: a survey. Proc IEEE 111(3):257–276. https://doi.org/10.1109/JPROC.2023.3238524

Vision Transformers for Remote Sensing (ViToRS) Image Scene Classification

Pankaj Kumar Gharai and Mogalla Shashi

Abstract Building deep learning models to perform computer vision tasks in remote sensing with higher accuracy using lesser computational resources and time has always been a challenge. Instead of using a single large-scale dataset of high spatial resolution for complete model training, this research divides the training into two parts. First part of training uses a single dataset of 'low spatial resolution and high volume', while second training part retrains the model on 'high spatial resolution and low volume' label-specific datasets. This paper proposes a method that comparatively evaluates 13 different vision transformer architectures and selects the best ones for building a composite model for remote sensing image scene classification. An efficient way to use transfer learning through training pretrained models and retraining fine-tuned models using the proposed method is discussed. Multiple label-specific vision transformer architecture models are run in parallel as a composite architecture to achieve higher accuracy for all class/labels, rather than running a single model for all class/labels with lower accuracy for some of the class/labels.

Keywords Deep learning (DL) · Remote sensing (RS) · Earth observation (EO) · Image classification · Transfer learning

1 Introduction

Recent interest of researchers in the field of machine learning (ML) using transformers has opened up new ways to solve issues in computer vision tasks that were present in earlier architectures [1]. Deep learning (DL) is slowly being used in the domain of remote sensing and earth observation (EO) for different types of computer vision tasks and applications. Large satellite image data has always been a major

P. K. Gharai (✉) · M. Shashi
Department of CS and SE, Andhra University College of Engineering, Visakhapatnam, India
e-mail: pankajgharai@gmail.com

M. Shashi
e-mail: prof.mshashi@andhrauniversity.edu.in

B. Annappa et al. (eds.), *High Performance Computing, Smart Devices and Networks*, Lecture Notes in Electrical Engineering 1262,
https://doi.org/10.1007/978-981-97-7794-5_8

challenge for researchers to process with self-supervised learning models. One of the basic tasks of computer vision problems is image classification, since accuracy obtained in classification will further qualify the machine learning architecture to venture in other tasks like object detection, image segmentation, etc.

Application of deep learning in remote sensing is rapidly emerging field of study at present. The interest of researchers in this field is partly due to availability of better datasets. But main contribution to this field is due to emergence of various techniques and pretrained models in past few months [1, 2]. Geospatial data is well known for its application in agriculture, urban planning, disaster management, climate change, and defense. Deep learning methods are suited for many remote sensing tasks due to its success in implementation in practical applications. However, due to complex, high-resolution and voluminous datasets, it becomes tedious to work with limited computing resources. In this paper, we will analyze if there is any method for building models of higher accuracy with limited resources.

In the past decade, convolution neural network (CNN) architectures were used primarily to process data from satellite images. These architectures produced models with highest accuracy and were used for different computer vision tasks in areas of remote sensing and earth observation (EO). But there was a limitation of attaining higher accuracy even when new architectures based on CNN were proposed [1]. Recent progress in transformer [3] architecture has shown that they have advantage in situation where global dependencies and contextual understanding are important. Transformer architecture have been successfully implemented in computer vision called vision transformers (ViT) [4].

RS image scene classifications using CNN architectures is still the most widely used method. Most of current research has over dependence on CNN which has few suitable architectures like 'ResNet', giving more emphasis to training methods and hyper-parameter tuning. Dataset remains the same for complete training/validation/testing that may have limited global variations of manmade structures within the same class/label like residential images. Other limitation like considering only overall accuracy of a single model having lower accuracy of some class/label is accepted.

It is evident that 'image scene classification' is complex than simple 'image classification'. RS images can have major features of one class and minor features of other classes. So sub-tasks like object detection, object, and texture/pattern classification is also to be taken into consideration. ViTs are more suitable for such scenarios due to their versatility to generalize well to unseen images [1].

A novel approach is implemented in this research which is different in terms of dataset, model selection, and training methodology than the existing ones. Overfitting of is avoided by minimizing noise during initial phase of complex model training, which are difficult to eliminate at later stage. Recall values of each class/label are considered for multiple model selection instead of overall accuracy. More focus is paid to learning rate of composite model as a whole to retain information of best performing models for each class/label in the dataset.

2 Related Works

2.1 Transfer Learning

Transfer learning is a way to reuse a pre-trained model for a new problem, where we start with a pre-trained model instead of a scratch and train it to our specific area of interest. The benefit is that the data required to train deep neural networks is reduced significantly. In this research the objective is to process real world data and improve the learning efficiency.

Transfer learning has been shown to reduce time and increase performance of a model on re-training. In this research we start with a benchmark pre-trained model and further train it with specific datasets to increase overall efficiency of the model. There is a growing interest to implement transfer learning in field of vision transformers recently. Further it has been observed that increasing data processing efficiency through hardware costly as it takes time and cost to develop such technology. Also, such hardware is not easily available to novice researchers who require it for open-source development. Modifying the model training methodology is the only way to solve this problem.

2.2 Vision Transformers

ViT has been showing encouraging results in computer vision tasks. It has better positional feature information of the whole image than CNN architectures which have limited localized feature information [4]. This gives ViT enough ability to classify very large single label EO images even if trained on a dataset having small-sized EO images. The size of the dataset images is increased to extract features for training a pretrained model and to produce a fine-tuned model. The test image is then reduced to the size of the fine-tuned model's processor for prediction. It has been recently established that ViT architectures outperform CNN architectures in field of remote sensing [1]. Various ViT architectures from different research papers will be used in this paper for comparative study in field of remote sensing [11–22].

2.3 HuggingFace Transformers

The HuggingFace Transformers Library [2] is a collection of Python programs that implement various ML architectures in PyTorch, TensorFlow, etc. In this paper, PyTorch is being used as it supports maximum ViT architecture implementation in Hugging Face (HF) site [Ref: https://huggingface.co/docs/transformers/index].

3 Dataset and Models

3.1 Datasets

A single benchmark dataset is used for initial training of all pretrained models considered in this research. For testing and further fine-tuning, various benchmark datasets were used after comparing similarity of images with training dataset. List of all datasets used in this paper for experiment is given in Table 1.

Dataset for Training: EuroSAT(RGB version) [3]: For research, a RGB RS dataset which has medium-sized images and maximum training images per label was needed. Out of all the RS datasets available, 'EuroSAT' was the most suitable to work with due to many advantages. First it has 27,000 images with 10 labels, and each label has 2000 to 3000 training images. Second, the size of images is 64 × 64 pixels, which is suitable to be used to fine-tune a pretrained ViT architecture model with less computational cost and time. Third, these images have enough positional features for classifying if used for training using ViT.

Further, this research is focused on fine tuning pretrained models using proposed method and to produce a composite model that can be used as a generalized model to classify any large-sized EO image. Hence, other datasets will be used either as a benchmark or for further fine-tuning of the already fine-tuned model.

This dataset will be called as 'train dataset' or DTR in this paper unless otherwise stated. A RS image will simply be called as image in this paper unless otherwise

Table 1 Summary of datasets considered for this research

Dataset	Type	Resolution (m)	Image size	EuroSAT label	Dataset label	Images used
EuroSAT [5]	Train	10	64x64	All	All	27000
EuroSAT [5]	Test	10	64x64	AnnualCrop	AnnualCrop	500
CLRS [6]	Test	0.26–8.85	256x256	Pasture	Meadow	500
AID [7]	Test	0.5–8	600x600	Residential	Dense residential	410
RSI-CB256 [8]	Test	0.3–3	256x256	PermanentCrop	Green farmland	644
NWPU-RESISC45 [9]	Test	0.2–30	256x256	Herba. vegetation	Mountain	700
				Industrial	Industrial area	700
				River	River	600
RSD46-WHU [10]	Test	0.5–2	256x256	Forest	Nat. dense forest	450
				Highway	Overpass	375
				SeaLake	Water	402

stated. EuroSAT dataset was downloaded directly from source recommended by original publisher for training to ensure that dataset images are not modified in any way by image preprocessors, etc. Link for the same can be found here: [https://github.com/phelber/EuroSAT, https://doi.org/10.5281/zenodo.7711810].

Datasets for Testing

For testing fine-tuned models and composite models, appropriate images from different benchmark datasets were collected. Images for specific classes are selected from multiple benchmark datasets to form a single test dataset for accessing generalization capability. It is ensured that the images from test datasets are similar to EuroSAT images for the given class. Hence, the test dataset contains multiple label-specific class, each selected from only one benchmark dataset, and others will be rejected. These label-specific datasets for each label are combined together to form a single new labeled dataset and will then be used for testing as shown in Table 1. This new dataset will be called as 'test dataset' or DTE in this paper unless otherwise stated.

All the images of EuroSAT have fixed resolution of 10mts per pixel, which is the upper resolution range for most of 'test datasets' considered in this research. Most of images of benchmark datasets considered for testing have resolution from 0.2 to 8 mts. This increases the difficulty for fine-tuned models to predict the correct label and requires further fine-tuning the models whose recall for a given label specific dataset is best.

It is possible to combine same class images from various benchmark datasets and reduce the image size to 64 $\times$ 64 to build a very large dataset like EuroSAT. The new dataset can then be used as a test dataset for initial training of pretrained models. But it has been avoided to combine images from different datasets due to lack of third-party benchmarking.

In RS image scene classification, the most challenging problem was the presence of two or more class in the same image. For example, a large pasture grass field image may be having crop, narrow road or few trees at a corner. Since a ViT gives equal importance to every position of the whole image, it may predict an image label incorrectly. Most of the images from benchmark datasets were having single class features in the whole image. A few images from 'pasture' and 'river' from original source which had multiple classes in major portion of the image were removed.

3.2 *Model Architectures and Pre-trained Models*

Each ViT architecture can produce many pretrained SOTA models by modifying training hyper-parameters to achieve higher accuracy. But in this research, aim is to evaluate performance among various ViT architectures and not variants within each architecture. Further, the research is focused on selecting and fine-tuning models

Table 2 Summary of the models considered for this research

Model	Publisher	Hosted by	Pretrained model file
BeiT [11]	Microsoft	Huggingface	microsoft/beit-base-patch16-224
CvT [12]	Microsoft	Huggingface	microsoft/cvt-13
DeiT [13]	Facebook	Huggingface	facebook/deit-base-distilled-patch16-224
FocalNet [14]	Microsoft	Huggingface	microsoft/focalnet-tiny
LeViT [15]	Facebook	Huggingface	facebook/levit-128S
PoolFormer [16]	Sea AI Lab	Huggingface	sail/poolformer_s12
SegFormer [17]	Nvidia	Huggingface	nvidia/mit-b0
SwiftFormer [18]	Amshaker	Huggingface	MBZUAI/swiftformer-xs
Swin [19]	Microsoft	Huggingface	microsoft/ swin-tiny-patch4-window7-224-in22K(Base)
SwinV2 [20]	Microsoft	Huggingface	microsoft/swinv2-tiny-patch4-window8-256
VAN [21]	VAN	Huggingface	Visual-Attention-Network/van-base
ViT [4]	Google	Huggingface	google/vit-base-patch16-224
ViTMSN [22]	Facebook	Huggingface	facebook/vit-msn-small

using a proposed method to produce composite model that can be used as a generalized processor. Hence, either tiny or base versions of each ViT architecture are considered for comparison. Source of all pretrained models is 'HuggingFace' hosted pretrained models (PTM) given in Table 2.

Feature maps obtained by images from a vertically downward pointing satellite camera when fed to an image preprocessor are more rotation and flip tolerant if they are more symmetrically cropped for training. Many features in RS scene classification images are evenly distributed from center to the edge and have geometrically symmetric features. Thus, RS images take less time to train/retrain and has lower training/validation/testing loss.

4 Methodology

Main purpose for this research is to increase image scene classification accuracy through proposed method which is a generalized algorithm to build a composite model. For this experiment, latest 'ViT architectures', best 'pretrained models' (PTM), and RS-specific datasets are selected as resources. The composite model generated is thus the best model for the existing resources at the time of experiment.

A composite model of classifiers is to be built, where each model/classifier is selected in such a way that the final fine-tuned models (FTM) inside the final composite model is the best classifier for predicting the corresponding class. This composite model will then be deployed as a generalized model for predicting a new image. For producing such classifiers with high accuracy, it has to be ensured that

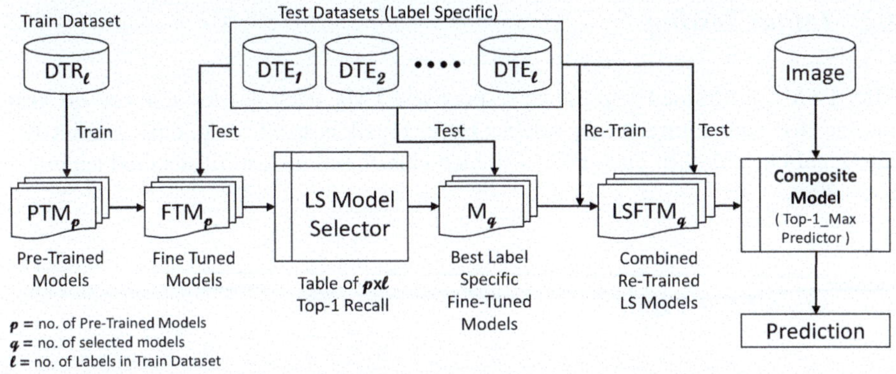

Fig. 1 Flowchart of proposed method (in this research $p = 13$, $q = 4$ and $l = 10$)

images of each class should have distinct features that separates them from each other by a huge boundary. For example, 'permanent crop' and 'industrial' images have more unique features than common ones. But a 'dense residential' and 'medium residential' images are having more common features than unique ones. It will be suitable that, such image classes with more common features are combined together as a single class and PTM are trained with both images. In this research, no such merging of dataset images was done. All training and testing were done on 'Google Colab' platform that supports multiple runs at the same time on GPUs specifically for ML. An online cloud-based environment for ML is affordable, interactive, and easy to learn for researchers. It speeds up training and saves all files and training parameters for reviewing later.

First the best-performing FTM available for the specific label is selected and then finally combined these label-specific (LS) models into a composite model for predicting new RS images. The proposed method is shown in Fig. 1.

4.1 Training

The method starts with PTMs that are trained on some large image datasets and hosted by reliable source online [2, Table 2]. Transfer learning is used for training rather than building each classifier/model from scratch.

To build FTMs out of each PTM, EuroSAT dataset is first shuffled, and then, training and validation splits of 90% and 10% fractions, respectively, generated. Here, maximum images in the training set are considered since ViT models require more number of images for producing models having better accuracy [4]. Validation set can be small since we are using state-of-art pretrained models that are already extensively trained on large image datasets. Test split from EuroSAT is not generated since separate test datasets will be used for determining classification recall/accuracy.

4.2 Initial Testing

After PTMs are trained to produce respective FTMs, each has to be tested on 'test dataset' for calculating recall values. Here, recall is used to quickly predict the classification ability of each FTM for each class/label instead of detailed accuracy evaluation.

4.3 Model Selection

The 'LS model selector' uses a two-pass strategy to select the best pretrained model for fine-tuning in order to identify images belonging to each class/label. These are explained below:

First Pass (10% Rule) Since images belonging to 10 labels are considered in this research, minimum qualifying criteria are set for all models based on the value of the recall to be 10% (*100/no. of labels*) on test dataset. So, all the models that have top-1 recall less than 10% for any class/label are dropped from further selection. Results for 'first pass' on 'test dataset' are given in Table 3. Each row provides recall values of different models for the specific class/label. Values shown in bold are predictions less than 10% i.e. the minimum qualifying criteria of a model hence, disqualified.

Second Pass (Label Best Top-1 Rule)

Now each label has to be assigned a model based on its performance on 'test dataset'. A model has to outperform other models in identifying new test images belonging to a given class/label. If there is a 'tie' where two or more models have the same recall for a label, the model which has a greater number of labels with 'highest recall' in selection table is preferred. For example, in Table 4, there is a tie between BEiT and ViT models for AnnualCrop, and since BEiT is the top scorer for 3 class which is higher than ViT model with only 2 top scores, BEiT is selected for AnnualCrop. When there is a tie between two or more models and there are the top scorers for equal number of labels, then model with overall better recall for all labels is selected. Results for 'second pass' on 'test dataset' are given in Table 4. Each row provides recall values of different models for the specific class/label. Values shown in bold is the highest recall for the given row, corresponding model(s) is shortlisted for further selection.

Finally, the 'LS model selector' assigns a single model for each class/label. These models are called M_q, which are the best label-specific models for the respective class. These models were combined into a composite model for further testing on 'test dataset'.

Table 3 Recall values of models for 'first pass' by 'LS model selector'

Label specific testing (top-1 recall)	BEiT	CvT	DeiT	FocalNet	LeViT	PoolFormer	SegFormer	SwiftFormer	Swinbase	SwinV2	VAN	ViT	ViTMSN
AnnualCrop	99.40	95.00	98.60	98.00	85.00	95.00	97.00	95.20	98.60	97.40	97.00	99.40	98.80
Forest	97.33	64.44	84.00	89.33	**0.88**	65.55	85.55	12.88	92.22	95.11	**0.20**	74.44	89.11
HerVeg	95.00	50.42	97.00	84.00	**0.00**	88.42	86.57	**7.71**	96.14	95.28	47.57	95.85	96.85
Highway	97.33	27.20	91.46	94.66	**0.80**	73.33	60.26	**0.00**	96.00	76.80	**1.60**	84.00	82.93
Industrial	93.28	98.28	98.00	93.57	99.71	86.57	88.42	**7.85**	96.57	97.71	40.57	98.00	89.85
Pasture	22.66	**2.16**	42.83	33.16	**0.00**	**6.33**	10.50	**7.00**	41.50	20.83	**3.00**	35.00	35.50
PermCrop	70.34	54.65	63.35	68.32	**0.00**	57.14	60.09	37.11	71.27	56.05	76.24	66.77	79.19
Residential	59.75	**2.19**	20.73	16.82	**0.00**	11.95	**0.00**	**8.04**	28.04	42.68	**5.36**	30.48	65.36
River	57.71	17.28	67.14	51.14	**0.14**	72.85	35.14	**2.14**	87.42	56.85	22.71	78.57	40.28
SeaLake	76.60	62.80	86.45	78.32	68.22	83.99	66.00	93.59	67.98	79.55	69.21	72.41	82.75
Overall	76.94	47.44	74.95	70.73	25.47	64.11	58.95	27.15	77.57	71.82	36.34	73.49	76.06

Table 4 Recall values of models for 'second pass' by 'LS model selector'

Label specific testing (top-1 recall)	BEiT	DeiT	FocalNet	SwinBase	SwinV2	ViT	ViTMSN	Final selected model
AnnualCrop	**99.40**	98.60	98.00	98.60	97.40	**99.40**	98.80	BEiT
Forest	**97.33**	84.00	89.33	92.22	95.11	74.44	89.11	BEiT
HerbaceousVegetation	95.00	**97.00**	84.00	96.14	95.28	95.85	96.85	DeiT
Highway	**97.33**	91.46	94.66	96.00	76.80	84.00	82.93	BEiT
Industrial	93.28	**98.00**	93.57	96.57	97.71	**98.00**	89.85	DeiT
Pasture	22.66	**42.83**	33.16	41.50	20.83	35.00	35.50	DeiT
PermanentCrop	70.34	63.35	68.32	71.27	56.05	66.77	**79.19**	ViTMSN
Residential	59.75	20.73	16.82	28.04	42.68	30.48	**65.36**	ViTMSN
River	57.71	67.14	51.14	**87.42**	56.85	78.57	40.28	SwinBase
SeaLake	76.60	**86.45**	78.32	67.98	79.55	72.41	82.75	DeiT

4.4 *Model Retraining*

All label-specific selected models have to be retrained for further fine-tuning on 'test datasets'. The retraining has to be limited until each M_q is more specific in predicting the label without oversaturation. Training with too many images and/or epochs will make the fine-tuned model to loose features of other class labels acquired during previous training. Hence, a single epoch with 60% increase in learning rate for 'AdamW' optimizer was found out to be most suitable during experiment. The aim for retraining was to increase the accuracy of the model/classifier for specific labels, but still able to predict other labels with similar accuracy as previous training. Hence, four models (BEiT, DeiT, ViTMSN, and SwinBase) were selected by 'LS model selector' and retrained to get label-specific fine-tuned models (LSFTM). Retraining of models on 'test datasets' not only took very less time than initial training of PTM on 'train dataset' but also produced very accurate classifiers that is discussed in results section in this paper.

4.5 *Composite Model of 'Final Classifiers'*

After model selection and retraining, it is evident that images from 'test datasets' can be identified into appropriate classes by the corresponding models with good recall. All label-specific selected and retrained models/classifiers called $LSFTM_q$ are combined to build a composite model. Each classifier in the composite model after selecting the top-1 recall for an image has equal voting weight in binary prediction. The class/label which gets the maximum votes is the prediction for the image for the composite model. It will result in increased accuracy of the model in predicting the label of a new image. The final composite model will again be tested on 'test dataset' for accuracy.

5 Results

In this research, the aim was to build a generalized model to classify any image (of any class) from around the globe. As per the method flowchart given in Fig. 1, testing was performed with 'test datasets' at three stages. First testing was carried out on each FTM after initial training of PTMs with 'train dataset'. The results of first testing are given in Table 3. Second testing was carried out before retraining on composite model made of M_l (selected models) on 'test dataset', and results are given in Fig. 2a. Third testing was carried out after retraining on final composite model made of $LSFTM_q$ (retrained selected models) on 'test dataset', and results are given in Fig. 2b. The results of testing before and after retraining in Fig. 2 have been shown as confusion matrix for detailed analysis. 'Confusion Matrix' was preferred over other evaluation methods due to need for visualizing the noise distribution across all the labels and compare it with the previous training.

If both the confusion matrix is studied, it can be observed that there was significant accuracy improvement for five labels, namely pasture, PermanentCrop, residential, river and SeaLake after retraining. In particular, FTM incorrectly classified mainly three labels, i.e., (residential as industrial, 257 nos.), (river as herbivorous vegetation, 139 nos.), and (pasture as AnnualCrop, 113 nos.). It can be easily explained that these errors were contributed mainly due to the difference in 'train dataset' and 'test datasets' for respective class images. The 'EuroSAT' images pixel size resolution was fixed at 10 mts, whereas images from other datasets were varying in resolutions and zoom levels. For example, a residential image when zoomed would look like large industrial sheds in 'EuroSAT'. Narrow rivers in mountains would look like Herbivorous Vegetation, and pasture images in 'EuroSAT' were mostly looking like crop fields. But these differences can be neglected at initial stage since a dataset

Predicted Images (columns) / Actual Images (rows)

	AnulCrop	Forest	HerbVeg	Highway	Industry	Pasture	PermCrop	Residential	River	SeaLake
AnulCrop	491	0	0	0	0	1	8	0	0	0
Forest	0	425	15	2	0	0	0	8	0	0
HerbVeg	0	16	682	1	0	0	1	0	0	0
Highway	1	0	0	367	2	0	0	0	5	0
Industry	0	0	0	11	681	0	1	6	1	0
Pasture	113	79	97	19	0	177	6	0	2	7
PermCrop	63	0	0	91	19	13	455	0	3	0
Residential	0	0	0	18	257	0	5	130	0	0
River	0	3	139	18	1	10	24	6	399	0
SeaLake	12	66	7	0	0	2	0	0	1	314

(a)

Predicted Images (columns) / Actual Images (rows)

	AnulCrop	Forest	HerbVeg	Highway	Industry	Pasture	PermCrop	Residential	River	SeaLake
AnulCrop	500	0	0	0	0	0	0	0	0	0
Forest	0	444	0	0	0	2	0	0	0	4
HerbVeg	0	0	700	0	0	0	0	0	0	0
Highway	0	0	0	375	0	0	0	0	0	0
Industry	0	0	0	3	691	0	0	5	1	0
Pasture	0	0	2	1	0	490	7	0	0	0
PermCrop	0	0	0	0	1	1	641	0	1	0
Residential	0	0	0	1	0	0	0	409	0	0
River	0	0	3	0	0	0	0	0	597	0
SeaLake	3	1	0	0	0	4	0	0	1	393

(b)

Fig. 2 Composite model's performance on 'test dataset' (**a**) before retraining, (**b**) after retraining

with large number of images per class is needed for initial training. Later these inaccuracies can be removed by slight retraining on smaller dataset having images of higher resolution and more specific to the class/label.

Minimizing Overfitting Noise

Most of the pretrained models based on architectures published recently are very complex and have overfitting problems due to presence of many hyper-parameters. It is observed that if a single model having best overall recall ('SwinBase' with 77.57% from Table 3) is selected as a standard practice, noise is persistent even after retraining. When individual class performance is not considered for selection, the single model ('SwinBase') will continue to have noise for some classes (residential, pasture, and SeaLake in case of 'SwinBase') even after better training methods. There is a need to compensate this poor performance for some classes by including models that are better than the single model ('SwinBase') in predicting those classes. But it has also to be ensured that the models compensating for the poor performance of a single model have to qualify for all class/labels with good performance as well. It is also observed in Table 3 that some rejected models (CvT, LeViT, SwiftFormer, and VAN) are showing too high and low recall values for some class/labels. Hence, a ViT architecture's performance overall types of data in dataset relating to the scene classification task has to be studied first.

Composite model having overall recall of 78.03% (Fig. 2a) minimizes noise easily with one time retraining on test dataset and produces recall of 99.22% (Fig. 2b) with noise distributed evenly across the confusion matrix. The retraining of selected models was as simple as changing just one parameter in complex ViT-based architecture. This shows that a training/retraining a single model extensively over the same data using various strategies may not minimize noise. Experimental results show that a composite model is more suitable for predicting new images than existing methods for RS image scene classification tasks.

6 Conclusion

In this paper, a novel approach is used to build a composite model for RS image scene classification that can also be applied to other image classification tasks. It has been shown how training dataset having images confined to a specific region like 'Europe' but having high volume of images could be used for initial model training. Thereafter, 'low volume high spatial resolution' class/label-specific images from around the globe can be used to retrain the model for improving the prediction accuracy. So, from the proposed method, it can be concluded that instead of training models with very large datasets of high spatial resolution images, we can apply transfer learning to easily build a composite model of classifiers to achieve higher prediction rates. The proposed method can be applied to 'Multi-Class' and 'Multi-Spectral' RS image scene classification for further research. The proposed method can also be used at hardware level in UAV by assigning dedicated GPU/FPGA for each composite model's classifier and feeding the camera input image for parallel processing. Finally, this method is still under active development, and better strategies are being implemented to reduce training time as well as improving accuracy without overfitting during real-time implementation.

References

1. Dimitrovski I, Kitanovski I, Kocev D, Simidjievski N (2023) Current trends in deep learning for earth observation: an open-source benchmark arena for image classification. ISPRS J Photogramm Remote Sens 197:18–35
2. Wolf T et al (2020) Transformers: state-of-the-art natural language processing. In: Proceedings of the 2020 conference on empirical methods in natural language processing: system demonstrations. Association for Computational Linguistics
3. Vaswani A, Shazeer N, Parmar N, Uszkoreit J, Jones L, Gomez A, Kaiser, Polosukhin I (2017) Attention is all you need. In: Advances in neural information processing systems, pp 5998–6008
4. Dosovitskiy A, Beyer L, Kolesnikov A, Weissenborn D, Zhai X, Unterthiner T, Dehghani M, Minderer M, Heigold G, Gelly S, Uszkoreit J, Houlsby N (2021) An image is worth 16×16 words: transformers for image recognition at scale. ICLR
5. Helber P, Bischke B, Dengel A, Borth D (2018) Introducing EuroSAT: a novel dataset and deep learning benchmark for land use and land cover classification. In: IGARSS 2018—2018 IEEE international geoscience and remote sensing symposium, pp 204–207
6. Li H, Jiang H, Gu X, Peng J, Li W, Hong L, Tao C (2020) CLRS: continual learning benchmark for remote sensing image scene classification. Sensors 20(4)
7. Xia G-S, Hu J, Hu F, Shi B, Bai X, Zhong Y, Zhang L, Lu X (2017) AID: a benchmark data set for performance evaluation of aerial scene classification. IEEE Trans Geosci Remote Sens 55(7):3965–3981
8. Li H, Dou X, Tao C, Hou Z, Chen J, Peng J, Deng M, Zhao L (2020) RSI-CB: a large scale remote sensing image classification benchmark via crowdsource data
9. Cheng G, Han J, Lu X (2017) Remote sensing image scene classification: benchmark and state of the art. Proc IEEE 105(10):1865–1883
10. Xiao Z, Long Y, Li D, Wei C, Tang G, Liu J (2017) High-resolution remote sensing image retrieval based on CNNs from a dimensional perspective. Remote Sens 9(7):725
11. Bao H, Dong L, Piao S, Wei F (2022) BEiT: BERT pre-training of image transformers
12. Wu H, Xiao B, Codella N, Liu M, Dai X, Yuan L, Zhang L (2021) CvT: introducing convolutions to vision transformers
13. Touvron H, Cord M, Douze M, Massa F, Sablayrolles A, Jégou H (2021) Training data-efficient image transformers & distillation through attention
14. Yang J, Li C, Dai X, Yuan L, Gao J (2022) Focal modulation networks
15. Graham B, El-Nouby A, Touvron H, Stock P, Joulin A, Jégou H, Douze M (2021) LeViT: a vision transformer in ConvNet's clothing for faster inference
16. Yu W, Luo M, Zhou P, Si C, Zhou Y, Wang X, Feng J, Yan S (2022) Meta former is actually what you need for vision
17. Xie E, Wang W, Yu Z, Anandkumar A, Alvarez JM, Luo P (2021) SegFormer: simple and efficient design for semantic segmentation with transformers

18. Shaker A, Maaz M, Rasheed H, Khan S, Yang M-H, Khan FS (2023) SwiftFormer: efficient additive attention for transformer-based real-time mobile vision applications
19. Liu Z, Lin Y, Cao Y, Hu H, Wei Y, Zhang Z, Lin S, Guo B (2021) Swin transformer: hierarchical vision transformer using shifted windows
20. Liu Z, Hu H, Lin Y, Yao Z, Xie Z, Wei Y, Ning J, Cao Y, Zhang Z, Dong L, Wei F, Guo B (2022) Swin transformer V2: scaling up capacity and resolution
21. Guo M-H, Lu C-Z, Liu Z-N, Cheng M-M, Hu S-M (2022) Visual attention network
22. Assran M, Caron M, Misra I, Bojanowski P, Bordes F, Vincent P, Joulin A, Rabbat M, Ballas N (2022) Masked Siamese networks for label-efficient learning

Examining an Effective Nearest Prime Pixels-Based Image Encryption Cryptography

G. Kalpana, Trishala Reddy, and Vaishnavi Reddy Chintala

Abstract Internet communication is absolutely necessary in the modern world. Data can be sent internationally using the Internet. The Internet was created as a result of scholarly attempts to disseminate knowledge; great security was never its goal. When sending or receiving data via a channel, the information being sent should not be taken by unknown individuals like hackers. In order to obfuscate secret information based on images that can be decoded by the human visual system, a unique type of encryption is used called image cryptography. Image encryption is a suitable process to protect image data. This paper presents an efficient nearest prime pixels-based image cryptography. The two types—image encrypted with key(s) and encryption of images without key(s)—are the major topics of the study. Additionally, it lists a number of qualities of an effective image encryption method. The ultimate decrypted image is created by modeling the pixels as a matrix and retrieving the data column by column.

Keywords Image cryptography · Nearest prime pixels · 2's complement · Secret sharing

1 Introduction

Technology has advanced quickly over the past twenty years, and the transmission and storage of vast amounts of multimedia information over the Internet has experienced a boom. The importance of digital images in multimedia technology makes

G. Kalpana (✉)
Department of CSE (AI&ML), Malla Reddy Engineering College for Women, Hyderabad, India
e-mail: dr.gkalpanamrecw@gmail.com

T. Reddy
University of Texas, Dallas, USA

V. R. Chintala
Department of CSE, Malla Reddy Engineering College for Women, Hyderabad, India

B. Annappa et al. (eds.), *High Performance Computing, Smart Devices and Networks*, Lecture Notes in Electrical Engineering 1262,
https://doi.org/10.1007/978-981-97-7794-5_9

it even more crucial for consumers to protect their privacy. On the Internet, private information is kept and sent, including exam papers from universities, business plans, nuclear calculations, and military and commercial secrets. Security concerns should be addressed while handling these secrets since an intrusive party could use a transmission channel flaw to steal this information [1]. Information confidentiality is a service to protect the information from unauthorized disclosure and is offered by encryption. In the context of text data, encryption is the process of converting the plaintext into ciphertext, and decryption is the process of restoring the plaintext from the ciphertext. The encryption process can be either symmetric or asymmetric key system [2].

Digital images are also less sensitive than information in comparison because an individual pixel change does not affect the complete image. In other words, although a digital image is more vulnerable to attackers than data, a tiny modification is acceptable. AES, DES, RSA, IDEA, and other cryptographic algorithms have all been proposed thus far [3]. The methods used to encrypt images are distinct from those used to encrypt data. Maintaining the confidentiality and reliability of the image is required due to the numerous security issues that arise during the processing of digital images and transmissions.

Different technologies, such as picture cryptography and image steganography, were created to manage the privacy of secret images. Images, texts, and other visual data are encrypted using the cryptographic technique known as image cryptography such that the human visual system can decrypt the data without the use of computers [4].

In order to limit the potential message sender(s) and/or receiver(s), cryptography is used. The foundation of cryptography are codes known as keys. The message can be encoded by the sender so that a proper key is required to decode it. While transmitting, an intruder may clearly see that something is encrypted, which causes a doubt, and an intruder attempts to decrypt the concealed secret. Image encryption masks the image so that it does not expose the concealed secret easily. Therefore, image steganography employs an innocent-looking cover image to conceal the transmission of the secret image. But in both situations, the whole secret is accessible in a single location. Therefore, the full hidden will be revealed if the intrusive party is successful in performing the secret reveal.

An image is partitioned into n shares in an image cryptography scheme such that a minimum of k shares is required to reconstruct the image. The random number generator divides the sum. Numerous cryptographic primitives, or fundamental building blocks, exist. Examples include block ciphers, stream ciphers, and hash functions. For thousands of years, military and diplomatic communications have been protected with cryptography. The National Institute of Standards' data encryption standard is the most widely used symmetrical encryption algorithm. Some of the hidden transformations in the algorithm include black-box transformations.

In a symmetric key system, both encryption and decryption are performed using the same key, while in an asymmetric key system, different keys like public and private keys are used for encryption and decryption of data. The image encryption methods proposed in this research use the symmetric key system. Diffusion and confusion are the desirable properties of encryption methods. Diffusion is achieved by repeatedly performing several permutations, and confusion is achieved by complex substitution methods.

Given that we are aware that the key is what determines how well the picture encryption method performs, the key should be sufficiently large to be difficult to guess. It should also be responsive to slight modifications. Even if there is only a one-bit change between two keys, the encryption method should produce a completely different cipher image. The cipher image may be impacted while transmission is occurring over a noisy channel. The encryption strategy should therefore be resistant to noise attacks. Recovering the original image should be possible for the receiver. The speed of encryption techniques is crucial in real-time applications. Being quick with the encryption process is always preferred.

2 Literature Survey

Rajaram et al. [5] suggested technique for safe bank check transmission over the network, and the embedded visual cryptography scheme (EVCS) utilizes gray threshold with morphology operation. In this method, the input image is first divided into a number of secret share images using a classic visual cryptography technique, followed by the application of a gray threshold with morphological operation. Include a fresh cover image for every share. At last, stack the embedded pictures to retrieve the original image data. Peak signal-to-noise ratio (PSNR) and mean square error (MSE) value are used to calculate the performance of the suggested approach. The suggested EVCS exhibits great performance in terms of PSNR and MSE for safe transmission of bank checks over networks.

Malakooti and Dobuneh [6] have put forth a method for pictures based on a cutting-edge lossless digital multimedia encryption system that uses orthogonal transformations to encrypt image data. This approach is based on symmetric key block cipher encryption. The creation of a revolutionary lossless digital encryption scheme for multimedia is emphasized by the authors. During the decryption process, they calculated the inverse of the orthogonal matrices using the symmetric features of the orthogonal transforms. They employed a number of traditional picture encryption techniques, including the Malakooti transform (MT), Hadamard transform (HT), and discrete cosine transform (OCT).

Yang et al. [7] introduced the idea of probabilistic VC and covered the creation of encrypting matrices for threshold structures of $(2, n)$, (k, n), and (n, n). The probabilistic VC, which is named in relation to the deterministic VC, allows the secret pixel to be deciphered probabilistically using statistical principles.

Wang et al. [8] offered a creative method for processing color and grayscale photos without pixel enlargement. The image can be produced by k or more shares, but details cannot be obtained by betting fewer shares than k. The secret cannot be generated by good shares or even rejected ones.

Hou and Quan et al. [9] proposed that the hidden image may be recovered with k shares or more, but not with fewer. Similar to this, progressive visual cryptography (VC) creates the hidden image piece by piece, while superimposing more shares produces the actual results. If fewer shares are staked, only the hidden image's outline can be obtained. The specifics of the information generated gradually expand along with the amount of shares.

Shyu et al. [10] developed a method of secret sharing in which shares are generated as rectangles. By superimposing the rectangular shares, the hidden image can be recreated and visibly displayed. In today's research, multiplies secret sharing has drawn increasing attention. More secrets can be incorporated into the shares by rotating them at various angles. Circular shares, however, require different encryption and decryption techniques than do square or rectangular shares.

3 Nearest Prime Pixels-Based Image Cryptography

The suggested approach creates an 8-bit random sequence that is used to create keys. By subtracting digital values from their closest prime pixel values, the key is obtained. Encryption and decryption take place over a number of steps. Every image's pixel has been transformed into pixels. Additionally, locate the closest prime. Determine the difference between each pixel in the image and the closest prime pixel in order to generate keys. Take into account the variation in value as the key; it is repeated for every single pixel in the image.

Sub-pixel alignment has a lot of problems, including pixel expansion. Reduce the quantity of sub-pixels used to replace the original image's pixels in order to solve this issue and lessen pixel expansion. In visual cryptography, the rgs share generating method was primarily created to lessen pixel expansion. Both the quantity and size of the sub-pixels have an impact on the dimension of the transparencies. The ratio of the sub-pixel size and the quantity of sub-pixels in the transparency determines the size of the transparency. A good VCS will possess shares that can be readily aligned and smaller transparency. For the original image to be recovered, the transparencies do not need to be perfectly aligned.

3.1 Encryption Algorithm

Round1: Between the key and the nearest prime pixel, do an XOR operation.

Round2: Round3: Save the result in a matrix (column wise), the opposite of round 1's outcome.

Round4: Retrieve pixels from rows of each n*4 matrix. Find equivalent pixel values and considered to be as final decrypted image.

3.2 Decryption Algorithm

Round 1: Find the binary equivalents of the decrypted image data, then reorder it into a matrix.

Round 2: Reverse the binary pixels that were retrieved row by row.

Round 3: To get the closest prime pixel values, reversed pixels are XORed with the key.

Round 4: Subtracting the key value from the closest prime pixel value will reveal the original image.

Every single pixel of the initial picture is represented by two pixels in each pixel share in the VCS scheme. Each pixel in the initial picture is read, and if a white pixel is found, one of the primary two rows is randomly selected, and each share is given a 2-pixel block as indicated in the third and fourth columns. In the event of a black pixel, one of the final two rows is picked at random with equal chance, and a sub-pixel is allocated to each share. Superimposition of shares operates during decryption in a manner akin to the Boolean OR function. The resultant pixel will be white if two white pixels with corresponding shares overlap. Black pixels will occur from the overlap of black pixels in one share with either white or black pixels in the other share. When the sub-pixels of both share in the third and fourth columns are superimposed, the final column displays the resulting sub-pixel. The use of a 22 sub-pixel block for representing each original image pixel was suggested as a solution to the problem of distorting of the width-to-height ratio of the initial picture. This produces a picture that is nearly four times the size of the original picture while maintaining the original image's aspect ratio, giving the final product a clearer and more natural-looking appearance.

4 Result and Analysis

The color image dataset for experiment is shown in Fig. 1. Their respective cipher images are represented in Fig. 3. The output or de-ciphered images are shown in Figs. 2 and 4.

Mean square error (MSE) and peak signal-to-noise ratio (PSNR) are measured for performance evaluation. The most popular error measures utilized in image processing are MSE and PSNR. The peak error is measured by PSNR, whereas the MSE is the total squared variance that exists between the restored hidden image and the initial secret image (Fig. 5).

$$\mathrm{PSNR} = 10.\log_{10}\left(\frac{B^2}{\mathrm{MSE}}\right). \tag{1}$$

where B is the maximum fluctuation in the input image.

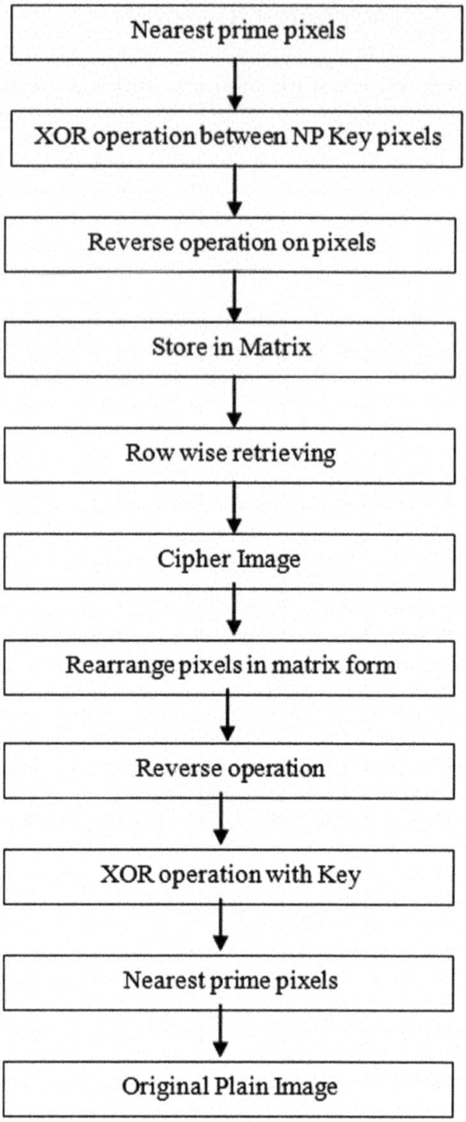

Fig. 1 Flow process of cryptography

$$\mathrm{MSE}(i_1, i_2) = \frac{1}{NM}\sum_{x=1}^{M}\sum_{y=1}^{M}\left[i_1(x, y) - i_2(x, y)\right]^2. \tag{2}$$

Therefore, nearest prime pixels-based image cryptography is achieved efficiently. Nearest prime pixels technique is used for the key generation.

Fig. 2 Color image dataset

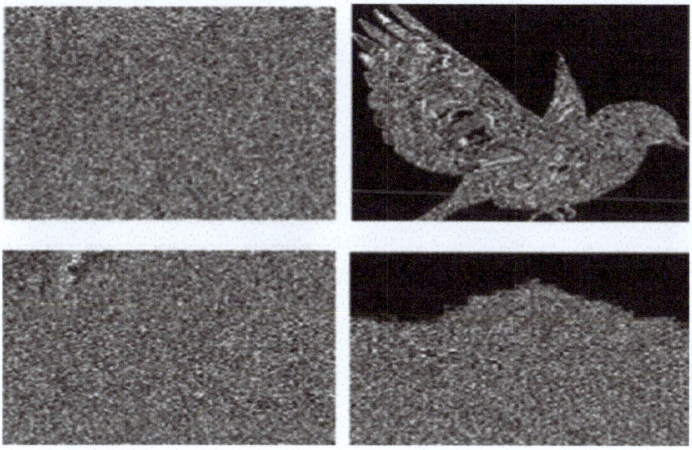

Fig. 3 Ciphered image

Fig. 4 De-ciphered image

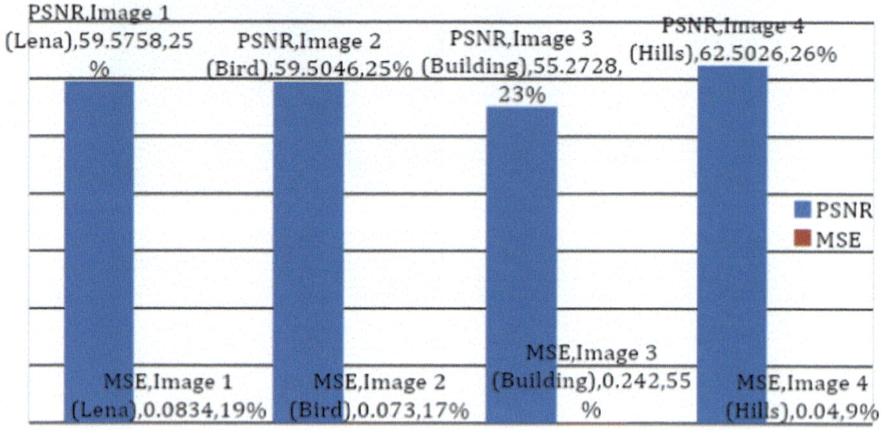

Fig. 5 Performance analysis

5 Conclusion

In this paper, an efficient nearest prime pixels-based image cryptography is described. Current trends heavily rely on image cryptography. This study concentrated on creating unique keys using the closest prime pixels. Mean square error (MSE) and peak signal-to-noise ratio (PSNR) are measured for performance evaluation. It was found that picture encryption methods demand great levels of confusion, a complete lack of association with the images that are entered, low levels of computing complexity, and high levels of resistance to cryptanalysis.

References

1. Kumar S, Sharma R, Rahman MA, Jha RK, Neha S, Verma AR, Singh Y (2019) Robust secret sharing of image encryption in fractional Fourier transform domain. In: 2019 international conference on computing, power and communication technologies (GUCON)
2. Singh VK, Singh PK, Rai KN (2018) Image encryption algorithm based on circular shift in pixel bit value by group modulo operation for medical images. In: 2018 4th international conference on computing communication and automation (ICCCA)
3. Singh SP, Bhatnagar G, Gurjar DK (2018) A secure Image encryption algorithm based on polar decomposition. In: 2018 IEEE 14th international colloquium on signal processing & its applications (CSPA)
4. Nipanikar SI, Hima Deepthi V (2017) Entropy based cost function for wavelet based medical image steganography. In: 2017 international conference on intelligent sustainable systems (ICISS)
5. Rajaram S, Suganya R (2017) Embedded visual cryptography for secure transmission of bank cheque. In: 2017 IEEE international conference on circuits and systems (ICCS)
6. Malakooti MV, Dobuneh MR (2012) A lossless digital encryption system for multimedia using orthogonal transforms. IEEE, Dobuneh Islamic Azad University, Dubai, UAE. 978-1-4673-0734-5/12/2012
7. Yang CN (2004) New visual secret sharing schemes using probabilistic method. Pattern Recogn Lett 25:481–494. https://doi.org/10.1016/j.patrec.2003.12.011
8. Wang D, Yi F, Li X (2011) Probabilistic visual secret sharing schemes for grey scale images and color images. Inf Sci 181
9. Hou Y-C, Quan Z-Y (2011) Progressive visual cryptography. J Electron Imag
10. Shyu SJ, Chen K (2011) Visual multiple secret sharing based upon turning and flipping. Inf Sci

Integrating RFID Technology with Student Information Systems for Enhanced Management of Attendance and Financial Records

D. R. Kumar Raja, ZuraidaAbal Abas, V. Eswari, G. Hemanth Kumar, and V. Kalpana

Abstract Radio Frequency Identification (RFID)-based student ID system for monitoring attendance records and fee dues. The proposed system utilizes RFID technology to track the movement of students as they enter and exit various locations on campus. The system also allows for the automated tracking of fee payments, providing real-time updates on student account balances. The paper describes the design and implementation of the system, including hardware and software components, as well as the integration with existing student information systems. Results from a pilot study are presented, demonstrating the effectiveness and efficiency of the system in accurately recording attendance and fee payment data. The proposed RFID-based student ID system has the potential to significantly improve the management of student attendance and financial records, leading to increased accountability and streamlined administrative processes.

Keywords RFID reader · RFID tags · ESP 32 · GSM module · Local server

1 Introduction

The traditional methods of monitoring student attendance and fee dues in educational institutions have become outdated and ineffective. The manual record-keeping systems are prone to errors, tampering, and fraudulent activities, resulting in inaccurate data and financial losses. To address these issues, radio frequency identification

D. R. Kumar Raja (✉) · Z. Abas
FTMK, UniversitiTeknikal Malaysia (UTeM), Melaka, Malaysia
e-mail: kumarrajadr@gmail.com

V. Eswari · G. Hemanth Kumar
School of ECE, REVA University, Bengaluru, KA, India

V. Kalpana
Mohan Babu University, Tirupathi, AP, India

B. Annappa et al. (eds.), *High Performance Computing, Smart Devices and Networks*,
Lecture Notes in Electrical Engineering 1262,
https://doi.org/10.1007/978-981-97-7794-5_10

(RFID) technology has been increasingly implemented in various sectors, including education. RFID-based student ID systems provide a convenient and efficient way of monitoring attendance records and fee dues, ensuring accuracy and transparency.

RFID technology is a wireless communication technology that uses radio waves to transmit data between a reader and a tag attached to an object. RFID tags are small, portable devices that can store and transmit data wirelessly. The tags can be embedded in various objects, such as student ID cards, allowing for easy tracking and monitoring.

RFID technology has several advantages over traditional methods, including faster data collection, improved accuracy, and real-time monitoring. The implementation of RFID-based student ID systems in educational institutions has several benefits. One of the primary benefits is the accurate and real-time monitoring of student attendance records. The traditional manual attendance record-keeping systems are prone to errors and often require significant amounts of time and resources to maintain. With RFID technology, attendance records can be automatically recorded when a student enters or leaves a classroom, allowing for quick and accurate tracking of attendance records. Another benefit of RFID-based student ID systems is the ability to monitor fee dues. Educational institutions often struggle to track fee payments and outstanding balances, leading to financial losses and operational inefficiencies. RFID-based student ID systems can help to monitor fee payments and track outstanding balances in real time, allowing educational institutions to efficiently manage their finances.

RFID-based student ID systems can also improve security in educational institutions. Traditional student ID cards can be easily duplicated, leading to security breaches and fraudulent activities. RFID tags are more secure and difficult to duplicate, providing an additional layer of security to educational institutions. In addition, RFID technology can be used to monitor student movements within the educational institution, helping to prevent unauthorized access and ensuring student safety. The implementation of RFID-based student ID systems can also improve communication between educational institutions and parents/guardians. With RFID technology, parents/guardians can receive real-time updates on their child's attendance and fee dues, improving communication and accountability. Additionally, RFID-based student ID systems can help to improve the overall efficiency of educational institutions, reducing administrative tasks and allowing staff to focus on other important tasks.

RFID-based student ID systems offer a convenient and efficient way of monitoring attendance records and fee dues in educational institutions. The technology provides several advantages over traditional methods, including faster data collection, improved accuracy, and real-time monitoring. Educational institutions can benefit from RFID-based student ID systems by improving their operational efficiency, enhancing security, and improving communication with parents/guardians.

2 Literature Survey

Radio frequency identification (RFID) technology is a popular method for identifying and tracking objects. In recent years, RFID-based systems have been used for attendance monitoring and fee dues in various educational institutions. The aim of this literature review is to investigate the different types of RFID-based student ID systems used for monitoring attendance records and fee dues.

The use of RFID technology for student attendance monitoring has been extensively researched in the last decade. Studies have reported that RFID-based attendance systems are more accurate and efficient than traditional attendance methods such as paper-based registers [1, 2]. Such systems have also been found to reduce administrative workload and enhance student engagement [3, 4]. However, the implementation of RFID-based student ID systems has also raised concerns about privacy and security. Researchers have highlighted the importance of ensuring that the data collected by RFID systems is secure and that the system is not used to infringe on students' privacy [5, 6]. In addition to attendance monitoring, RFID-based student ID systems have also been used for fee dues management. Studies have reported that such systems can reduce the time and effort required for fee collection and enhance financial transparency [7, 8]. Despite the advantages of RFID-based student ID systems, there are still research gaps in the area. Specifically, there is a need to investigate the factors that influence the adoption and effectiveness of RFID-based student ID systems. Further research is also needed to examine the ethical and legal implications of using such systems in educational institutions. While there have been several studies on the use of RFID-based student ID systems for monitoring attendance records and fee dues, there are still research gaps that need to be addressed. There is a need for research on the factors that influence the adoption and effectiveness of RFID-based systems in educational institutions.

The existing literature has identified several factors that may impact the adoption of RFID-based student ID systems, including cost, ease of use, and perceived usefulness [9, 10]. The use of RFID technology in student identification has been explored in several studies. RFID-based student ID systems offer several advantages, including increased efficiency, reduced costs, and improved accuracy in attendance record-keeping. An RFID-based attendance system was implemented in a university in Saudi Arabia. The system was found to be accurate, reliable, and easy to use [11–17]. An RFID-based student attendance system for a Malaysian primary school. The system used RFID tags and readers to identify students as they entered or exited the school premises. The data collected was transmitted wirelessly to a central database for processing and analysis. The system was found to be effective in reducing absenteeism and improving student attendance [12–23].

However, there is a need for more research to explore these factors in greater depth and to investigate other factors that may be relevant.

3 Methodology

Consider a university with a large student population of over 10,000 students. The institution is facing challenges with manual attendance monitoring, fee collection, and exam hall access control. These issues result in a lot of time and effort spent on administrative tasks and create inconveniences for students and staff alike.

To address these challenges, the university decides to implement an RFID-based student ID system. Each student is issued an RFID-enabled ID card that contains their personal information, photograph, and a unique identification number. The card is also linked to the student's academic records, fee payment status, and attendance records (Fig. 1).

Attendance and fee dues improve communication and accountability. Additionally, RFID-based student ID systems can help to improve the overall efficiency of educational institutions, reducing administrative tasks and allowing staff to focus on other important tasks.

RFID-based student ID systems offer a convenient and efficient way of monitoring attendance records and fee dues in educational institutions. The technology provides several advantages over traditional methods, including faster data collection, improved accuracy, and real-time monitoring. Educational institutions can benefit from RFID-based student ID systems by improving their operational efficiency, enhancing security, and improving communication with parents/guardians.

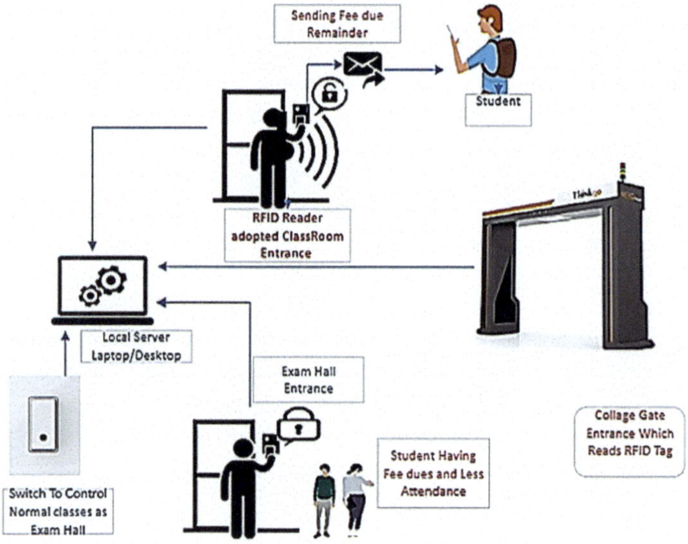

Fig. 1 Sample implementation of RFID system

Attendance Monitoring: The RFID-based system allows for automated attendance monitoring in lectures, tutorials, and other academic activities. Each classroom is equipped with RFID readers that can detect the presence of students with RFID-enabled ID cards. The data collected from these readers is sent to a central database, which generates attendance reports for each class. Faculty members can access these reports and monitor the attendance of their students in real time.

Fee Dues Reminder: The system also sends automated reminders to students who have pending fee dues. When a student enters the campus or a classroom, the RFID reader checks their fee payment status. If a student has unpaid fees, the system sends an alert to the student's registered email or mobile number, reminding them to clear their dues. This helps reduce the number of students with outstanding fee dues and improves the financial management of the institution.

Exam Hall Access: During exams, the RFID-based student ID system provides access control based on fee due and attendance as parameters. Students with unpaid fees or poor attendance records are not allowed to enter the exam hall. This ensures that only eligible students can take the exam and reduces instances of cheating and malpractice.

To set up a system for monitoring attendance, fee status, and exam hall access using an RFID card reader, buzzer, LCD screen, and Firebase database are outlined in the following description:

First, the system needs to be initialized and connected to a Wi-Fi network to enable communication with the Firebase database. Once connected, the system stores information about students' attendance, fee status, and exam eligibility in the database. Next, the RFID reader and buzzer are set up to read the student's RFID card, and the LCD screen is set up to display relevant information about the student's attendance and eligibility. A button is also set up to mark the end of the semester, and variables are initialized to keep track of attendance, fee status, and exam hall access for each student. When a student scans their RFID card, the system reads the card's UID and retrieves the corresponding information from the Firebase database. If the UID matches, the student's name, SRN, and attendance status are displayed on the LCD screen. If the student has fees due, the system sends an alert message reminding them to clear the fees before the due date. If the fees are not cleared, the student's access to the exam hall is denied. The system marks the attendance of each student who scans their RFID card while entering the class, incrementing the attendance count in the database for every class attended. If the switch is turned on, the system displays the student's semester exam and senses their attendance percentage, calculates the average attendance of all students, and checks their eligibility for the exam. If the student's attendance is less than 75%, the system sends a warning message to their registered mobile number. If the student has fees due, the system displays a message reminding them to clear the fees before the exam. Finally, the system displays the student's name, SRN, and exam eligibility status on the LCD screen, and this process is repeated for each RFID card read until the end of the program.

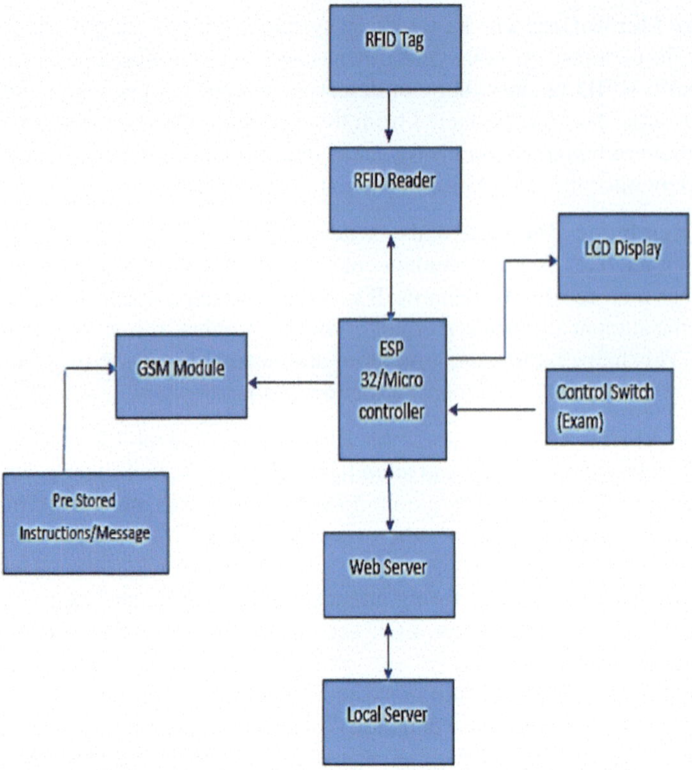

Fig. 2 Block diagram

The RFID-based STUDENT ID system, such as the number of students, locations where the RFID readers will be placed, and desired features of the system. Figure 2 shows to identify the requirements and specifications of RFID-based STUDENT ID system.

RFID Tag: Each student should be assigned a unique RFID tag that can be attached to their ID card. The tag should be compatible with the RFID reader and must have enough storage to store student information.

RFID Reader: Install the RFID readers at strategic locations, such as entry and exit points, library, and labs. Configure the readers to communicate with the ESP32 microcontroller.

ESP32: The ESP32 microcontroller will be the main controller of the system. It will receive information from the RFID readers, process it, and communicate with other components of the system, such as the GSM module and web server.

GSM Module: The GSM module will be used to send SMS notifications to parents or guardians when a student enters or exits a particular area or is absent. The GSM module will receive pre-stored messages from the ESP32 to send SMS notifications.

Pre-stored Message: Pre-stored messages will be created for different events, such as entry, exit, or absence. The messages will be stored in the ESP32 memory and sent to the GSM module when required.

Local Server: The local server will be used to store the data collected from the RFID readers, such as student ID, time, and location. The server will also communicate with the ESP32 to receive data and send commands to the control switch.

Web Server: The web server will be used to provide access to the student ID and attendance data to authorized personnel, such as teachers and school administrators. The server will communicate with the local server to retrieve data.

Control Switch: The control switch will be used to trigger alarms or lock/unlock doors when a student enters or exits a particular area. The ESP32 will send commands to the control switch when required.

LCD Display: The LCD display will be used to display information, such as student ID, time, and location. The display will be connected to the ESP32 and will receive information from it.

Testing: After assembling the components, the system should be tested thoroughly to ensure that it meets the requirements and specifications. Testing should be done in different environments and scenarios to identify and fix any issues.

Deployment: Once the system has been tested and approved, it can be deployed in the school or educational institution. Training should be provided to the staff who will be using the system, and regular maintenance and updates should be performed to keep the system running smoothly.

Algorithm

Step 1: Initialize the system and connect to the Wi-Fi network.
Step 2: Connect to the Firebase database.
Step 3: Set up the RFID reader and buzzer.
Step 4: Set up the LCD screen.
Step 5: Set up the button for the end of semester.
Step 6: Initialize variables for attendance, fee status, and exam hall access.
Step 7: Read RFID card and get the card UID.
Step 8: If (UID == Match_Found), Display Name, SRN, and Attendance from the database.
If (Fee_Due != Cleared),Send Alert Message to clear the due before the due date.
Else: Access Denied.

Step 9: If std1 scan Tag while Enter Sense data when std1 scan tag when Exit, Mark std1 attendance Present and update the database.
Step 10: Sense Data from Database and increment the attendance by 1 for every class.
Step 11: If (Switch == ON), Display Semester Exam, Sense Attendance/Total Classes * 100, Get the average of Student attendance,
Step 12: If (Average < 75%), Not Eligible ("Send Warning Message to registered mobile number").
Else IF (Fee due != cleared), Display Not Eligible ("Send Warning Message").
Else: Eligible, Send Message ("All the Best for Your Examination").
Step 13: Display the student's name, SRN, and Eligibility on the LCD screen.
Step 14: Repeat steps 7–13 for each RFID card read.
Step 15: End the program.

4 Results and Discussion

The suggested system accomplishes two goals: the first is to register, track, and monitor a student's attendance using an RFID tag; the second is to provide student information services, such as timetables, lecture times, classroom numbers, and other student-related data shown on screens or LCDs. The standard procedure for gathering student absence data on the advertisement wall typically involves paperwork and handwriting. As a result, the paperwork method necessitates a workforce, doubles up on efforts, and is time-consuming and ineffective.

Table 1 compares the suggested system with the traditional attendance system (paperwork) based on several characteristics (Figs. 3, 4, 5 and 6).

Table 1 Comparison between the systems

1. **Parameters**	2. **Traditional system**	3. **Proposed system**
4. **Human interference**	5. Yes	6. No
7. **Time-consuming**	8. Over 5 min	9. Maximum of 2 min
10. **Efforts spend**	11. Yes	12. No
13. **Speed**	14. Slow (human)	15. High (computer)
16. **System security**	17. More vulnerable	18. Only authorized people
I. **Resources (documents)**	19. More paper work	20. one electronic record alone
21. **Data reliability**	22. Low	23. High accuracy
24. **Time for registration**	25. More than 8 min for each student	26. 1–2 min
27. **User friendly**	28. No	29. Yes

Fig. 3 Image with scanned RFID card

Fig. 4 Image when switch pressed

Fig. 5 Image showing eligibility

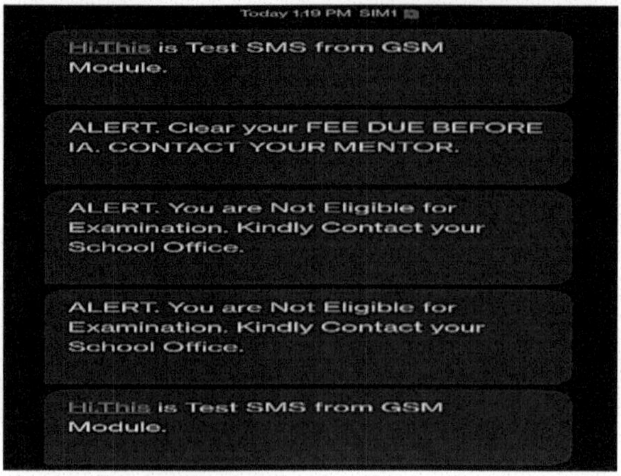

Fig. 6 Image with received alert SMSs

5 Conclusion

The use of RFID-based student ID systems has surged due to their convenient and secure method of identifying and tracking students on school and college campuses. This technology has several advantages, including improved efficiency, accuracy, and security. Real-time access to information stored on the tags allows for quick and easy identification of students. Additionally, incorporating a GSM module and an ESP32 LCD display into the system enables fee due alerts to be sent directly to students, aiding them in avoiding late payment penalties. This feature makes the RFID-based student ID system with fee due alerts a valuable solution for schools and colleges seeking to enhance their student identification and fee tracking capabilities.

References

1. Alshehri S, Drew S (2017) RFID-based student attendance system. Int J Adv Comput Sci Appl 8(10):271–277
2. Kaur M, Singh J (2019) RFID-based attendance system for educational institutes: a review. Int J Eng Adv Technol 9(1):253–257
3. Houta S, Abdel-Naby M (2017) An RFID-based attendance management system in higher education institution. Int J Adv Comput Sci Appl 8(11):221–225
4. Akram R, Alshahrani S (2020) An enhanced RFID-based attendance system for higher education institutions. Int J Adv Sci Technol 29(2):1–8
5. Raghunathan V, Kuppuswamy V (2017) RFID in schools: balancing privacy with security. IEEE Technol Soc Mag 36(3):63–68
6. Mohanty SP, Patnaik S (2019) RFID-based student tracking: an ethical dilemma. In: Handbook of research on digital learning. IGI Global, pp 274–293
7. Abuhejleh A, Al-Zoubi M (2017) An RFID-based system for fee dues management in universities. Int J Adv Comput Sci Appl 8(8):345–349
8. Oladimeji R, Jide-Ojo A (2020) Development of RFID-based fee payment system for Nigerian universities. J Appl Res Higher Educ 12(5):1115–1129

9. Alzahrani AI, Alshahrani SM, Akram R (2020) Investigating the factors influencing the adoption of RFID-based student attendance systems in higher education institutions: a case study of Saudi Arabia. IEEE Access 8:158114–158123
10. Zhang H, Li S (2019) A study of the adoption of RFID technology in the educational industry. Int J Emerg Technol Learn 14(5):160–173
11. Alshaikh F, Alkahtani M (2018) A review of RFID-based student attendance systems. Int J Adv Comput Sci Appl 9(3):232–237
12. Alsubaie A, Almulhim M, Alsubaie S, Alrasheed W, Aljohani F, Alghamdi H (2018) A smart attendance system using RFID technology. IEEE Access 6:76663–76672
13. Al-Sulaimani HM, Al-Shahri MA (2022) Smart student ID card system using RFID and magnetic stripe technologies
14. Mohapatra MK, Behera SK, Panda M (2020) Smart student ID card system using RFID and GPS technologies
15. Bharathwaj A, Krishnan N, Sundaram S, Harish S, Prashanth V, Ganesh CS (2018) RFID-based intelligent security system for school children. 7:337–339
16. Zhang Y, Cheng L, Hu W (2019) Design and implementation of "*RFID-based attendance system. 2019 14th IEEE conference on industrial electronics and applications*" (ICIEA), 1922–1926
17. Dalal YM, Kumar Raja DR, Ashwin Kumar UM (2023) Eco park: a low-cost and sustainable approach for smart parking systems. In: World conference on communication & computing (WCONF-2023), 14–16 July 2023
18. Mohanraj R, Gomathi R (2019) An "*RFID-based attendance system using GSM for academic purposes*". In: 2019 IEEE 5th international conference for convergence in technology (I2CT), 787–791
19. Olaniyi OM, Adebayo AO (2020) A real-time student attendance monitoring system using RFID technology. Int J Adv Trends Comput Sci Eng 9(3):6403–6409
20. Almansor M, Ali FZ (2021) Student attendance using rfid system. Shanghai LigongDaxueXuebao/J Univ Shanghai Sci Technol 22:253–264
21. Biswas P, Ghosh S (2021) An integrated system for attendance monitoring, fee due monitoring based on RFID technology. IEEE Trans Consum Electron 67(1):23–29. https://doi.org/10.1109/TCE.2020.3049065
22. Niveditha J, Kumar Raja DR, Ashwin Kumar UM (2023) Smart cities: transforming urban living with IoT. In: 5th international conference on inventive research in computing applications (ICIRCA-2023) on 3–5 Aug 2023
23. Hemanth Kumar G, Kumar Raja DR (2019) Automation of aerators based system on water quality parameters in aquaculture. In: Kumar A, Paprzycki M, Gunjan V (eds) ICDSMLA 2019. Lecture notes in electrical engineering, vol 601. Springer, Singapore

Heart Disease Prediction System Using BIRCH Clustering

G. V. S. Rajkumar, K. Naveen Kumar, Addagarla Monish, Sujith Maddikonda, and M. Ramu

Abstract Diseases of all types that affect the heart or blood vessels, which lead to heart attacks, strokes and other heart defects known as heart diseases; nearly 630000 deaths occur in America due to heart diseases each year and 11.7% of adults in America have been diagnosed with heart disease which indicates more than one of every ten was affected with heart disease. Early recognition of heart disease will reduce the severe impact of the disease on the person. Efficient and more accurate disease prediction has been an important topic in the prediction system using machine learning. This paper aims to develop an Improved Heart Disease Prediction Model (IHDPM) with a new approach using Unsupervised and Supervised Learning techniques. The results showed that the combination of BIRCH clustering and decision tree classifier can give more accuracy, i.e. 83.33%, and the proposed model, i.e. IHDPM is fast and accurate compared to the other existing model.

Keywords Unsupervised learning · Supervised learning · Heart disease · BIRCH clustering · Decision tree classifier

1 Introduction

Heart disease comprises of conditions affecting the circulatory system, including coronary heart disease, angina, congestive heart failure, cardiomyopathy, hereditary cardiovascular disease, and congenital heart disease involving structural defects. Coronary heart disease involves the narrowing of arteries that supply the heart with

G. V. S. Rajkumar · K. N. Kumar (✉) · A. Monish · S. Maddikonda · M. Ramu
GITAM School of Technology, GITAM, Visakhapatnam, India
e-mail: nkuppili@gitam.edu

G. V. S. Rajkumar
e-mail: gganapav@gitam.edu

M. Ramu
e-mail: rmutyala@gitam.edu

B. Annappa et al. (eds.), *High Performance Computing, Smart Devices and Networks*, Lecture Notes in Electrical Engineering 1262,
https://doi.org/10.1007/978-981-97-7794-5_11

oxygen and blood, potentially leading to fatal outcomes. Elevated cholesterol levels, along with other factors such as thallium levels, high blood pressure, anomalous cholesterol, and obesity, contribute to heart disease. Family history is also a potential risk factor, though not used for forecasting due to non-availability. Age, gender, stress, and an unhealthy diet are additional risk factors.

As a person's age rises, which is, as one gets older, the odds of developing a heart attack rise. For men, the risk of heart disease is higher. Females often bear the same risk after menopause. A stressful life can cause arterial damage, and the risk of coronary heart disease increases. This unique thing depends upon the above-listed factors. An attempt is made to assess the danger of heart disease. The techniques used are very much dependent upon deep learning, machine learning, data mining, etc. The heart disease prediction model is created with new approach using both unsupervised and supervised learning techniques by considering different attributes.

2 Related Works

Samson Agboola et al. [1] proposed a technique by combining HAC and Naive Bayes classification algorithms. After collecting data, they normalized the data and applied clustering to group data to compare group values and actual to filter followed by classification to predict the diabetics. Their results show the proposed model improved accuracy compared to other models. Bhatt et al. [11] proposed a technique by using classifiers and comparing the heart disease dataset by which they can benefit from reliable feature selection so that they don't have to consider all features available in the dataset. Decreasing the number of features required to analyze and predict would inadvertently reduce the execution time and make a more optimal solution. The algorithms used are K-Nearest Neighbour (KNN), Naive Bayes, J-Rip, SVM, Ada Boost, Stochastic Gradient Descent (SGD), and Decision Table (DT).

Due to the complex nature of heart disease, the suggested tests had to be prioritized, and feature sets will be obtained. Monika et al. [2] have described how each model used works in predicting the type of heart disease, for example, how Naive Bayes classifier uses the pre-assumption of class conditional independence that the values are conditionally independent of each other and its advantages being how easily it can handle different amounts of data and d-merits being that accuracy has to be sacrificed. So, by using these methods of experimentation and prediction of hidden patterns in the dataset by usage of data mining methods, they were able to make effective predictions on the type of heart disease and the demerits of the data classification that they utilized. Tanvir et al. [3] have proposed techniques where they have opted for the K-means clustering, which performed relatively well with regard to other clustering techniques.

They have also used normalization to enhance the Euclidean distance between data points by calculating a number of centres, which will reduce the number of iterations, thereby reducing the model's computation time. The results show that the proposed model improved accuracy compared to other models.

3 Methods Used in IHDPM

The proposed model encompasses the following viz., Principal Component Analysis for feature extraction, BIRCH clustering and Decision Tree Classifier. The details are as follows.

3.1 *Principal Component Analysis*

PCA is a method which is used to achieve reduction of dimensionality in the data [4]. PCA is defined as the orthogonal linear transformation which converts data to a new coordination system and the greater variance represented by the scalar vector known as eigen vector lies on first principal component. Higher dimensionality leads to overfitting which reduces the ability to generalize the data as it increases. So, PCA can be used to reduce dimensions.

PCA plots data on to the line drawn from origin. This line should be adjusted in such a way that sum of squares of distances from the projected points to the origin is maximum. Line with the largest sum of squared distances is considered as best fit and Unit vector with loading scores is considered as eigen vectors. By projection eigen vectors will form the axes, and determine the direction of new axis which reduces the original dimensionality feature. After finding all the eigen vectors with eigen values, the eigen value with less value can be dropped as it bears less information.

$$\text{Covariance}(x, y) = \frac{1}{n-1}\sum_{i=1}^{n}(xi - \overline{x})(yi - \overline{y}) \quad (1)$$

Covariance Matrix is formed and relation between the data attributes is identified. Then Eigen values and Eigen Vectors are found. Each Eigen vector represents one new axis and Eigen vector with less Eigen Values is eliminated as it contains less information resulting in reduction in dimension which leads to reduce in noise and less overfitting.

3.2 *BIRCH Clustering*

Compact and small part of the large dataset is first generated which retains much information. CF tree of BIRCH clustering frameworks the clustering of some suitable data, which improves the algorithm in clustering larger data sets. Clustering feature is a node in a BIRCH tree, which provides the level of abstraction [5].

3.3 Decision Tree Classifier

This unique algorithm falls under the Supervised Learning community. It solves both problems related to regression and classification. To solve the problem tree representation is used, where every leaf mode is related to class label whereas features are displayed on the tree inner node. By this algorithm, any Boolean function on discrete attributes can be represented by using it. Initially, we consider root as the whole training set. If the values are continuous then they are disconnected until the model is constructed. Records are disseminated regularly according to the meanings of the attributes. For the purpose of ordering attributes. The key challenge faced in the algorithm is the identification of root node attributes in every stage, such process is known as selection of attributes. The two famous attribute selection tests are Information Gain and Gini Index [4, 6].

4 Existing System

Diverse works in literature connected to the heart disease prediction system using data mining techniques have encouraged this work. [7–12] have described various techniques for diabetes and heart prediction using machine learning models. Accordingly, a model intelligent Heart Disease System was built using data mining techniques and classification techniques namely Naive Bayes and decision tree, yield results such that each technique has its uncommon and rare strength in realizing the objectives of the defined mining.

5 Proposed System

To improve the accuracy of the prediction, a new approach is proposed which includes dimensionality reduction using PCA followed by clustering technique and classification technique. PCA is used to reduce dimensions and clustering for filtering the data points. With the proposed approach, high accuracy of the prediction model can be achieved.

The architecture diagram listed in Fig. 1 of IHDPM describes the proposed model. After data scaling, clustering technique is applied to filter the data points by comparing the group value and class value and clustering is followed by classification to predict whether the heart is normal or abnormal. Data is clustered into groups, i.e. heart presence or absence by using spectral clustering, Hierarchical ward linkage clustering and BIRCH clustering and the best clustering technique is taken by using external validation technique. Prior to this PCA is useful for the data for reducing dimensions of the data. The filtered data is classified using supervised learning by selecting the best classification technique.

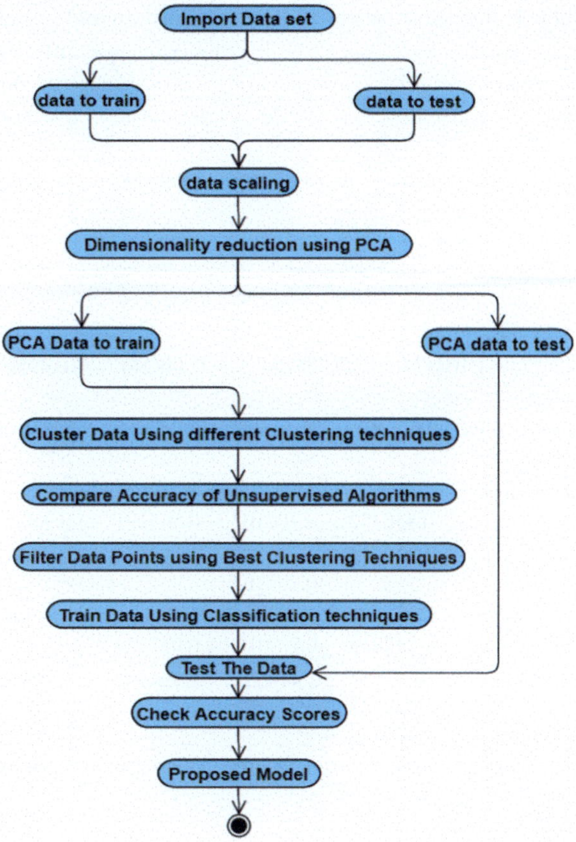

Fig. 1 Architecture of proposed system

6 Requirement Analysis

6.1 Data Set Description

Heart disease dataset taken from UCI repository [13] contains 303 records and 13 attributes [.]. Based on the data analysis with proposed methods, the class can be labelled with either heart disease presence or absence (Table 1).

7 Data Attributes Description

- Age: age of the patient in years.
- Sex: Whether the patient is male or female. Male indicates 1 and female indicates 0 (in Boolean).

Table 1 Represents various attributes used in the model

Seq	Age	Sex	Chest pain type	BP	Cholesterol	FBS Over 120
0	1.210112	0.670570	0.888858	– 0.667134	– 1.372978	– 0.40935
1	– 1.215216	0.670570	0.888858	– 0.957894	1.050616	– 0.40935
2	0.438417	– 1.491268	0.888858	– 1.830176	– 0.007302	– 0.40935
3	0.107690	– 1.491268	0.888858	2.821992	1.512253	– 0.40935
4	– 1.104974	0.67570	– 0.148944	0.495908	– 0.257355	– 0.40935

EKG results	Max HR	Exercise angina	ST depression	Slope of ST	Number of vessels fluro	Thallium
-0.977135	– 0.440352	– 0.707107	– 0.545191	– 0.950366	– 0.692425	1.214592
-0.977135	1.330656	– 0.707107	0.132084	0.705110	– 0.692425	– 0.863442
1.032972	– 1.217868	– 0.707107	– 0.037234	0.705110	– 0.692425	– 0.863442
0.027918	– 1.433845	1.414214	1.994592	0.705110	– 0.692425	– 0.863442
1.032972	1.287460	– 0.707107	– 0.883829	– 0.950366	– 0.692425	– 0.863442

- Chest pain: On a scale of 1–4 how much pain is the patient suffering in his chest area?
- BP: The mmHg of blood pressure the patient has. (80–120 is normal).
- Cholesterol: The amount of cholesterol the patient has measured in deciliters. (Age of 20 or above is 125–200 dL. Ages below 19 < 170 dL).
- FBS over 120: Fasting Blood Sugar level of patient measured in millimoles/liter. Variable in Boolean.
- EKG result: EKG or ECG is test that measures the electrical activity of the heartbeat. It is indicated with 3 values 1,2,3 meaning normal, having ST-T wave abnormally, probable or definite left ventricular hypertrophy.
- Max HR: Heart rate of patient. Normal heart rate is 60–100 heartbeats/minute.
- Exercise angina: It is Boolean variable. One indicates stable angina and other is unstable angina.
- ST depression: Trace in ST segment is abnormal or normal.
- Slope of ST: Helps in identifying the deviation.
- Number of vessels fluro: no. of major vessels (0–3) coloured by fluoroscopy.
- Thallium: The amount present in blood of patient.

8 Methodology of Proposed Model-IHDPM

This proposed model encompasses the algorithms viz., Principal Component Analysis (To reduce dimensions of data), BIRCH clustering algorithm [5] (to filter the data points), Decision Tree Classifier (for classification) [14–16].

Methodology consists of the following steps:

- Step-1: Involves collecting the data set of heart disease diagnosis lab reports. Data is split for training and testing (80% for training and 20% for testing).
- Step-2: To scale the data or normalize and also finding if any missed values exist. Scaling is done as the values differ high range. Scaling the data is very important because it is used to get features with maximum variance, and also changes the direction of PCA towards high-magnitude features as PCA get the features with maximum variance.
- Step-3: Feature extraction of important data from the unstructured, high dimensional, noisy data can be done through principal component analysis. By PCA variance within the data is preserved by projecting data into lower dimensional subspace.
- Step 4: Excluding the class variables and considering only the class variables data points are clustered and divided into two groups—one with normal heart and other with abnormal heart using BIRCH clustering.
- Step-5: Class value is compared with value in the group for each record and only the records which have the same value for both are considered for training using classification technique.
- Step-6: Applying the decision tree classifier to train the data and evaluate model to find accuracy score using test data.

Correlation matrix is an important part of the exploratory data analysis process. This is used to know whether features affect the target variable, and it is a commonly used method for feature selection. The correlation matrix for the data generated by using the seaborn library revealed that MAX HR is negatively correlated and highly sensitive to the data as shown in Fig. 2.

8.1 Applying Principal Component Analysis

Using PCA, 13 eigen values and eigen vectors of covariance matrix are calculated. 13 eigen vectors determine directors of new space, and 13 eigen values determine the variance of the data. Eigen values and vectors are computed using the eigen decomposition.

By using equation,

$$|A - \lambda I| = 0(2)$$

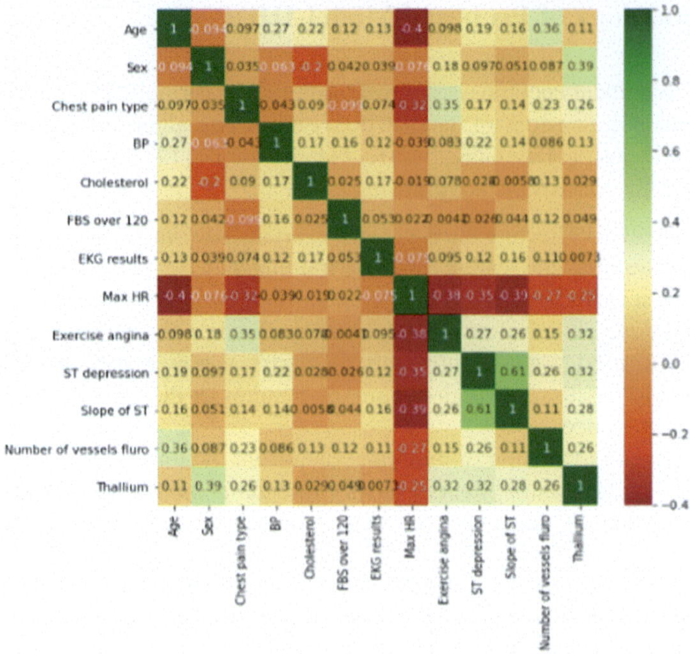

Fig. 2 Correlation matrix

Eigen values (λ) and eigen vectors are computed.

After finding all the 13 eigenvectors with eigen values, the eigen value with less value can be dropped as it contains less information. It is found that singular or eigen vector representing axis PC 8–12 with least eigen values are to be dropped as it contains less information and dimension can be reduced from 13 to 7. Dimensions are reduced from 13 to 7 as shown in Fig. 3, using PCA which avoids overfitting and noise, thereby PCA accuracy can be increased as dimensions are reduced.

8.2 Applying BIRCH Clustering

By using BIRCH clustering, data is clustered into two groups. Accuracy score for BIRCH clustering is 0.68 (Fowlkes mallows score), and 0.336 (rand index score), 0.273 (normalized mutual info score). BIRCH clustering is considered as, the best technique and data points are filtered using this technique. Rand index score for BIRCH clustering is 0.336 which is acceptable and that means the similarity measure between two clustering, i.e. that heart disease presence or absence and by taking into account all training data samples with pair. BRICH clustering is done by clustering Feature tree which is a highly balanced tree, and it contains clustering feature entries. Cluster of these data points (consider here, cluster of data points in cluster which

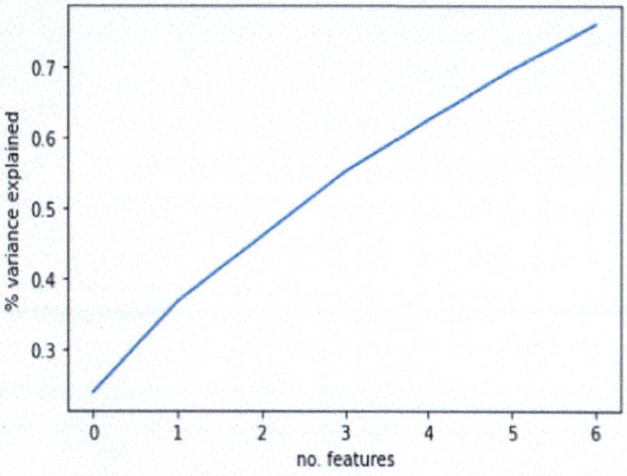

Fig. 3 PCA graph with 7 dimensions

represents heart disease presence or (1) represented by CF entries, i.e. total items in subcluster (*N*), Linear sum of the points (LS), Sum of square of points (SS). Using this diameter is calculated which is threshold and each leaf node has maximum CF entries which reach to threshold.

In Fig. 4, data points are clustered into two groups, heart disease presence or absence. *X*-axis represents principal component 1 and *Y*-axis represents principal component 2. Data is clustered to divide in to two groups and compared to the actual class labels to filter the data points. Data points are filtered using BIRCH clustering. Out of 216 data points 171 are correctly clustered. To this 171 data points which are subjected to PCA classification algorithm is applied to train the data and testing is performed for the prediction.

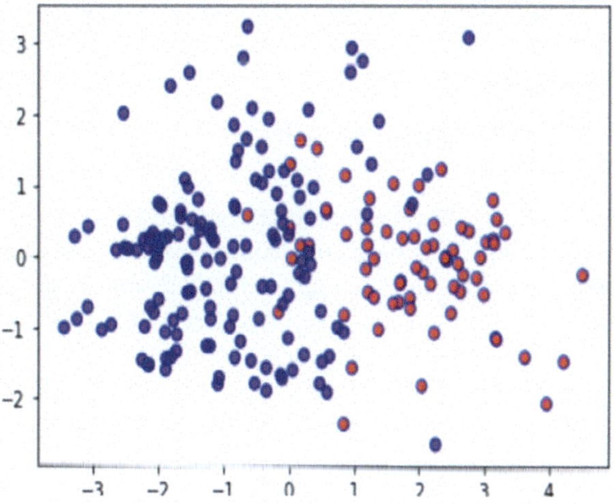

Fig. 4 Graphical representation of clustering using BIRCH

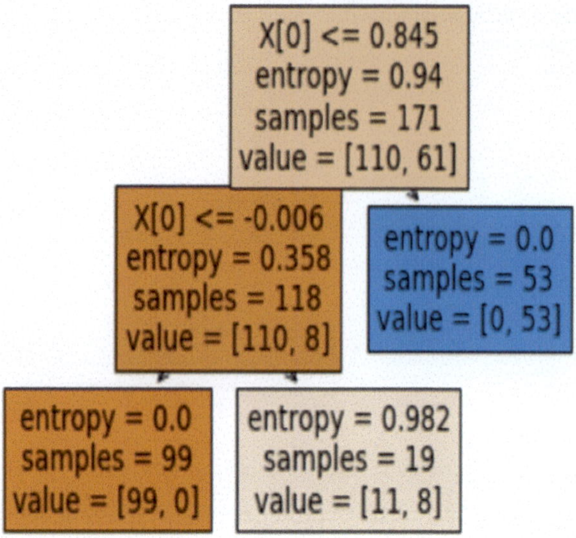

Fig. 5 Decision tree

8.3 Applying Decision Tree Classifier

In implementation, better attribute is selected by Attribute Selection Measures (ASM) for splitting the data points. Make feature as a node for decision and split data into the reduced subsets. To every child node tree should be made by recapping the method recursively till one condition is matched.

Initially, whole data is at the root node and based on the condition in the root node splitting takes place and this condition is whether x [0] $< =0.845$ as shown in Fig. 5 and that points are placed in the right children of the root and the points that will not meet the condition will go to left node and right child is the pure as the entropy is 0.

9 Experimental Results

The results of the proposed model are shown below in Table 2. It is observed that the proposed model is based on, using clustering and classification techniques, dimensionality reduction. The highest predictive accuracy score for this proposed model was given by Decision Tree Classifier with 83.33% accuracy while SGD classifier scored least predictive accuracy score with 74.07% (Tables 3 and 4).

Table 2 Results of the classifier

	Decision Tree classifier	Logistic Regression	SVM	Random Forest Classifier	Gaussian Naive Bayes	SGD Classifier
Accuracy Score	0.8333	0.7962	0.777	0.7777	0.7592	0.7407
Confusion Matrix	28 2 7 17	26 4 7 17	26 4 8 16	25 5 7 17	24 6 7 17	24 6 8 16

Table 3 Results of the classifier for model I

	Decision Tree classifier	Logistic Regression	SVM	Random Forest Classifier	Gaussian Naive Bayes	SGD Classifier
Accuracy Score	0.7037	0.7037	0.6296	0.7222	0.5740	0.7037
Confusion Matrix	24 6 10 14	28 2 14 10	30 0 20 4	25 5 7 17	22 8 7 17	30 0 23 1

Table 4 Results of the classifier for model II

	Decision Tree classifier	Logistic Regression	SVM	Random Forest Classifier	Gaussian Naive Bayes	SGD Classifier
Accuracy Score	0.74074	0.7037	0.6296	0.7037	0.7222	0.4814
Confusion Matrix	25 5 9 15	28 2 14 10	30 0 20 4	25 6 10 14	22 8 7 17	3 27 1 23

10 Comparing with Other Models

The model which is constructed using only classification technique with normalization is considered as existing model I as in Fig. 6. The highest predictive accuracy score for this model was given by Gaussian Naive Bayes classifier with 72.22% accuracy while SGD classifier scored least predictive accuracy score with 57.40%.

Model using only classification technique and dimensionality reduction with normalization is considered as existing model II, shown in Fig. 7. The highest predictive accuracy score was given by Decision Tree Classifier with 74.07% accuracy while SGD classifier scored least predictive accuracy score with 48.14%.

Fig. 6 Existing model I

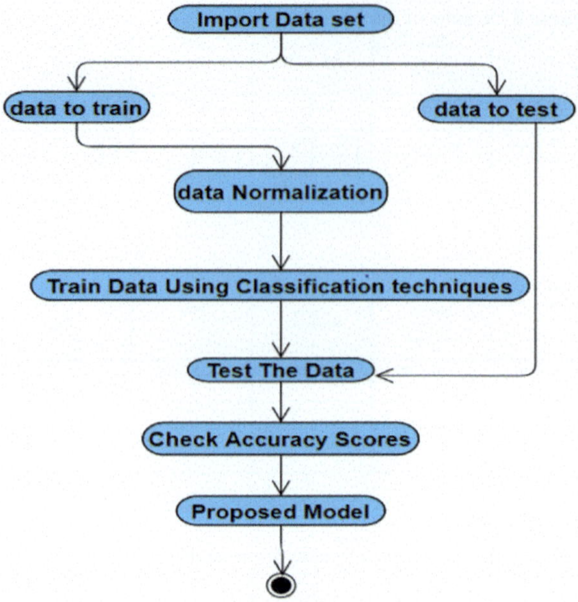

Fig. 7 Existing model II

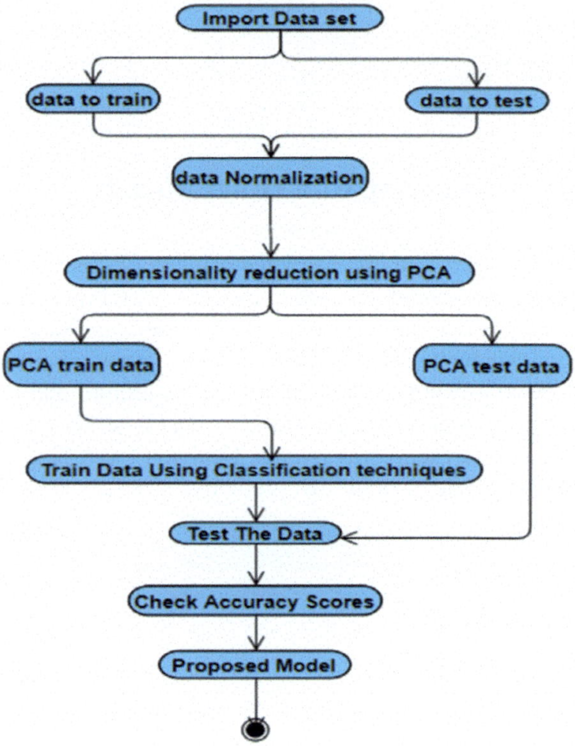

11 Conclusion

Using this new approach, i.e. IHDPM accuracy improved when compared to the existing models. Combination of Decision Tree classifier and BIRCH clustering by Using Principal Component Analysis gave good accuracies. The highest predictive accuracy score was given by decision tree classifier with 83.33% while SGD classifier scored least predictive accuracy score with 74.07%. The proposed improvement leads to improvement of accuracy of classification.

Model using only classification technique and dimensionality reduction with normalization is considered as existing model II. The highest predictive accuracy score was given by Decision Tree Classifier with 74.07% accuracy while SGD classifier scored least predictive accuracy score with 48.14%.

References

1. Agboola S, Mataimaki B (2017) Classification of some seasonal diseases: a hierarchical clustering approach. Biomed Statist Inform 2:122–127
2. Gandhi M, Singh S (2015) Predictions in heart disease using techniques of data mining. 2015 1st International conference on futuristic trends in computational analysis and knowledge management, ABLAZE. pp 520–525. https://doi.org/10.1109/ABLAZE.2015.7154917
3. Ahmad T, Munir A, Bhatti SH, Aftab M, Raza MA (2017) Survival analysis of heart failure patients: a case study. PLoS ONE 12(7):e0181001. https://doi.org/10.1371/journal.pone.0181001
4. Sehgal S, Singh H, Agarwal MV, Bhasker, Shantanu (2014) Data analysis using principal component analysis, 2014 International conference on medical imaging, m-health and emerging communication systems (MedCom), Greater Noida, India, pp 45–48, https://doi.org/10.1109/MedCom.2014.7005973
5. Fontanini AD, Abreu J (2018) A data-driven BIRCH clustering method for extracting typical load profiles for big data, 2018 IEEE Power & energy society general meeting (PESGM), Portland, OR, USA, pp 1-5, https://doi.org/10.1109/PESGM.2018.8586542
6. Saurav Z, Mitu MM, Ritu NS, Hasan MA, Arefin S, Farid DM (2023) A new method for learning decision tree classifier, 2023 International conference on electrical, computer and communication engineering (ECCE), Chittagong, Bangladesh, pp 1–6, https://doi.org/10.1109/ECCE57851.2023.10101557
7. Singh R, Rajesh E (2019) Prediction of heart disease by clustering and classification techniques. Int J Comput Sci Eng 7(5):861–866
8. Zhu Z, Fan P (2019) Machine learning based prediction and classification of computational jobs in cloud computing centers, 15th International wireless communications & mobile computing conference (IWCMC), Tangier, Morocco, pp 1482–1487, https://doi.org/10.1109/IWCMC.2019.8766558
9. Yeruva S, Nimmagadda SP, Siempu R (2019) Improved diabetes prediction model for predicting type-II diabetes. Int J Innov Technol Explor Eng (IJITEE) 8:12
10. Kadhm MS, Ghindawi IW, Mhawi DE (2018) An accurate diabetes prediction system based on K-means clustering and proposed classification approach. Int J Appl Eng Res ISSN 13(6):4038–4041
11. Bhatt CM, Patel P, Ghetia T, Mazzeo PL (2023) Effective heart disease prediction using machine learning techniques. Algorithms 16(2):88

12. Hassan CAU, Iqbal J, Irfan R, Hussain S, Algarni AD, Bukhari SSH, Alturki N, Ullah SS (2022) Effectively predicting the presence of coronary heart disease using machine learning classifiers. Sens J 22:7227
13. Cleveland, Hungary, Switzerland, and the VA Long Beach, UCI machine learning repository, Available: https://archive.ics.uci.edu/dataset/45/heart+disease
14. Almustafa KM (2020) Prediction of heart disease and classifiers' sensitivity analysis. BMC Bioinform 21:278. https://doi.org/10.1186/s12859-020-03626-y
15. Prasad R, Anjali P, Adil S, Deepa N (2019) Heart disease prediction using logistic regression algorithm using machine learning. Int J Eng Adv Technol IJEAT 8(3):659–662
16. Tangirala S (2020) Evaluating the impact of GINI index and information gain on classification using decision tree classifier algorithm. Int J Adv Comput Sci Appl. https://doi.org/10.14569/IJACSA.2020.0110277

Estimation of Channel Availability in Cognitive Radio Networks

B. Surya Samantha, D. Nitul, and Sanjeev Kumar

Abstract Cognitive Radio Networks (CRNs) improve spectrum usage by opportunistically sharing licensed spectrum with cognitive capable devices when the licensed users are idle. However, the constraint in such opportunistic communication is that the cognitive users need to evacuate the channel as and when the licensed user becomes active. Consequently, it leads to sudden termination of ongoing sessions of cognitive devices which in turn yields undesirable communication overhead and degraded service quality. So, channel availability estimation is an important aspect of the acceptability of CRN routing protocols. In this paper, we have performed an analysis of channel estimation in CRN in a scenario of mobile licensed as well as cognitive users. *First*, some mathematical propositions and theorems are stated for determining the availability of a channel, possibility of successful communication, and suitability of the channel to select cognitive users at specific instant in time. The stated propositions are subsequently proved in support of the channel estimation. *Second*, the mathematical analysis is with *ns-3* simulation. Both mathematical and simulated observations are in line and reveal the benefits of adopting the proposed strategy in cognitive radio networks.

Keywords Cognitive radio networks · Channel estimation · Mathematical analysis · Ns-3 Simulation

1 Introduction

The Cognitive Radio Network (CRN) has come into existence in order to resolve two major issues in wireless communications. First, the scarcity of spectrum in unlicensed frequency bands to meet the user demand. Second, unbalanced utilization of licensed band spectrum assigned to various network services, primarily, TV broadcasting. This fact is in accordance with the report produced by FCC, stated in [1].

B. S. Samantha (✉) · D. Nitul · S. Kumar
Department of Computer Science and Engineering, SRM University—AP, Andhra Pradesh, India
e-mail: beri.suryasamantha@gmail.com

B. Annappa et al. (eds.), *High Performance Computing, Smart Devices and Networks*, Lecture Notes in Electrical Engineering 1262,
https://doi.org/10.1007/978-981-97-7794-5_12

According to the report, ISM band is spectrally congested by devices of WLAN, PAN, Sensor Network, etc., whereas, licensed bands (e.g., the TV spectrum) are utilized only within range of 15–85%, both spectrally and temporally. As a consequence, Cognitive Radio Networking (CRN) is designed and proposed to allow cognitive radio capable devices to opportunistically use unused portion of the spectrum (or spectrum hole) of licensed band [2]. In CRN, the cognitive radio-capable devices are not assigned a fixed spectrum for communication, rather they sense environment for existence of spectrally and temporally unused spectrum and tune their transceivers to communicate accordingly when licensed user is not active.

In CRN, availability of licensed channels (also called primary channels) for use of cognitive devices (also called secondary users) determines success of the network. The term channel availability signifies, probability of a particular set of licensed channels (used by primary nodes) being available for use of cognitive radio devices. It is a key parameter for the success of CRN technology. In a scenario, where the PUs are highly active (say 85%), there is little space left for CU for concurrent communication. So, a new concept of communication in CRN is coming up recently where CUs are allowed to communicate concurrently [3, 4] with PU devices. In order to enable concurrent communication, CU device must control their transmission power to avoid transmission of PU nodes. For both opportunistic as well as concurrent communication, estimation of channel availability plays a significant role in successful design of CRN protocols. Because, in opportunistic communication, if a channel is selected for CU communication with high probability of availability, then CRN communication will sustain for longer period and hence will improve efficiency. Likewise, in concurrent communication transmission range is determined based on distance of CU from PU nodes. However, in either case, the channel availability estimation is not an easy task because of the unpredictable PU appearance. Furthermore, in mobile scenario of PU and CU nodes, the prediction of channel availability is much difficult compared to static scenario due to unpredictable positions of CR nodes.

This paper aims at mathematical analysis of channel availability estimation followed by simulation verification. We explore both opportunistic as well as concurrent transmission of CU nodes. It is assumed that when the owner of a channel is inactive, that channel is available to CU with probability 1. If PU of channel is active then availability probability of that channel is determined by the protection range of PU. The protection range of PU defines circular region around PU within which the transmission of CU is not allowed. If CU is outside protection range, then concurrent communication of CU is possible provided the transmission power of the CU is controlled so as to avoid interference with PU communication. Throughout this discussion, we have proposed some mathematical propositions and theorems for determining availability of a channel, possibility of successful communication among CR devices using the selected channel, and suitability of the channel to select cognitive users at specific instant of time. These propositions are based on the characteristics and principles of CRN. The stated propositions are theoretically proved in order to support the proposed channel estimation mechanism. The mathematical claims are also verified by simulated results. Rest of the paper is organized as follows. Section 2 states some related work with motivation of this research. Section 3 describes the network mode followed by analysis in Sect. 4. Paper is concluded in Sect. 5.

2 Related Work and Motivation

Channel state information significantly influences network performance in opportunistic communication. Based on such state information the availability of the channel for future use may be estimated. In [5], by Ning et al. have described a method of such opportunistic schemes taking into account the channel gains over time. The same authors Ning et al. in [6] have proposed another mechanism of channel estimation based on network coding and time division broadcasting schemes centered on outage probability. The proposal of [6] tries to resolve some limitations of their proposal given in [5]. However, none of the two protocols are suitable for CRN as they have not considered uncertain predictability of channel in cognitive environment. Few more channel estimation schemes, [7] authored by Torkestani and Meybodi, [8] authored by Kharraz et al. that take into account the low layer performance for channel estimation. They have suggested to use the received signal strength (RSS) in order to estimate the availability of the channel for CU devices. In [9], the authors Salameh et al. have proposed a probabilistic channel quality and availability estimation-based MAC protocol for CRN. They presume the existence of spectrum sensing mechanism to determine the available free channels for CR nodes. Talay et al. [10] have proposed similar algorithm for dynamic spectrum access on cognitive radio networks. However, the difficulties in such schemes are the complexity of collecting and analyzing low layer data. It is always easier to examine the link connectivity estimation within the network layer compared to other lower layers Bamis et al. [11].

There exist few proposals that estimate channel availability based only on the network layer activities. Chowdhury et al. in [12], have proposed a routing metric with minimized interference of the CUs against the PU communications. The proposal suggests a channel estimation mechanism by observing the history of channel availability. In [13] the authors Avokh et al. have proposed an interference-aware routing solution which is also based on estimated channel availability. All these researches have considered the channel estimations in order to support the protocol designed for the purpose. They have not investigated the channel availability issues only from the routing or MAC point of view. Consequently, few uncertainties in channel estimation, like location of CUs in protection range of PUs, reachability of CUs are overlooked. However, these parameters are vital in channel estimation [14]. A recent work by Angela Sara Cacciapuoti et al. [15] demonstrates explicit accounting of channel availability estimation for mobile CU devices. However, they have assumed static PUs in the scenario.

Importance of channel availability estimation for use of cognitive devices and availability of comparatively less work focused on this area, motivates us to carry out this research. Here, an exclusive channel availability estimation is carried out through mathematical modeling and simulation. In the first part of the research, some mathematical propositions and theorems are stated with subsequent proofs derived from the characteristics and principles of CRN. This theoretical basis helps to estimate the channel availability before the use of the channel in CR communication in an environment of mobile licensed (or primary) and cognitive (or secondary) users. We have assumed a concurrent communication of CU with PU if the CU is located outside the protection range of the PU. Otherwise, opportunistic communication to CU devices. Furthermore, we have also assumed that the PU and CU devices independently move with different angles of movement with respect to a base point. Both types of devices have a random speed with a min–max range. In the later part of the work, a set of simulation experiments in *ns-3* is shown for the verification of mathematical claims.

3 Network Modeling

PUs are mobile with single channel per PU. Their activities are modeled as two-state ON–OFF process with on duration T_{ON} and off duration T_{OFF}. The channel occupancy probability of channel *i(assigned to* PU *m)* is $\alpha_i^{\mathrm{ON}} = \frac{T_{\mathrm{ON}}}{T_{\mathrm{ON}}+T_{\mathrm{OFF}}}$ and the probability of the channel being idle is denoted by $\beta_i^{\mathrm{OFF}} = 1 - \alpha_i^{\mathrm{ON}}$. In opportunistic communication, the idle period of the primary channel is used by CU devices to communicate data among fellow CU devices. It is assumed that the PU nodes move with speed v_{pi} in direction θ_{pi} and these parameters are known to both PU itself and nearby CU sin the topology (Fig. 1).

CU nodes form adhoc network and are mobile with known velocity $\left(v_{cj}\right)$ and direction of movement (θ_{cj}) for the node CU_j. A CU can use the channel i under two scenarios. *First,* if the owner PU is idle, *second*, if the CU is outside the protection range (R_i) of PU_i when the owner of the channel is active. The duration of time for which any of the above two criteria is true, the CU can communicate using channel i. The transmission duration of CU devices is considered as T_d, and is segmented into N equal slots of duration T. During T_d, either N CUs can communicate with one packet each or one CU can communicate N packets. The time slot T is again subdivided into sensing period (T_s) and data transmission period (T_t). The CU node

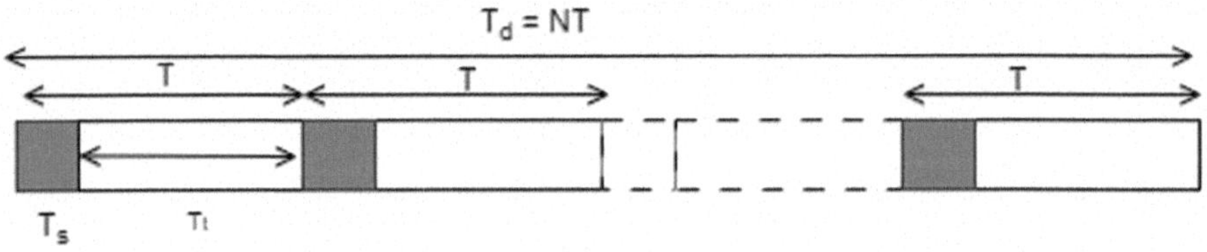

Fig. 1 Transmission duration of CU nodes

updates the PU location after every T_d period. Since the position of a PU in its OFF period does not affect the transmission of CU, hence location update period of T_d is sufficient to model the CU activities.

Update period of PU location information at CU database plays a substantial role in the performance of the channel availability estimation. A careful selection of the location update may enhance the performance.

4 Mathematical Analysis

Analysis is made to estimate the availability of a channel to CU nodes exploiting the node's position information at a specific instance of time. It is assumed that every PU is assigned a single channel exclusively (Section III) and there may be one or more CUs competing to acquire the same channel. In this section, some propositions and theorems are presented to describe the CU node connectivity and channel availability.

Proposition 1 A cognitive node CU_j estimates the probability of availability of channel m (assigned to primary user PU_i) as.

$$p_{j\to i} = \begin{cases} 1\ if d_{(i,j)}(t) > R_i \text{ for } \forall t \in T \\ \alpha_i^{\text{ON}} \text{otherwise} \end{cases} \quad (1)$$

R_i is the protection range of PU_i, α_i^{ON} is as defined in α_i the Section III, $d_{(i,j)}(t)$ denotes the distance between PU_i and CU_j estimated at sensing period of a time slot T.

Proof Once the estimated distance ∂_{ij} (see Appendix) between PU_i and CU_j at time t is available with a CU the availability of the channel may be estimated. If $d_{(i,j)}(t) > R_i$, then PU_i can freely use the channel since it is outside of CU_j protection range, i.e., the channel is available with probability $p_{j\to \text{i}} = 1$. Differently, if $d_{(i,j)}(t) < R_i$ then PU_i is inside CU_j protection range and, hence, the channel is available with probability $p_{j\to i} = \alpha_i^{\text{ON}}$.

As per the discussion above, a channel is assumed fully available for a node CU_i in an arbitrary time t_i if and only the CU is outside the protection range during that time slot or the owner PU is idle. This fact ensures the interference less communication of CU in a CRN environment. This assumption also ensures that both the activity and the position of PU nodes determine the channel availability to CU nodes.

Proposition 2 Any two cognitive nodes CU_j and CU_k can communicate if the following two conditions are satisfied.

(i) $d_{(j,k)} \leq r$, where r is the communication range of CU node that satisfies Proposition 1.
(ii) $\exists u \in \{N_{j\to \text{pu}} \cup N_{k\to \text{pu}}\}$ that satisfies Proposition 1.

Proof If two nodes are within the transmission range of each other then they can communicate provided they have a common free channel available. In absence of common channel, the nodes would not be able to communicate.

Proposition 3 A primary channel m owned by PU_i is considered not suitable for selection for the cognitive node CU_j if the $\underset{\Delta v}{\rightarrow} i \rightarrow j$ tends to zero. Where, the $\underset{\Delta v}{\rightarrow} i \rightarrow j$ is the relative velocity of PU_i and CU_j.

Proof lower relative velocity indicates that the two nodes under consideration would be at same distance apart over longer time. Under such scenario the use of the channel m by node CU_j may lead to reclaim by the owner PU and forced to termination of CU connectivity. Hence, channel m is not suitable for node CU_j.

Theorem Assume that there are m primary channels available in the CRN environment. There are n number of CR nodes that compete for the same set of channels. Assuming there are m' channels available with probability *1* and m'' channels available with probability α_i^{ON}. A CR node computes the total probability of channel availability probability as.

$$p_{t(i)} = \frac{1}{|N_j(u)|}\left(m' + \sum_{k=1}^{m''} \alpha_k^{\text{ON}}\right)$$

$N_i(u)$ is the close neighbors of CU_j.

Proof The channels for which the CUs are not in their protection range would be available to CU with probability 1 and for other channel, it is decided by the ON–OFF probability. The available channels will be accessed by all the CU nodes in the neighborhood. So, total probability would be divided among them. If the number of neighbors in the area is more, then channel acquiring probability for a node is less. We assume that the MAC layer protocol is decided by the underlying network architecture and either CSMA/CA or any suitable protocol for the purpose is used.

5 Performance Evaluation

(a) Statistical evaluation

Let $f(x)$ represent the availability of a channel for the use of secondary user at any instant of time. Either, $f(x) = 0$ (i.e., the channel is not available) or $f(x) = 1$ (i.e., channel is available). We state that $f(x) = 1$ for $x \in A$ where, A is the set of all CUs that have a channel available and $f(x) = 0$ otherwise. Let us suppose the random variable X has probability function as defined in Eq. (1) then it can be stated that $E(f(x)) = p(X \in A)$. A Deployment scenario with $N (\geq 3)$ number of CU nodes and $M (\geq 1)$ PU nodes is considered for analysis. In this analysis, we want to observe the probability of successful one hop communication between any two source and destination S and D. Both the nodes can communicate as per the *proposition 2* above. The probability of one hop connectivity is distributed over n dimensional space (assuming n number of CR nodes) of binary function, *i.e.,* either is a one hop connectivity exit or does not exist between any two randomly selected nodes. As we have modeled the probability of connectivity as binary variable $Y \in [0, 1]$, hence the variable is completely determined by its mean. If we assume $Y = 1$ with probability as stated in Eq. (1), i.e., $p_x \in \left[0, \alpha_i^{ON}\right]$, then $E(Y) = p_x$ and $E\left(Y^2\right) = E(Y) = p_x$ because $Y^2 = Y$ for binary variables. So, $\sigma^2 = p_x - p_x^2 = p(1-p)$. Taking *99%* confidence interval, p_x is recomputed as $p_x = p\hat{x} \pm 2.58\sqrt{\frac{p\hat{x}\left(1-p\hat{x}\right)}{m}}$.

For experimental analysis, we have considered an observation of PU and CU activities in a duration of 60 secs. An average activity of the PU is considered 50%. In such situation, a PU will remain off for 30 s and that period shall be utilized by CU nodes. So, $T_d = 30$ s and divided into 10 (=N) slots each occupying 3 secs (=T). Further, the slot time T is divided into *1* s (=T_s) sensing and 2 s (=T_x) period.

Example 1 For experiments, we have created 100 topologies of ($m =$) 70–320 nodes and searched the successful one hop connectivity between any two randomly selected nodes. The success probability in each case is observed and the upper and lower limits of probability are calculated using a 99% confidence interval as (for m = 70 shown below).

$$p_x = 0.141 \pm 2.58\sqrt{\frac{0.141(1 - 0.141)}{m}}$$

The observed data is shown in Table 1. Other parameters for this experiment are as follows. Transmission range of the PU node is 120 m, protection range is 70 m and PU activity is 45%.

(b) Simulation evaluation

Simulation in *ns-3* [16] is carried for the self-assessment of the proposed channel estimation scheme. Three parameters, *channel availability, successful End-to-end (E2E) connectivity, and duration of path survivability* are examined through the

Table 1 Probability distribution

Sample size (CU Nodes)	Simulated probability	With 99% confidence	
		Min Prob	Max Prob
70	0.141	0.034	0.248
120	0.146	0.036	0.229
170	0.152	0.081	0.223
220	0.158	0.095	0.221
270	0.265	0.196	0.334
320	0.471	0.399	0.543

simulation. The *ns-3* with *CRE-ns-3* patch [17] for CRN is used as a simulation tool. Around 300 CUs collocated with 11 PUs (hence 11 primary channels) are deployed in a region of 500 × 500 m area. Both CUs and PUs are allowed to move following Random Walk model [18].

During the simulation, number of available channels after every 60 s are examined and average ratio of available channel is recorded. Channel availability ratio (Fig. 2) decreases with increasing PU activities. Moreover, as and when protection range of PU is more, the channel availability is less. Increased PU activities and larger protection range make a narrow spectrum hole for CU devices and hence less available channel for CU nodes. If the result is compared with the mathematical results of Table 1 then it is seen that result of Fig. 2 is in line with the table. For example, with node 300, PU activity 45%, and protection range 70 m, the availability ratio is within 40–55%.

With the increase of protection range, successful connectivity decreases (Fig. 3). It is due to the less opportunity for CU to communicate. Likewise, as and when hops in the path increases, the success rate of *E2E* connectivity falls off due to the same reason. Similarity with the mathematical results of the Table 1 is also noticed here.

During a simulation, we have observed period of connection survival with varying movement angles of PU and CU. Results of 1-, 2- and 3-hops connections are shown in Fig. 4. Since our intention is to observe the impact of CU-PU angle of movement deviation, we keep the direction and speed of all CUs same so that minimum disconnection occurs due to out of range CUs. In such cases, disconnections are caused due to protection range of PUs only. Increased deviation in angle between CU and PU leads to higher survivability of the connection because the CUs go out of PU protection range quickly and hence increased availability of the channel for CUs. Connection survivability probability decreases with the increase in path length.

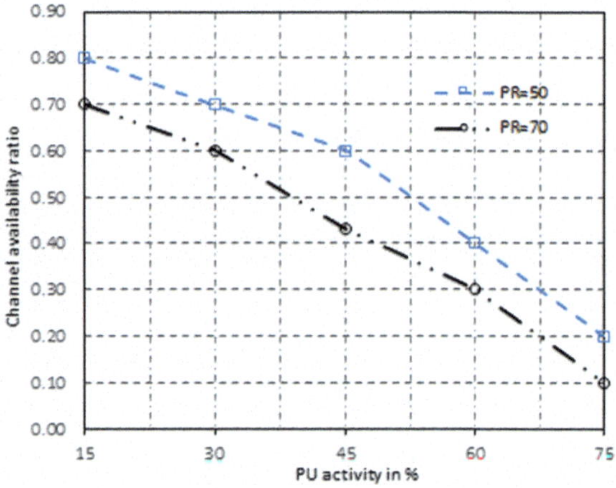

Fig. 2 Channel availability ratio

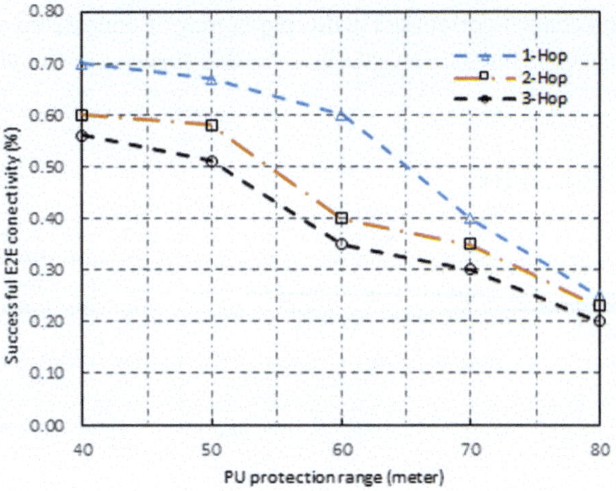

Fig. 3 Successful E2E connectivity

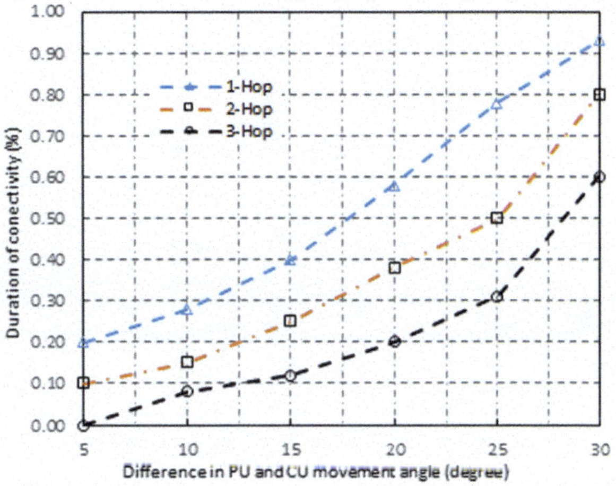

Fig. 4 Duration of connectivity

6 Conclusion

In this paper, we have performed an analysis of channel estimation in CRN network along with their subsequent benefits in data communication. In a scenario of mobile PU and CU nodes, channel availability is estimated through mathematical propositions and theorems and later these findings are verified with *ns-3* simulation. Mathematical claims are proved theoretically and correctness is shown through simulation by examining few important parameters like *channel availability, successful End-to-end (E2E) connectivity, and duration of path survivability*. These parameters are vital to examine the efficiency of a network in terms of user satisfaction. Both analytical and simulated results are agreed to each other and hence the channel estimation mechanism discussed in this paper may be considered valid. However, more rigorous analysis maybe carried out to further investigate the proposed mechanism.

Appendix

A primary user p_i and a secondary user c_j are initially located at position (x_1, y_1) and (x_2, y_2) at time instant t_0. Node p_i travels at a speed v_{pi} and v_{cj} in the direction of θ_{pi} and θ_{cj} with the base. At time instant t_1 node p_i and c_j is located at position $\left(x_1', y_1'\right)$ and c_j located at position $\left(x_2', y_2'\right)$. the values of these Co-ordinates are computed as

$$x_1' = v_{pi}t \cos\theta_{pi},\ y_1' = v_{pi}t \sin\theta_{pi} \tag{2}$$

$$x_2' = v_{cj}t\ \cos\ \theta_{ci},\ y_2' = v_{cj}t\ \sin\ \theta_{cj} \tag{3}$$

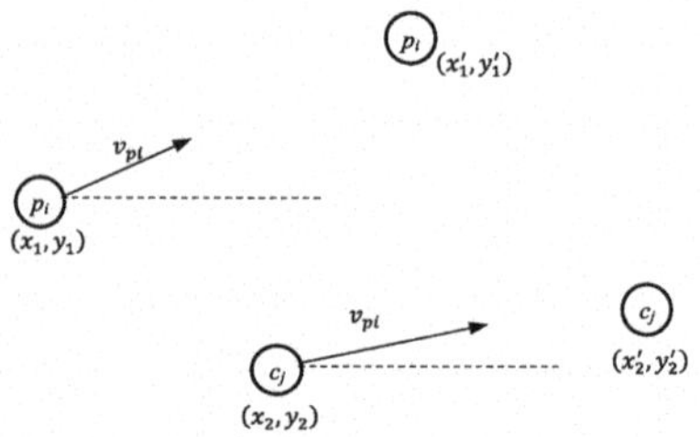

After a time of $t = t_1 - t_0$, the distance between two nodes p_i and c_i is computed as

$$\partial_{ij} = \sqrt{(x_2' - x_1')^2 + (y_2' - y_1')^2}$$

Substituting the values of x and y co-ordinates in terms of angle of movement and speed, it may be expressed as

$$\partial_{ij} = \sqrt{\left(v_{cj}t\cos\theta_{cj} - v_{pi}t\cos\theta_{pi}\right)^2 + \left(v_{cj}t\sin\theta_{ci} - v_{pi}t\sin\theta_{pi}\right)^2} \tag{4}$$

Expanding the Eq. (3)

$$= \sqrt{(v_{cj}t)^2\cos^2\theta_{cj} + \left(v_{pi}t\right)^2\cos^2\theta_{pi} - 2v_{ci}v_{pj}t^2\cos\theta_{cj}\cos\theta_{pi} + (v_{cj}t)^2 + \left(v_{pi}t\right)^2\sin^2\theta_{pi} - 2v_{cj}v_{pi}t^2\sin\theta_{cj}\sin\theta_{pi}}$$

$$= \sqrt{(v_{cj}t)^2\cos^2\theta_{cj} + (v_{cj}t)^2\sin^2\theta_{cj} + \left(v_{pi}t\right)^2\cos^2\theta_{pi} + \left(v_{pi}t\right)^2\sin^2\theta_{pi} - 2v_{cj}v_{pi}t^2(\cos\theta_{cj}\cos\theta_{pi} + \sin\theta_{cj}\sin\theta_{pi})}$$

$$= \sqrt{(v_{cj}t)^2\{\cos^2\theta_{cj} + \sin^2\theta_{cj}\} + \left(v_{pi}t\right)^2\{\cos^2\theta_{pi} + \sin^2\theta_{pi}\} - 2v_{cj}v_{pi}t^2(\cos\theta_{cj}\cos\theta_{pi} + \sin\theta_{cj}\sin\theta_{pi})}$$

$$= \sqrt{(v_{cj}t)^2 + \left(v_{pi}t\right)^2 - 2v_{cj}v_{pi}t^2(\cos\theta_{cj}\cos\theta_{pi} + \sin\theta_{cj}\sin\theta_{pi})}$$

So, the distance ∂_{ij} between node p_i and c_j is computed as

$$\partial_{ij} = \sqrt{(v_{cj}t)^2 + \left(v_{pi}t\right)^2 - 2v_{cj}v_{pi}t^2(\cos\theta_{cj}\cos\theta_{pi} + \sin\theta_{cj}\sin\theta_{pi})}. \quad (5)$$

Cognitive Radio Networks (CRNs) emerged to address two critical issues in wireless communications. First, there is a shortage of spectrum in unlicensed frequency bands to meet increasing user demands. Second, there is uneven utilization of the spectrum in licensed bands, particularly those assigned to services like TV broadcasting. According to a report by the FCC, devices such as WLAN, PAN, and Sensor Networks heavily congest the ISM band. In contrast, licensed bands, including the TV spectrum, are only used between 15 and 85% of the time, both spectrally and temporally. To tackle these issues, the Cognitive Radio Networking model was proposed. It enables cognitive radio devices to opportunistically utilize the unused portions (or spectrum holes) of licensed bands.

In CRNs, devices do not have fixed spectrums for communication. Instead, they dynamically sense the environment for available spectrum and adjust their transceivers to communicate when licensed users are inactive. The availability of licensed channels (or primary channels) for cognitive devices (or secondary users) is crucial for the network's success. Channel availability indicates the probability of licensed channels being free for cognitive radio use, which is vital for CRN technology.

When primary users (PUs) are highly active, there is limited space for secondary users (CUs) to communicate concurrently. Recently, a new concept allows CUs to communicate simultaneously with PUs by controlling their transmission power to avoid interfering with PU transmissions. Estimating channel availability is essential for both opportunistic and concurrent communication to ensure efficient CRN protocol design. In opportunistic communication, selecting channels with a high probability of availability extends communication duration and improves efficiency. For concurrent communication, the transmission range is determined by the distance between CU and PU nodes. However, predicting channel availability is challenging due to the unpredictable nature of PU activity, and it becomes even more complex in mobile scenarios with moving PU and CU nodes.

References

1. FCC (2002) Spectrum policy task force report, ET docket no. 02–155
2. Ghosh C, Roy S, Dave C (2011) Coexistence challenges for heterogeneous cognitive wireless networks in TV white spaces. IEEE Wirel Commun 18(4).22–31
3. Raikel BL, Sanchez SM, Fernandez EM, Souza RD, Alves H (2014) Genetic algorithm aided transmit power control in cognitive radio networks, 9th IEEEInter. Conference on cognitive radio oriented wireless networks and communications (CROWNCOM), Oulu, pp 61–66

4. Dutta N, Polkowski Z, Savulescu C (2016) Grid based routing in cognitive radio networks for concurrent communication, IEEE 9th International Conference on contemporary computing (IC3), Noida, India
5. Ning Z, Song Q, Yao Y (2013) A novel scheduling algorithm for physical-layer network coding under Markov model in wireless multi-hop network. Comput Elect Eng 39(6):1625–1636
6. Ning Z, Song Q, Huang Y, Guo L (2014) A channel estimation based opportunistic scheduling scheme in wireless bidirectional networks. J Netw Comput Appl 39:61–69
7. Torkestani JA, Meybodi MR (2011) A link stability-based multicast routing protocol for wireless mobile ad hoc networks. J Netw Comput Appl 34(4):1429–1440
8. Kharraz MA, Sarbazi Azad H, Zomaya AY (2012) On-demand multicast routing protocol with efficient route discovery. J Netw Comput Appl 35(3):942–950
9. BanySalameh H, Badarneh OS (2013) Opportunistic medium access control for maximizing packet delivery rate in dynamic access networks. J Netw Comput Appl 36(1):523–532
10. Talay AC, Altilar DT (2013) Self-adaptive routing for dynamic spectrum access in cognitive radio networks. J Netw Comput Appl 36(1):1140–1151
11. Bamis A, Boukerche A, Chatzigiannakis I, Nikoletseas S (2008) A mobility aware protocol synthesis for efficient routing in ad hoc mobile networks. Comput Netw 52(1):130–154
12. Chowdhury KR, Akyildiz IF (2011) CRP: a routing protocol for cognitive radio ad hoc networks. IEEE J Select Areas Commun 29(4):794–804
13. Mirjalily AAG (2014) Interference-aware multicast and broadcast routing in wireless mesh networks using both rate and channel diversity. Comput Elect Eng J 40:614–640
14. Parvin S, Hussain FK, Hussain OK, Han S, Tian B, Chang E et al (2012) Cognitive radio network security: a survey. J Netw Comput Appl 35(6):1691–1708
15. Cacciapuoti AS, Caleffi M, Paura L, Rahman MA (2015) Channel availability for mobile cognitive radio networks. J Netw Comput Appl 47:131–136
16. The Network Simulator—ns-3. [Online]. Available: http://www.nsnam.org
17. CRE-NS3: Cognitive radio extension for NS-3. [Online]. Available: https://www.coe.neu.edu/Research/krclab/crens3/index.html.
18. Zhang Q, Das SK, Rodriguez Carrion A (2016) A 2-D random walk mobility model for WiMAX location update. Comput Commun 78:86–96

Machine Reading Comprehension Using NLP

A. Vineela, G. Lavanya Devi, M. Poojitha, and K. Deepika

Abstract Machine Reading Comprehension (MRC) is a type of question-answering mechanism that involves reading documents and responding to questions. The challenge of enabling machines to effectively interpret paragraphs, comprehend related inquiries, and generate appropriate responses has long persisted. Unlike humans, who possess the ability to grasp the logic and intent behind a question and formulate a fitting reply, machines struggle with this task. Thus, the primary goal of this project is to leverage Natural Language Processing (NLP) to enable machines to read paragraphs, understand questions, and provide coherent answers. The input for the reading comprehension model is the question and the context. The passage's answer is the model's output. A sort of writing and reading comprehension (RC) is becoming more and more popular in the study of natural language processing (NLP). Datasets like SQUAD (Stanford question answer dataset) are used. Various models like BERT, and ELECTRA, are used to predict the answers to respective questions. There are a few models that can answer the questions in the Telugu language. In this analysis, the MRC is implemented in the Telugu language using BERT model. Although existing MRC models surpass humans on numerous benchmark datasets in English language comprehension, there remains a significant disparity between these models and real-world human-level reading comprehension.

Keywords Machine reading comprehension (MRC) · Question-answering mechanism · Reading comprehension · Model evaluation · BERT (Bidirectional Encoder Representations from Transformers)

A. Vineela (✉) · M. Poojitha · K. Deepika
Department of Computer Science and Engineering, GMR Institute of Technology, Rajam, India
e-mail: alladavineela4@gmail.com

G. L. Devi
Department of Computer Science and System Engineering, AU College of Engineering (A), Andhra University, Visakhapatnam, India

B. Annappa et al. (eds.), *High Performance Computing, Smart Devices and Networks*, Lecture Notes in Electrical Engineering 1262,
https://doi.org/10.1007/978-981-97-7794-5_13

1 Introduction

The ability of machines to comprehend human languages has sparked a growing interest over recent decades. In addition to that machine reading comprehension (MRC) has made significant progress in recent years. From a certain perspective, it is possible to consider the latest assignments referred to as MRC as an expansion of question-answering (QA) tasks. The survey conducted by Hirschman and Gaizauskas divides these QA models into three groups: question-answering and story comprehension, natural language database frontends, and dialogue interactive advisory systems. The final category might be thought of as the source of the current MRC tasks. Wendy Lehnert suggested that the Question-Answering systems should consider both the passage and the question, and then answer the question after making any necessary inferences and interpretations. Under her theory, Lehnert also developed a system called QUALM. In terms of the MRC corpus, numerous datasets covering a variety of fields and styles have been made available in recent years. MCTest was released in 2013 as a high-quality multiple-choice reading comprehension dataset that was insufficient to train neural models. CBT and CNN/Daily Mail were released in 2015. These two datasets were significantly larger than the previous ones because they were generated automatically from distinct domains. In 2016, the SQuAD dataset emerged as the initial extensive collection of questions and answers crafted by humans. The initial MRC task was reduced to requiring systems to return a sentence containing the correct response. Rule-based heuristic approaches, such as bag-of-words approaches and manually generated rules, serve as the foundation for the systems. Since around 2015, the MRC's research interests have grown significantly due to the introduction of effective architectures like attention mechanisms in NLP and deep neural networks. Better passage and question interactions and fine-grained text encoding were the main topics. The benchmark results of MRC were significantly improved by The Role of Contextualized Language Models and Beyond, which encouraged progress toward more complex reading, comprehension, and reasoning systems by strengthening the capacity of language encoders. As a result, the MRC's research gets closer to human cognition and applications in the real world. On the other hand, a growing number of researchers are interested in examining and deciphering how the MRC models operate, as well as investigating the actual capability beyond the datasets, such as performance in the adversarial attack and the benchmark capacity of the MRC datasets. The goal of this project is to conduct a comprehensive examination of the latest MRC datasets and techniques, offering a detailed analysis.

2 Literature Survey

Machine Reading comprehension is a type of question-answering mechanism where the machine can answer the respective questions according to the passage given. The paper proposed by Yao Cong proposed that the existing works only focus on the natural language and the performance of the model and not on the hierarchical information which represents words, sentences, and paragraphs [1, 2]. Furthermore, addressing the persistent noise issue, the PH-Model introduces a novel approach emphasizing hierarchical information. In comparative analysis, the DuReader 2.0 dataset has demonstrated superior performance over the PH-Model, boasting a remarkable 18.24% improvement in ROUGE-L and a substantial 24.17% enhancement in BLEU-4 baseline scores [3]. In the paper was proposed by Mucheng Ren proposed that the existing approaches mainly focus on complicated knowledge and not on world knowledge. The main focus of this analysis is reasoning which will contain five modules that are learned in an end-to-end manner and the dataset that was used is the ROPES dataset which is best suited for the reasoning of paragraphs. This model follows the sequential approach which consists of three modules named Contextual Encoding, Interpretable Reasoning, and Answer Prediction. The ROPES dataset is required for reasoning and the Winogrande dataset is required for common sense-related reasoning [4]. Generally, machine reading comprehension is limited to only some text spans and the model that was proposed by Ronghan Li was called multi-choice MRC which is more flexible to answer the questions by utilizing some extra commonsense knowledge [5]. This analysis focuses on the multi-hop MRC model. It involves providing a question in the form of a triplet with an absent entity and a set of supporting documents. The aim is to choose a response to the query from a range of potential answers. [6]. There is a model called AP-BERT which improves the performance through average pooling and shows better results in feature extraction [7]. There are many approaches that introduce hierarchical conversation flow transitions to enhance machine reading comprehension. This analysis focuses on the reasoning of answers with respect to the questions and advancements for further machine reading comprehension [8]. There is a frame-based neural network which is a novel approach that utilizes frames for the understanding of textual data and advancements in MRC [9]. In the early years of the 1950s and 1960s, research about Machine reading comprehension began. However, due to a lack of knowledge in computation and limited resources, the research has been stopped. In the 1970s and 1980s, rule-based approaches have a significant impact on Natural Language Processing. In 2015, the Stanford Question Answer dataset (SQUAD) showed a significant impact on Machine reading comprehension that has a 91% F1-Score with the BERT model [10, 11]. There are many MRC models that are implemented in different languages such as Bangla which uses transformer-based learning techniques [12]. Another language is Arabic which implemented the MRC on the Holy-Qur'an [13]. Some models in MRC improve the decision-making process for respective questions with answers [14]. With the advancement of deep learning techniques, the Recurrent Neural Network (RNN), Transformer models, and LSTM have developed more accurate MRC models. Various models like BERT, GPT, and Roberta has attained top-of-the-line abilities to produce findings that are more accurate.

The introduction of attention mechanisms like transformer models has shown a significant impact to provide accurate results. Later, self-tuning and fine-tuning mechanisms have also been in advancing MRC. There are several challenges.

Regarding MRC which includes noise, complex reasoning, and common sense-related questions. There are also very few models regarding machine reading comprehension that were implemented in the Telugu language.

3 Methodology

MRC models take both the article and the question as input. First, the encoding process is performed, which represents the text in a numerical format. A dataset that is used for the machine reading comprehension model is the TyDi QA (Typical Diverse Question Answering) dataset which contains the questions and their respective answers. The model is Multilingual Bert which would give the answers to the respective questions in Telugu language. After acquiring the information from the dataset, divide it into smaller parts called tokens. For the Telugu language, the tokenizer compatible is used to segment the text. The article and the question are combined to form the input sequence. To maintain the proper contextual coherence, the proper alignment of the input sequence is a must. There are many embeddings but for Telugu language, the word2vec and Glove embeddings are used. These are used as initial embeddings of words in the encoding layer. To enhance the encoding layer, the input sequence is incorporated with character embeddings. Character-level convolutional neural network (char-CNN) is used to represent each character as a vector capturing information. The Multilingual BERT model is used to encode the input sequence. Combine the multiple encoding layer which consists of a multi-head self-attention mechanism and a feed-forward neural network for every layer. The attention layer is used to generate the fixed-length representation of the input sequence. Then, the pointer network layer would be used to check whether the output consists of discrete tokens and whether the task involves sequence-to-sequence models. The linear transformation layer is included in the output layer to generate probabilities for question-answering tasks in Telugu. Then Finetune the model to adapt to the parameters for the TyDiQA Dataset. To maintain the explanation of the original text, it is essential for this encoding to encompass contextual information for every word, expression, and sentence. The encoding layer is the subject of this. Next, the model must comprehend the connection between the question and the article to provide accurate responses. The MRC model conducts a comprehensive joint analysis of the article's and the question's semantics during this procedure. The interaction layer is the subject of this. In the interaction layer, once the relationship between the article and the question has been established, the model captures clues to answer the question. The output layer is the final module that provides the answer prediction. Furthermore, this layer establishes the loss function and evaluation protocol for optimization purposes. Within the encoding layer, pre-trained word embeddings such as word2vec and Glove are employed, remaining static throughout training. These embeddings serve as trainable parameters alongside other network parameters. To effectively incorporate character-level information, the encoding layer can integrate character embeddings, representing each character with a vector, commonly achieved through methods like character convolutional neural networks (char-CNN). The interaction layer utilizes attention mechanisms to capture the semantic relationship between the query and the passage. Following the encoding and interaction layers, the model accesses necessary data from both the query and the passage. Consequently, the output layer gathers the analyzed information to generate the response in the appropriate format.

4 System Analysis and Design

The methodology we used for the Machine Reading Comprehension are as follows:

1. Input sequence: In this layer, the data will be taken from the datasets which are given by the model.
2. Token embedding layer: This layer converts the input tokens into dense vectors, called embeddings. The input embeddings are learned during the pretraining phase of BERT.
3. Encoder layers: The main section of the BERT model is the encoder layer, which is made up of multiple identical parts of the transformer architecture. Within each block, there is a combination of a multi-head self-attention mechanism and a feed-forward neural network. The self-attention mechanism allows the encoder to weigh the importance of each input token in the context of the other tokens in the sequence.
4. Attention layer: This layer is responsible for generating the fixed-length representation of the input sequence, which is then used for downstream tasks such as text classification or question answering.
5. Pointer network layer: This layer is applied to sequence-to-sequence models, in which the output consists of a few discrete tokens that correspond to the input rather than an entire phrase. These work well for sorting, word ordering, and other computational language issues.
6. Output layer: The final layer is a linear transformation layer that generates the output probabilities for the specific task at hand (e.g., sentiment classification or question answering).

5 Algorithm (MultilingualTeluguBERTQA)

The used BERT model is in this system. These are the main steps involved in using a BERT model for machine reading comprehension.

Step-1: Pre-processing—To pre-process the input data, including text from the context and the question

Step-2: Input Encoding—Convert the pre-processed data into numerical values, which can be processed by the BERT model. This is done by adding special tokens

[CLS] and [SEP] which are appended to the start and finish of the input sequence, respectively.

Step-3: Attention Mechanisms—A multi-head self-attention mechanism is used to consider all the tokens in the input sequence simultaneously.

Step-4: Fine-Tuning—This Fine-Tuning adds a task-specific layer to the top of the model.

Step-5: Prediction—The prediction is made by taking the hidden state of the [CLS] token and using it as the representation of the entire input sequence. This representation can then be used to generate a prediction.

Step-6: Evaluation—Evaluate the performance of the model (Fig. 1).

Existed System

There are many systems for machine reading comprehension (MRC) that have been proposed in recent years. Here are some examples of MRC systems along with reference papers:

BERT: Devlin et al.—Bidirectional Encoder Representations from Transformers BERT is a language model that has already been trained and may be adjusted for a

Fig. 1 Process flowchart

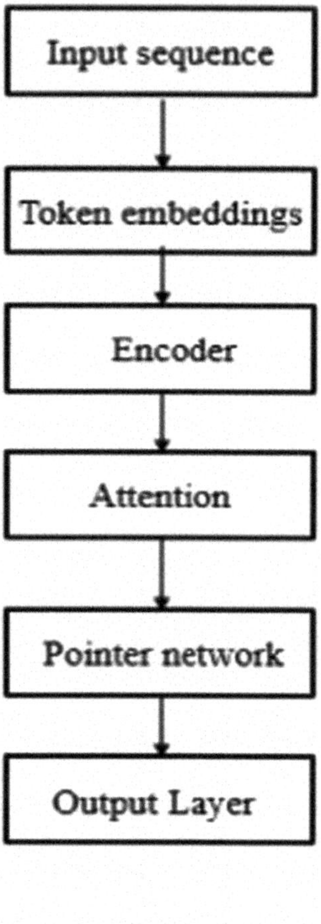

variety of natural language processing tasks, such as MRC. When tested against the SQuAD v1.1 and v2.0 benchmarks, it produced cutting-edge performance.

RoBERTa: Liu et al. present a robustly optimized BERT pretraining approach. A longer training period and a larger training corpus are used in the RoB-ERTa form of BERT. On a number of MRC benchmarks, including SQuAD v2.0 and Natural Questions, it produced exceptional results.

ALBERT: ALBERT, or A Lite BERT for Self-Supervised Learning of Language Representations, is a faster and more compact version of BERT that uses parameter-sharing strategies to minimize the amount of parameters while maintaining functionality. It produced novel results on a number of MRC benchmarks, notably SQuAD v2.0.

SpanBERT: Joshi et al. Improve pretraining by Representing and Predicting Spans a BERT variant called SpanBERT is trained to predict text spans as opposed to individual tokens. It produced better outcomes on the QuAC and CoQA datasets, among other MRC benchmarks.

XLNet: Yang et al. Generalized Autoregressive Pretraining for Language Understanding a permutation-based method is used by the pre-trained language model XLNet to capture bidirectional dependencies between all input positions.

6.T5: Raffel et al.—Examining the Boundaries of Transfer Learning via a Consolidated Text-to-Text Transformer The T5 language model is pre-trained and may be adjusted for a variety of natural language processing tasks, such as MRC. It is trained using a text-to-text format. On several MRC tasks in the SuperGLUE benchmark.

6 Proposed System

The pre-trained BERT model has been fine-tuned for machine reading comprehension (MRC) tasks. It was trained on a combination of several MRC datasets, including SQuAD 1.1, SQuAD 2.0, XQuAD, TyDiQA-GoldP, and MLQA. The model is available for use through the Hugging Face Transformers library.

The model uses a multilingual cased BERT architecture, which means that it can handle text in multiple languages and retains case information. It has been fine-tuned on a combination of MRC tasks to improve its performance on a variety of question-answering tasks. The fine-tuning process entails training the model for the particular task and fine-tuning the weights of the pre-trained BERT model to enhance performance on that task.

The BERT model achieves advanced results on multiple MRC benchmarks, such as SQuAD 1.1, SQuAD 2.0, XQuAD, TyDiQA-GoldP, and MLQA, according to the Hugging Face model card. It has also been evaluated on several other datasets and has shown strong performance across a range of MRC tasks. In summary, mrm8488/Bert-multi-cased-finetuned-xquad-tydiqa-goldp is a pre-trained BERT model that has been fine-tuned for MRC tasks. It uses a multilingual cased BERT architecture and achieves good results on several MRC benchmarks. It is available for use through the Hugging Face Transformers library.

7 Equations

The main performance metrics are Precision, Recall, F1 Score. The formulas for respective are as follows:
Recall: Mathematically, it can be represented as ratio of true positive and sum of true positive and false negative.

$$\textbf{Recall} = \frac{\textbf{True Positive}}{\textbf{True Positive} + \textbf{False negative}} \tag{1}$$

Precision: It can be expressed mathematically as the ratio of True Positive to the ratio of True Positive to False Positive.

$$\textbf{Precision} = \frac{\textbf{True Positive}}{\textbf{True Positive} + \textbf{False Positive}} \tag{2}$$

F1 Score: Mathematically, it is defined as the harmonic mean of precision and recall.

$$F1\textbf{score} = 2 * \frac{\textbf{Precision} * \textbf{Recall}}{\textbf{Precision} + \textbf{Recall}} \tag{3}$$

8 Results

The Bidirectional Encoder Representational Transformer (BERT) model achieves state-of-the-art performance on several machine reading comprehension (MRC) benchmarks, including SQuAD 1.1, SQuAD 2.0, XQuAD, TyDiQA-GoldP.

- On the SQuAD 1.1 benchmark, the model achieves an F1 score of 92.14% and an EM score of 88.55%. On the more challenging SQuAD 2.0 benchmark, the model achieves an F1 score of 81.89% and an EM score of 77.14%.
- On the XQuAD benchmark, which evaluates cross-lingual MRC performance, the model achieves an F1 score of 86.4% and an EM score of 73.5%.
- On the TyDiQA-GoldP benchmark, which contains questions and answers in 11 typologically diverse languages, the model achieves an F1 score of 45.2% and an EM score of 33.7%.
- On the dynamic dataset, which evaluates dynamic passages and questions, the model achieves precision about 61.3% and F1 score about 61.2% (Table 1).

Table 1 F1 score of various models applied on different datasets

Model	SQuAD 1.1	SQuAD 2.0	XQuAD
ALBERT xxlarge	93.16	89.44	–
RoBERTa large	93.18	89.97	–
mBERT	–	–	83.7
XLM-R base	–	–	86.5
MultilingualTeluguBERTQA	92.14	81.89	86.4

9 Comparison Results

9.1 Discussions

Machine reading comprehension (MRC) is an active area of research and development in natural language processing (NLP), and there has been some work done on implementing MRC systems in Telugu language.

One recent study in this area was conducted by researchers at the Indian Institute of Technology (IIT) Kharagpur, who developed a Telugu language MRC system based on the BERT model. They trained their system on a dataset of Telugu passages and questions and evaluated its performance using a Telugu version of the SQuAD benchmark. They achieved an F1 score of 65.35% and an EM score of 49.78%, which is a promising result for a relatively low-resource language like Telugu.

Another study by researchers at the University of Hyderabad focused on developing a Telugu MRC system for medical domain. They used a pre-trained BERT model and fine-tuned it on a dataset of Telugu medical passages and questions. They evaluated their system using a dataset of Telugu medical questions and achieved an F1 score of 66.29%.

While there has been some progress in developing MRC systems for Telugu language, there are still some challenges to be addressed. One major challenge is the lack of large, high-quality datasets for Telugu MRC, which can make it difficult to train accurate models. Additionally, the limited availability of Telugu language resources, such as pre-trained language models and NLP tools, can make it challenging to develop high-quality Telugu MRC.

10 Conclusion and Future Scope

The Proposed BERT (Bidirectional Encoder Representation Transformer) model is a state-of-the-art machine reading comprehension model that has been pre-trained on a large corpus of text data and fine-tuned on several benchmarks, including SQuAD and XQuAD. While this model has not been specifically implemented for Telugu language, its strong performance on a range of natural language processing tasks suggests that it could be a promising starting point for developing an MRC system in Telugu.

It's crucial to remember, though, that this model was trained using a vast corpus of text data in multiple languages, and it's possible that its performance may not generalize well to Telugu language, which has its own unique characteristics and challenges. Additionally, developing an MRC system in Telugu would require the creation of high-quality Telugu language resources, including large, annotated datasets and pre-trained language models specifically tailored to Telugu language.

Overall, while this model is a tool for natural language processing, developing an MRC system specifically for Telugu language would require additional research and development to address the specific challenges and opportunities presented by Telugu language.

References

1. Cong Y, Wu Y, Liang X, Pei J, Qin Z (2021) PH-model: enhancing multi-passage machine reading comprehension with passage reranking and hierarchical information. Appl Intell 51(8):5440–5452
2. Allen LK, Creer SD, Poulos MC (2021) Natural language processing as a technique for conducting text-based research. Lang Linguist Compass 15(7):e12433
3. Wu Z, Xu H (2022) Trustworthy machine reading comprehension with conditional adversarial calibration. Appl Intell 53(11):14298–14315
4. Ren M, Huang H, Gao Y (2022) Interpretable modular knowledge reasoning for machine reading comprehension. Neural Comput Appl 34(12):9901–9918
5. Li R, Wang L, Jiang Z, Liu D, Zhao M, Lu X (2021) Incremental BERT with commonsense representations for multi-choice reading comprehension. Multimed Tools Appl 80(21):32311–32333
6. Liao J, Zhao X, Li X, Tang J, Ge B (2022) Contrastive heterogeneous graphs learning for multi-hop machine reading comprehension. World Wide Web 25(3):1469–1487
7. Zhao S, Zhang T, Hu M, Chang W, You F (2022) AP-BERT: enhanced pre-trained model through average pooling. Appl Intell 52(14):15929–15937
8. Liu X, Yang M, Lyu Z, Lin D, Li P, Xu R (2022) Hierarchical conversation flow transition and reasoning for conversational machine comprehension. Neural Comput Appl 35(3):2413–2428
9. Guo S, Guan Y, Tan H, Li R, Li X (2021) Frame-based neural network for machine reading comprehension. Knowl-Based Syst 219:106889
10. Singh U, Kedas S, Prasanth S, Kumar A, Semwal VB, Tikkiwal VA (2020) Design of a recurrent neural network model for machine reading comprehension. Proced Comput Sci 167:1791–1800
11. Feng J, Sun J, Shao D, Cui J (2022) Improving the robustness of machine reading comprehension via contrastive learning. Appl Intell 53(8):9103–9114
12. Aurpa TT, Rifat RK, Ahmed MS, Anwar MM, Ali AS (2022) Reading comprehension-based question answering system in Bangla language with transformer-based learning. Heliyon 8(10):e11052
13. Malhas R, Elsayed T (2022) Arabic machine reading comprehension on the Holy Qur'an using CL-AraBERT. Inf Process Manage 59(6):103068
14. Cui Y, Liu T, Che W, Chen Z, Wang S (2022) ExpMRC: explainability evaluation for machine reading comprehension. Heliyon 8(4):e09290

Fog Enabled Systematic Car Parking System

Gundaveni Shylaja, Prashanth Ragam, and B. Naga Jagadesh

Abstract As we know the population in the cities of our country is increasing day by day due to migration of the people from one state to another state because of their business purpose and their children's education purpose for a quality education. As the population is increasing daily in the cities even the vehicle usage of the people is also increasing. As we know according to the population of the city and according to their financial status every family has a vehicle for their traveling in the city. Here the problem arises when many vehicles in the city the car or bike vehicles are unable to find the parking area to park their vehicles. The smart car parking system became an essential need for the above-mentioned problem. Traffic caused by multiple cars vying for the same garage place results in unnecessary time consumption, more oil usage, and eco-friendly damage. Whereas outcome concerns our setup shows that fog-based performance of the well-organized car parking system keeps down latency crucially. From our implementation, it is noticed that cloud-based smart car parking is not suited for stressful full network usage whereas the proposed fog-based model collectively brings down the network traffic which saves several resources such as waiting time, and reduces fuel usage, and pollution.

Keywords Fog node · Cloud server · Sensor layer

1 Introduction

The automobile industry is getting bigger to bigger according to the current advanced technology standards. And personal conveyance has become a powerful transport resource now [1]. Because of this rapid increment of private transport vehicles in today's world, availability of the free parking space in the required place or region for parking a car or bike became a very difficult task, especially in the most populated cities like Mumbai, Delhi, has not only become difficult it also results in increasing

G. Shylaja · P. Ragam (✉) · B. N. Jagadesh
School of Computer Science and Engineering, VIT-AP University, Amaravati 522237, India
e-mail: prashanth.rajam429@gmail.com

B. Annappa et al. (eds.), *High Performance Computing, Smart Devices and Networks*, Lecture Notes in Electrical Engineering 1262,
https://doi.org/10.1007/978-981-97-7794-5_14

of the pollution in the city like unnecessary fuel consumption. Because of these types of situations in those cities in today's world, the most polluted city is Delhi. Due to this lack of availability of car parking places in highly populated cities one hand of the city is polluted [2], with harmful smoke pollution, and on the other hand, due to the lack of availability of parking slots people started parking their vehicles on the roads and on the streets of the cities which will cause a high problem of the traffic jam attacks [3]. Taking all the scenarios above mentioned. Therefore, a smart car parking system has come into an idea in order to deal with the problems listed above. Usage of cloud servers and (IoT) [4], researchers have created a smart car parking system that considers all the considerations. So that drivers may park their cars in the designated spaces, and it will aid them in making the right choice. This will cut down on time wasted and excess carbon dioxide produced. Although it will also broadcast information about parking spaces not being available to users, the combination of fog nodes allows it to transmit information about nearby vacant parking spaces [5].

2 Background

There are several problems with cloud computing, including longer reaction times and increased communication costs [6]. Azam et al. [7] have found that fog nodule greatly decreases information transmission delay. This prime alternative is to address their needs for multiple links with less response time services which transfer from cloud server to fog node with quick transfers along with fast operations [8]. Fog computing is a new paradigm that can help to address these challenges. Fog computing brings computing and storage resources closer to the edge of the network, which can reduce latency and improve network utilization. [9]. Fog computing's ability to reduce the frequency of data transmissions to the cloud is one of its main advantages. This reduces the amount of network traffic that the apps using the cloud must use, thus improving response times. Numerous studies have demonstrated that fog has lower latency than clouds. For instance, the authors showed in [10, 11], and [12] that fog has lower latency than the systems' cloud-based implementation. Fog computing also improves scalability and minimizes network traffic, making it ideal for IoT frameworks [13]. For real-time applications, network use is always an important factor that can be substantially reduced with fog computing [14]. At the time a car approaches the garage place there by the proposed car parking architecture plays a major with the help of fog computing and taking pictures of vacancies of the parking lots in the garage using the camera's sensors after taking the inputs of the slots thereby it displays the vacancy in the car parking screen which is placed in the entry of the garage. Fog computing is necessary for the suggested technique since the data on available parking spaces must be updated often, making it desirable to minimize latency and network bandwidth usage. Fog computing, which stresses executing calculations near the network edges, can be used to resolve problems in an efficient manner, preventing the need for frequent data transports to the cloud [15].

3 Methodology

For this, the fog-three-tiered design has been proposed for a successful automobile parking system. Parking lot data will be displayed on the entry of the screen of the garage once the parking lot photos have been captured by the camera sensors there after the implemented fog nodes in the different areas of the garage places take up the role and start computing those captured images by the camera sensors [16]. Thereafter, it shows the data of empty parking lots which will display the data on the parking screen like wherever the specific parking lot is empty (Fig. 1).

Explanation of the architecture of the smart car parking system. Generally, without using the cloud server we cannot directly set the architecture of the fog node in our proposed smart garage technology. The intermediate layer between the fog node and the cloud server is called a forward proxy. This forward proxy helps gather all the fog nodes' data and sends it to the cloud server after a certain time when the data is not required by the fog node. The suggested fog-based architecture, as we are aware, comprises a variety of components, including cameras, microcontrollers, fog nodes, chips, smart LEDs, and cloud servers. The parking spaces will be photographed by the cameras. Once the photos have been taken. The image processing method is

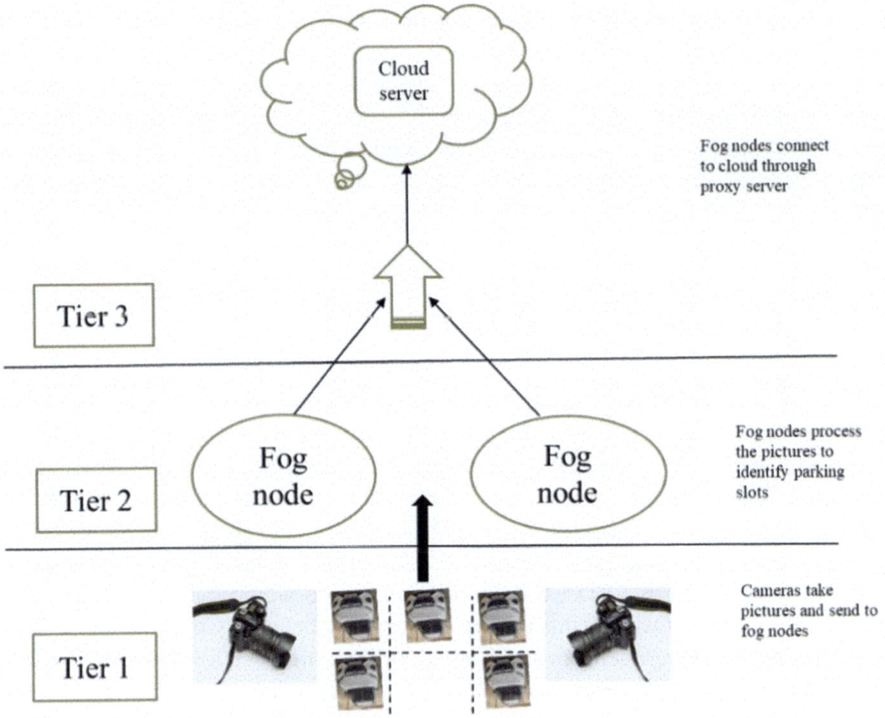

Fig. 1 Fog-based smart parking system [11]

used to determine whether parking spaces are available [17]. The fog node enters the scene after images have been taken using cameras placed in open parking spaces; it acquires the images and refreshes the parking status display every five seconds. There will be plenty of parking spaces available, so in order to cover every parking space, we must deploy many cameras that are positioned across the parking area. In this architecture, even the microcontroller will play a major role. The microcontroller will serve as the link between the cameras and the fog node, which will communicate through the cloud server and transfer information to the cloud server in the wake of a predetermined amount of time span [18]. We are aware that we cannot have a fog node without a cloud server.

Different fog nodes may be deployed at various parking spaces depending on the situation. For suppose let us consider a scenario if the car owner needs a specific parking lot in the required location, he needs to send the required data to the owner with the help of the internet selection option. There by the fog nodes come into play role it computes the input data given by the car owner and checks the vacancy of the parking lots in the required area. In the worst-case scenario if the required parking area is no parking lot vacancy there by the fog node computes again and fetches the next-door parking lot vacancy to the car owner. One design suggested a fog-based parking system, which can accommodate both a single parking space and several spaces. If there is a scenario where we can only set up one fog node for a parking garage based on its parking lots, it will transfer data to the cloud server for long-term storage if it is the only fog node deployed in that area. On the other hand, shows how a single cloud server is connected to several parking lots and fog nodes. In this scenario, network utilization and latency will be constant for every fog node, but the amount of duration and connectivity usage need to pass and download information from a centralized remote server which grows the number of components in the suggested architecture. Our project is essentially computerized.

3.1 Camera and Microcontroller Layer

To manage how our setup operates, we will utilize a microcontroller. The IR SENSOR installed at the entry will emit a high level of light whenever a vehicle enters the parking lot. The microcontroller then receives this output and feeds it to its interrupt pin. After running some code in the interrupt service routine, the microcontroller decreases the available capacity and outputs an intelligent signal to direct the car to its parking location. The LCD displays "Parking Full!" when every parking space is filled. Initially green LED LIGHT will come on to signal that a car may enter the parking lot [19]. Overall, this picture capturing of the empty parking lots will be done through the camera sensors which are designed with specialized microcontroller chips and each camera sensor uses 12 V. Cameras have been set up; therefore, an image of the parking spaces must be given in the parking space by making a camera angle of the top view with camera sensors.

After the picture capture phase, digital image processing comes into the play role using features of the acquired image's pixels, it divides the image into various portions using the features of the acquired image's pixels, it divides the image into various portions, and this transforms the acquired image into an RGB image. A grayscale version of the resulting RGB image is then created, which is then threshold-optimized before the last binary picture is captured for the segmentation process. After the binary image has been enhanced to eliminate unwanted noise, it is detected to see if the parking place is open. The following are the procedures for taking the photo and locating the parking space: Initializing the System the Position the cameras still, For Image Acquisition, It Takes a photo of the parking lot. For Image segmentation, it takes an RGB picture and creates a gray. The final image required for segmentation is taken from a scaled-down RGB image using thresholding techniques. For image enhancement technique it eliminates noisy image noise. Finally, for image deletion, it extrapolates the image's features.

3.2 *Fog Node*

Fog nodes are used to gather data in order to determine the condition of the parking space more accurately. They are the layer between the cameras and the cloud. Parking lots are designed in such a way that each parking space has its own index ID, like parking vacancy (PV1), which denotes row 1 and the vacancy of the parking lot lane. Parking spaces and their current status are updated via fog to smart LEDs that may be seen at the parking area's entrance. The image data is momentarily stored by this fog node before being sent to the cloud [20]. This proposed architecture of the fog node computing plays a major play role which is set up between edge nodes and the cloud server which helps in gathering the information about the parking lots by it starts computing the information provided by the cameras. These parking spots at the edge are shown in real-time data.

3.3 *Layer of Clouds*

The cloud's function in this framework is to keep all picture data in storage and to delete it when the fog node no longer requires it. Only proxy servers allow for communication between this fog and the cloud. There will be some exception cases in which sometimes the fog node needs to retrieve the information from the cloud server to know the past parking lots data here without any fraction amount of delay the cloud server sends the data to the fog node. Here we observe and declare that interoperability is one of the important characteristics of computing it also has a significant impact when we consider the natural variety of these edge nodes. Fog nodes work together by sharing their resources to help each other with computing and storing data. This way, they can meet the needs of all nearby fog nodes. In this scenario, we assume that fog nodes may always interact with one another via proxy servers and via nearby nodes to exchange crucial information. Since this interoperability is an intrinsic quality, the communication between fog nodes will not account for the delay. If there is no vacancy in the parking lots at the parking area where the requesting node will show the vacancy of all the neighboring parking areas.

3.4 Requirements and Specifications for Support

(a) **Software requirement analysis and planning**

User characteristics in the development iFogsim smart car parking system we are having three layers: In the first-tier layer we require Cameras (CCTV)**,** Parking slots, Cars (vehicles)**,** Parking lanes**,** Android mobile application (3G/4G, and good WIFI services)**,** RFID, laser, infrared, radar, ultrasonic, acoustic.**,** and twelve sensors are connected are connected to raspberry. In the second-tier layer, we require a Fog node, Microcontroller device**,** Chip, and routers. In the third-tier layer, we require a Cloud server and LED screens. By using Microsoft Azure Web Services, the raw, unprocessed data from different application sources are transferred to the SQL (structured query databases). The General constraint for parking system it processes the data before utilizing it. Users can obtain parking system data and find out more about the available parking spaces by using the admin portal. The gateway can be used to locate the problem sensor. The data is sent to the mobile app, allowing the user to access and check for available parking spaces.

Constraint 1: The 0.02-to-0.4-m infrared sensor range is one of this architecture's drawbacks.

Constraint 2: To access the data, consumers need a smartphone. The system is useless for users if they do not possess smartphones.

For instance, while using a cloud server, latency issues and network utilization are more prevalent. When a lot of people access the data at once, the transmission of the data is delayed and too late. Fog has been added to smart transportation in order to improve computational power and decrease delay rate.

Assumption: Since we are aware that the sensors and cameras would identify vacant parking spaces, let us assume that this car parking system will operate with ultrasonic sensors. Without using the fog, the cloud server plays a large part in the game. Cloud computing has several drawbacks, including slower reaction times and more expensive communication. However, if a fog node is used between the cameras and the cloud server, the latency will be reduced and the data will be transmitted much more quickly than it would have been otherwise. The delay and network usage issues related to cloud services are proven to be significantly minimized through fog computing.

Dependency: It is completely dependent on the sensor layer is the third layer and includes CCTV cameras, lasers, infrared radar, ultrasonic, and ultrasonic radar. The communication layer is linked to the sensor layer. The communication layer acts as a bridge between the application layer and the sensor layer [21]. For the communication layer to work, 3G/4G or WIFI services must be accessible and even a microcontroller then uses a WIFI connection to send the photographs it has taken to the fog. To be considered as a component of the system, the device (android smartphone) needs to be registered using the web API. The sensor number allows for the detection of the slot's value.

4 Experimental Set Up

We designed a setup where high-definition smart cameras capture images of parking spaces, and these images are then sent to the fog node as soon as they are available. After analyzing the images to identify the vacancy of parking slots, the fog node displays the images of the slots on a smart LED connected to it via Wifi. To implement the connection between the fog node and the cloud server, a proxy server is deployed. We used iFogSim, an IoT device toolbox, to imitate various scenarios [22]. Through iFogSim, we can evaluate the delay and how efficiently the network is utilized for the simulation, we defined the variables for parking spaces and camera counts. During our experiment, we established four parking lots. At first four cameras were installed in each parking lot to capture images of the area. Our fog nodes are connected to a central cloud device that we built utilizing a proxy server. It is important to note that we made one fog node for each location. We constructed using the guidelines of the simulation environment, the cameras were used as sensors. Since they are equipped with Wifi capabilities and connected via a microcontroller, they possess intelligent connectivity. To examine the results for different scenarios, we augmented the number of camera sensors. To compute the effects on the response time and overall network bandwidth usage by a fog node, we increased the number of cameras for a particular scenario.

5 Outcomes of the Experiment

The following outcomes of fog-based network systems in relation to delay and network usage, along with a comparison to the cloud server implementation. As per the outcomes of experiments, the fog-based application exhibits reduced delay and lower network bandwidth usage compared to the cloud server version. Table 1 shows the consequences of the cloud-based implementation as well as the network utilizations and latency results in the fog environment.

Table 1 Simulation findings for the suggested parking structure for clouds and fog

Cameras	Fog latency (ms)	Cloud latency (ms)	Fog network usage	Cloud network usage (Kb)
14	6.54	7.425	2563.12	25,632.14
22	7.26	10.2	4000.25	38,069.50
26	9.48	1.073	4952.16	48,224.65
40	10.02	292.6	7996	638,802.54
44	10.3	1968.6	8795.6	64,557.22
48	10.73	2886.4	9595.2	64,557.22
32	9.30	292.6	6396.8	62,398.28

5.1 Analysis of Latency

Reduced latency is necessary in situations when excellent real-time performance is required. Fog computing offers the benefit of decreased delay and rapid response times for customer devices by performing computations at the network's edge, instead of frequently relying on the cloud for processing. Utilizing the delay, which is determined from [10].

$$\text{Latency} = x + y + z$$

where while taking photographs and video, the triple CPU execution delays y. The time to upload pictures for processing and storing on the fog node. And finally, z processing at the fog node.

5.2 Assessment of Network

As the server's traffic volume increases, only the cloud's resources are used, leading to a rise in network usage. This increased traffic on the cloud server results in reduced data rates on the network. For geographically dispersed servers, one fog node is set aside to handle requests from one specific region. Here, the formula is:

$$\text{Network Usage} = \text{Latency} * \partial$$

5.3 *Key Findings in Experiment*

The usefulness of the suggested fog-based smart parking technology solution is displayed by experimental results. As can be seen, multiple fog and cloud-based setup scenarios were created, and it was possible to realize varying numbers of cameras connected to fog nodes and cloud servers. Here, created four fog nodes to examine the effects of fog computing. Every fog node is linked to sixteen cameras. Furthermore, scaling up the system requires adding more fog nodes and cloud servers. In the case of twenty cameras, each fog node has five cameras connected to it. For the sake of performance measurement, the total number of fog nodes remains constant throughout all situations. We used a router to connect the cameras to the cloud in case of a cloud emergency. In the cloud example, we used a router to connect the cameras to the cloud.

6 Conclusion

Fog computing has recently taken on a crucial role, particularly in application domains where time is of the essence. The demand for faster responses has increased along with the number of data-generating devices. To achieve that, we suggested a computer vision-based fog-based smart parking architecture to detect parking spaces. Vacant parking spaces, allow cars to identify a parking lot in a particular period span which helps the car to consume less amount of gasoline. The testing results show that, in comparison to the cloud, the suggested fog design requires less network bandwidth while simultaneously reducing latency. The proposed research has some limitations, including the use of cameras to identify parking spaces. Since the images of the cars like car number, company name of the car, and owner image of the car will be stored in the cloud server, this data should be maintained in private for the car owners because as we know hackers are everywhere in cyberspace if once the information is leaked, they start to misuse the data. Here by keeping this a main point of the security information we have used proper encryption techniques, it is possible to protect the privacy of data stored in the cloud, and this can be a key topic for future research. It is also very important to note that the increasing number of parking places will require widespread application of the suggested framework to preserve efficiency. As a result, In the future, we plan to look into the load-balancing problems in fog nodes and provide a workable solution.

References

1. Attaran M (2023) The impact of 5G on the evolution of intelligent automation and industry digitization. J Ambient Intell Humaniz Comput 14(5):5977–5993
2. Demir S, Basaraner M, Gumus AT (2021) Selection of suitable parking lot sites in megacities: A case study for four districts of Istanbul. Land Use Policy 1(111):105731
3. Aljohani M, Olariu S, Alali A, Jain S (2021) A survey of parking solutions for smart cities. IEEE Trans Intell Transp Syst 23(8):10012–10029
4. Sarker VK, Gia TN, Ben Dhaou I, Westerlund T (2020) Smart parking system with dynamic pricing, edge-cloud computing and Lora. Sensors 20(17):4669
5. Celaya-Echarri M, Froiz-Míguez I, Azpilicueta L, Fraga-Lamas P, Lopez-Iturri P, Falcone F, Fernández-Caramés TM (2020) Building decentralized fog computing-based smart parking systems: From deterministic propagation modeling to practical deployment. IEEE Access 8:117666–117688
6. Mukherjee M, Shu L, Wang D (2018) Survey of fog computing: fundamental, network applications, and research challenges. IEEE Commun Surv Tutorials 20(3):1826–1857
7. Aazam M, Zeadally S, Harras KA (2018) Fog computing architecture, evaluation, and future research directions. IEEE Commun Mag 56(5):46–52
8. Chang YCP, Chen S, Wang TJ, Lee Y (2016) Fog computing node system software architecture and potential applications for nb-iot industry, In: 2016 International computer symposium (ICS). IEEE, pp 727–730
9. Alrawais A, Alhothaily A, Hu C, Cheng X (2017) Fog computing for the internet of things: security and privacy issues. IEEE Int Comput 21(2):34–42
10. Dar BK, Shah MA, Shahid H, Naseem A (2018) Fog computing based automated accident detection and emergency response
11. Saharan K, Kumar A (2015) Fog in comparison to cloud: a survey. Int J Comput Appl 122:3
12. Atlam H, Walters R, Wills G (2018) Fog computing and the internet of things: a review. Big Data Cognitive Comput 2(2):10
13. Dastjerdi AV, Gupta H, Calheiros RN, Ghosh SK, Buyya R (2016) Fog computing: principles, architectures, and applications, In: Internet of things. Elsevier, pp 61–75
14. Dastjerdi AV, Buyya R (2016) Fog computing: Helping the internet of things realize its potential. Computer 49(8):112–116
15. Khattak HA, Islam SU, Din IU, Guizani M (2019) Integrating fog computing with vanets: a consumer perspective. IEEE Commun Stan Mag 3(1):19–25
16. Wang H, Liu T, Kim B, Lin CW, Shiraishi S, Xie J, Han Z (2020) Architectural design alternatives based on cloud/edge/fog computing for connected vehicles. IEEE Commun Surv Tutorials 22(4):2349–2377
17. Koushika B, PraveenKumar G, Priyadharshini ST, Balamurugan A (2021) An efficient car parking slot identification system using image processing. Annal Rom Soc Cell Biol 9:3400–3414
18. Benrazek AE, Farou B, Kouahla Z, Ferrag MA, Seridi H (2023) IoVT-based efficient solution for optimal active smart camera selection in a tracking mission. Int Things 1(24):100941
19. Bakal SS, Gokhale AP, Smart parking system for marriage halls
20. Ahmed A, Mohan A, Gene C, Guillaume P (2020) Docker container deployment in distributed fog infrastructures with checkpoint/restart. In: 2020 8th IEEE International conference on mobile cloud computing, Services, and Engineering (MobileCloud), IEEE pp 55–62
21. Lamonaca F, Sciammarella PF, Scuro C, Carnì DL, Olivito RS (2018) Synchronization of IoT layers for structural health monitoring. In: 2018 Workshop on metrology for industry 4.0 and IoT, IEEE, pp 89–94
22. Lera I, Guerrero C, Juiz C (2019) YAFS: a simulator for IoT scenarios in fog computing. IEEE Access 7:91745–91758

Generalized Procedure for Cryptanalysis of ARX-Based Block Ciphers

Praveen Kumar Gundaram, Appala Naidu Tentu, and Neelima Guntupalli

Abstract Satisfiability Modulo Theories (SMT) investigate the functionality of first-order formulas that utilize operations from various theories, including Booleans, bit-vectors, arithmetic, arrays, and recursive data forms. SMT solvers can efficiently resolve a wide range of logic, evaluate hardware and software security, and determine solutions in polynomial time. This paper introduces an algebraic cryptanalysis algorithm that effectively and precisely detects vulnerabilities in bit-level operations by utilizing the SMT solver. The proposed method changes a cryptographic algorithm into a format that an SMT solver can understand. The solver then finds keys for plaintext–ciphertext pairs. This method can also be used to develop new cryptanalytic methods to find cryptographic algorithm weaknesses. Furthermore, this approach is also useful for automated verification, testing of cryptographic algorithms and developing new cryptographic algorithms that are more immune to cryptanalytic attacks. The case study shows the SPECK block cipher cryptanalysis results of the proposed method. For instance, the case study showed how the SPECK block cipher could recover a partial round (up to 4 rounds) key in less than 1214 minutes using the proposed approach, which would have taken more than a day to get a key using traditional cryptanalytic techniques. Overall, this method provides an effective way to analyze cryptographic algorithms and is a powerful tool for analysts.

Keywords Cryptanalysis · Block cipher · Satisfiability checking · SMT solver · System of equations · Symbolic computation · Speck cipher

P. K. Gundaram (✉) · N. Guntupalli
Department of Computer Science and Engineering, Acharya Nagarjuna University, Guntur, Andhra Pradesh, India
e-mail: praveenkumar@crraoaimscs.res.in

P. K. Gundaram · A. N. Tentu
C R Rao AIMSCS, UoH Campus, Gachibowli, Hyderabad, Telangana, India

B. Annappa et al. (eds.), *High Performance Computing, Smart Devices and Networks*, Lecture Notes in Electrical Engineering 1262,
https://doi.org/10.1007/978-981-97-7794-5_15

1 Introduction

SAT (satisfiability) solvers can be used to analyze algorithms and get solutions. We only consider SAT solvers [1] that generate satisfiable values of the algorithm constraints and variables, if any exist, in order to solve the SAT problem. In SAT solver-based cryptoanalysis, the calculations of a cryptographic algorithm are stored in secret key variables, like key bits, which is similar to algebraic attacks. A system of multivariate equations derived from a finite field is utilized in algebraic attacks to encode bits. The equations containing the temporary variables are subsequently solved using algebraic methods. Automatically determining whether a logical formula complies with the constraints imposed by a theory (e.g., arithmetic, bit vectors, arrays, or sets) is satisfiability. When it comes to the validation, optimization, and synthesis of software and hardware, SMT [2] solvers are utilized. In addition to this, it enhances the automation of complex security systems, artificial intelligence, and cyber-physical systems.

Contribution: This study investigates the impacts of attacks on the ARX cipher. We present the complexity of SMT solver-based cryptanalysis and propose a cryptanalysis method for the ARX cipher using the SMT solver. Develop a user interface that facilitates the conversion of cipher encryption functions to the SMT solver interface format. Input plaintext and ciphertext to the interface tool, and subsequently impose encryption constraints in order to extract probable keys. The proposed procedure is implemented in the SPECK [3] cryptographic algorithm as a case study of the methodology and evaluates the time complexity of each possible key. The results are tabulated at the same time to validate the possible key. The results of the tabulation must be further analyzed and compared to the master key. The tool should generate a report that summarizes the results and recommends the best possible key. The interface should also be configured to allow users to manually adjust the encryption constraints.

This paper is formatted as described below: In Sect. 2, we introduce the SMT solver tool, its preliminary steps, and cryptanalysis application of SMT solvers is discussed. its methodology. In Sect. 3, we discussed about cryptanalysis techniques as well as existing result of various ciphers. In Sect. 4, an ARX cipher and cryptanalysis with the SMT solver algorithm are suggested, and the results are presented. In Sect. 5, the conclusion is presented.

2 SMT Solver

Satisfiability Modulo Theories (SMT) [4] solvers are specialized software designed to resolve logical formulas composed of various mathematical structures or theories. The solvers are utilized across various domains, including but not limited to hardware design, automated reasoning, software verification, and cryptography. SMT solvers can be used in many hard problems across domains because they can accommodate multiple logical theories and provide customizable components.

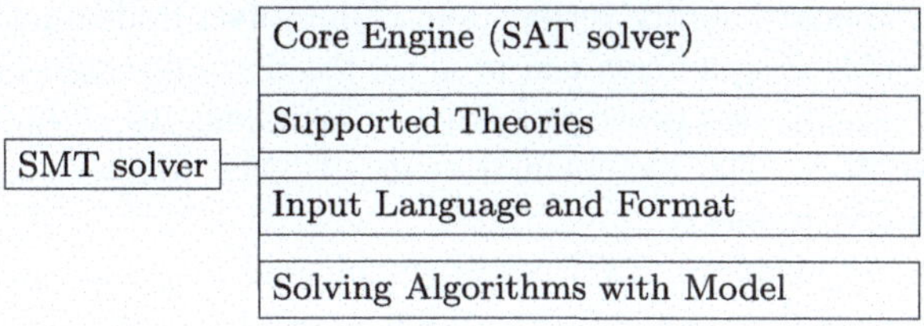

Fig. 1 Components of SMT solver

The following are the primary components of an SMT solver, as shown in Fig. 1: The core engine is comprised of SAT [5] and a logical constraint solution. The solver provides support for arrays, arithmetic, bit vectors, and custom theories. Standardized formats are utilized to represent logical formulas in order to ensure compatibility with a wide range of problem descriptions and tools. Algorithms and methods that assist solvers in exploring the solution space in order to determine the satisfiability of complex problems and generate models.

2.1 Fundamental Elements of SMT Solvers

SMT solvers are a combination of satisfiability solvers and theory solvers. SMT solvers are used to find the most efficient solution space for cryptographic algorithms. SMT solvers are used to verify the correctness of algorithms and identify vulnerabilities. They are also used to verify the security and robustness of cryptographic algorithms. SMT solvers are an important tool in the development of secure cryptographic systems and cryptanalysis frameworks. Support is provided for multiple input formats, including SMT-LIB and Z3 [4]. SMT solvers have been enhanced by technological advances in data structures, algorithms, and optimization techniques. SMT provides evidence for the resolution of complex and large-scale problems [6]. These are descriptions of SMT solvers and their operations.

- Logical Formulas: SMT solvers receive first-order logic formulas. The formula can include propositional variables, quantifiers, arithmetic expressions, bit-vector operations, or mathematical theory relations.
- Theories: SMT solvers support multiple math theories. Theories like array, string, arithmetic, and bit vector formulate axioms.
- Satisfiability checking: SMT solvers [4] check input formulas.
- Decision procedure: SMT solvers use formula-specific algorithms and procedures. Bit-vector theory solves problems using bit-wise algorithms instead of integer and real-number algorithms in arithmetic.
- Solution Search: If the formula is met, the solver can search. The variable assignments that produce a model that meets constraints or validates the formula are identified in this procedure.

- Unsatisfiability verification: If the solver finds the formula unsatisfiable, no variable assignments can produce the desired result. Formal methods and software verification prove properties or invariants for all inputs.
- Mode: The incremental mode of SMT solvers allows satisfiability queries and constraints.
- Applications: SMT solvers go beyond software and hardware verification, model checking, symbolic execution, constraint solving, and optimization.

SMT-LIB, a widely used standard among SMT solvers including CVC4, Boolector, Solar, Yices, and Z3, is used to define logical formulas that are compatible with multiple SMT solvers [7]. SMT solvers generalize Boolean satisfiability through the utilization of quantifiers and first-order theories [6]. SMT solvers evaluate the formulas of these theories to make sure they are valid and satisfying. Applications of SMT solvers include bound model verification over many domains, predicate abstraction, test case generation, and extended static verification. SMT solvers are generally beneficial automated analytical tools that have the potential to enhance the accuracy, performance, and security of a wide range of systems and applications.

2.2 Z3 Solver

Microsoft Research built the Z3 solver [4], which is a high-performance theorem prover. The Z3 solver possesses the ability to efficiently and precisely prove mathematical statements. It has also been implemented to resolve practical issues, including the validation of software integrity and security, as shown in Fig. 2. Z3 provides an extensive library of predefined functions and a robust API for resolving logical constraints. It develops models and uses the solver to solve application and software testing problems. The models assign values to input equations and create partial function graphs for predicates and function symbols. Since Z3 can solve many difficult logical and mathematical problems, it is useful in many applications.

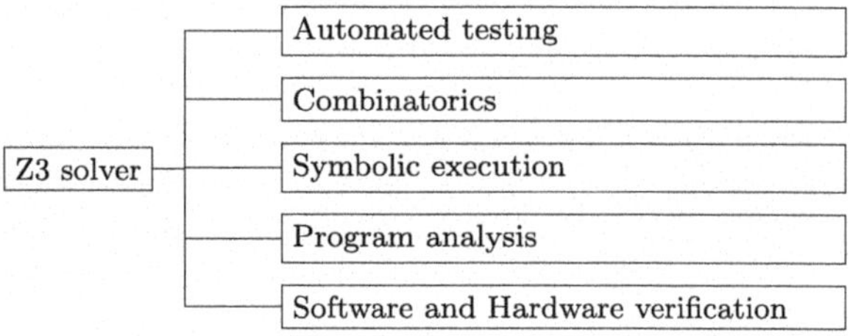

Fig. 2 Components of Z3 solver a model checker

2.3 Security Applications Based on SMT Solvers

The SMT solver is a robust general-purpose search tool [6]. Satisfiability Modulo Theories solvers are powerful tools used in various fields, including cryptography, to automate the process of verifying logical formulas involving complex theories or mathematical structures. SMT solvers are commonly used as black-box solvers for reduced equations. In the cryptographic field, SMT solvers are particularly useful for analyzing security properties, finding vulnerabilities, and verifying the correctness of cryptographic protocols and algorithms. SMT solvers pursue cryptanalysis for many reasons. It examines various cryptographic algorithm techniques and compares them to those already in existence. Finally, it can analyze algorithm flaws to understand cryptanalysis. In general cryptanalysis techniques, an SMT tool [4] is utilized to locate the master secret key.

- Finding cryptographic keys
- Modular root finding
- A collision attack
- Pre-image attacks
- Differential cryptanalysis
- Verification of cryptographic primitives

In order to perform cryptanalysis, the following strategy is used in SMT frameworks [8]:

i. Input constraints: Algebraic cryptanalysis solves finite field equations and inequalities. Input constraints consist of equations and inequalities that describe the cryptographic algorithm, its properties, and the desired result.
ii. SMT Solver Core: Using algorithms, the SMT solver core solves and verifies the satisfiability of input constraints. Some SMT solvers are Z3, CVC4, and Yices.
iii. Solver Interface: This component manages communication between the user and the SMT solver. The SMT solver accepts user-supplied constraints and returns the solution to the user.
iv. Finite Field Arithmetic: Algebraic cryptanalysis frequently uses finite fields including multiplication, and inverses in finite fields.
v. Solving Strategies: Depending on the nature of the algebraic constraints, various solving strategies may be used. Strategies may consist of linearization (converting nonlinear equations to linear ones), Gaussian elimination, and Grobner basis computations, among others.
vi. Theory Solvers: Algebraic cryptanalysis may involve theories from diverse mathematical fields (e.g., linear algebra, number theory, algebraic geometry). The SMT solver uses specialized theory solvers to handle these mathematical domains.
vii. Optimization Techniques: Cryptanalysis frequently necessitates the selection of the most probable solution from a pool of possibilities. Optimization techniques can be used to direct the solver in the direction of solutions with particular properties (e.g., maximize probability, and minimize key space). As soon as

the SMT solver determines satisfiability or provides a solution, the results must be analyzed. The solution may reveal cryptographic secrets, such as encryption keys or security vulnerabilities.

viii. Output: The solver must provide the user or the system with clear and comprehensible output. This may include equation solutions, statistical information, and vulnerabilities discovered.

ix. Performance Enhancements: Cryptanalysis SMT solvers frequently require high performance and optimization. Parallelization, heuristics, and hardware acceleration are techniques that can be used to accelerate problem-solving.

x. Integration with Cryptanalysis Tools: In order to provide a comprehensive solution for analyzing cryptographic algorithms, the SMT solver may need to be integrated with other cryptanalysis tools and frameworks.

3 Cryptanalysis Technique Using SMT Solver

Cryptanalysis of ARX ciphers evaluates their security by identifying exploitable vulnerabilities that allow recovery of the secret key or plaintext. SMT solvers automate reasoning over logical formulas and are used in formal verification and security analysis. Cryptographic properties and constraints can be encoded as logical formulas [4]. They can be used to search for solutions and validate the absence of particular properties. SMT solvers can be used in different ways for cryptanalysis, depending on the attack scenario and cipher. These methods are common:

i. Differential Cryptanalysis: SMT solvers can analyze block cipher differential characteristics. Differential cryptanalysis examines cipher round differences to find weaknesses. By encoding the operations and constraints of a cipher as logical formulas, SMT solvers can search for differential characteristics that lead to successful attacks.

ii. Linear Cryptanalysis: Linear cryptanalysis focuses on exploiting linear approximations in the cipher to deduce key bits or other security-related information. Encoding the cipher operations as linear equations enables SMT solvers to search for linear approximations.

iii. Algebraic Cryptanalysis: Algebraic cryptanalysis solves cipher equations algebraically to recover the secret key or other pertinent data. Encoding these algebraic equations as logical formulas allows SMT solvers to help. Additionally, they can search for assignments that reveal the key or other targeted properties.

For cryptanalysis, the SMT solver has not been extensively investigated. SMT solvers accept more generic First Order Logic (FOL) formulas, while SAT only accepts CNF Boolean formulas. Thus, SMT solver can model block cipher cryptanalysis problems like SIMON, SPECK, IDEA [9], and TEA [10] as shown in Table 1.

Table 1 Cryptanalysis comparison on ARX cipher

Cipher	Rounds	Attacked	Time	Data	Memory
SIMON [11]	24/44	Differential	$2^{58.427}$	$2^{62.012}$	2^{32}
SIMON [12]	10/44	Algebraic	$2^{29.8}$	$2^{2.59}$	Negl.
SIMON [9]	31/44	Linear-hull	2^{120}	$2^{63.5}$	$c.2^{31}$
SPECK [13]	4/44	Key recovery	2^{16}	2^{36}	Negl.
IDEA [8]	2/8.5	SMT solver based	2	Negl.	Negl.

4 A Proposed Method for Cryptanalysis

Adding, rotating, and XOR (ARX) block ciphers are symmetric-key cryptographic algorithms that can encrypt and decrypt data blocks of a certain size by performing basic operations like addition, rotation, and XOR. These ciphers are widely acknowledged for their effectiveness, simplicity, and resistance to attacks. Input blocks are subdivided into smaller sub-blocks by ARX ciphers, which operate in rounds. Each round uses addition, rotation, and exclusive operations. The master key controls all rounds, with each round utilizing a round key. The number of rounds in an ARX cipher depends on the block size and security level desired. In general, more rounds of ciphers provide higher security but are more challenging to discover. ARX ciphers are compatible with numerous platforms, including low-power devices, by virtue of their simplicity. Consequently, ARX ciphers remain a popular choice for secure data transmission.

4.1 Preliminaries

ARX ciphers use many operators to encrypt and decrypt, such as $\oplus$, $\boxplus$, $\odot$, ... Our proposed algorithm makes use of these operations, and for enhanced optimization, we provide customized functions for modulo operations, as illustrated below. The SMT solver for ARX ciphers consists of both legacy and custom functions with 16-bit and 32-bit operations.

Operation	*Example*: $fun(x, y)$ where x, y are SMT variables
Multiplication	$Z3mul = simplify((ZeroExt(16, x) * ZeroExt(16, y))$
Addition	$Z3add = simplify((ZeroExt(16, x) + ZeroExt(16, y))$
Exclusive OR	$Z3xor = simplify(x \oplus y)$
Extract Bits	$Z3ext = Extract(15, 0, Z3mul)$
Concat Bits	$Z3cat = Concat(x, y)$
Rotate right	$Z3rr = simplify((RotateRight(x, 7))$
Rotate left	$Z3rl = simplify(RotateLeft(y, 2))$

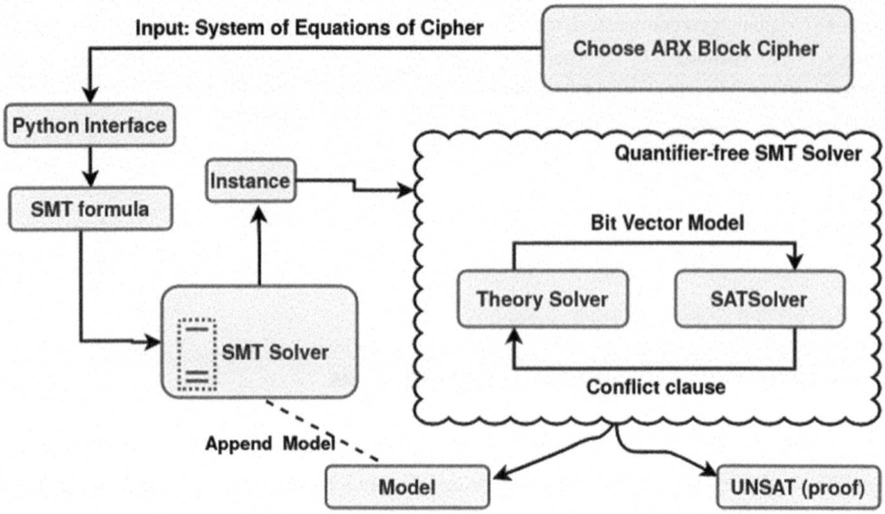

Fig. 3 Schematic view of an SMT solver

4.2 Procedure for Cryptanalysis of ARX Block Ciphers

In general, the cryptographic algorithm and the key space determine the cryptanalysis procedure. By solving multivariate polynomial equations with finite fields, algebraic cryptanalysis compromises cryptosystems. The proposed algorithm illustrates the utilization of an SMT solver for the purpose of key recovery. In the Z3py system of equations, cryptanalytic patterns are identified using the SMT solver. The algorithm-1 demonstrates that SMT solvers can process this system as a set of equations. SMT solvers need HPC machines to calculate algorithm complexity [8, 10]. Consequently, SMT solvers serve as efficient tools for the detection and decryption of ARX cipher algorithms or partial keys.

The preparation of a schematic representation of a solver customized for algebraic cryptanalysis employing SMT would be laborious and time-intensive. Provide the following summary of the processes and components of a solver: A solver has a language for representing mathematical formulas, algorithms for solving them, and a system for verifying its output. A solver must be capable of efficiently solving complicated formulas. It must handle Boolean, linear, and polynomial formulas. A SMT solver-based ARX block cipher cryptanalysis procedure as illustrated in Fig. 3.

5 SPECK Block Cipher

The building blocks for the SPECK block cipher [15] is a Feistel network, a structure that numerous other block ciphers, such as [16] also use. It usually uses a 128-bit key

Algorithm 1 Cryptanalysis of ARX ciphers using SMT solver [14]

Input: Pair of plaintext and ciphertext (PT, CT)

Output: Key or possible $keys$

```
function Z3bvARXcipherOneRoundEqu(X, Y, Key, rounds)
    return EncEq(X, Y)
end function
function ARXcipherEncAlg(PT, Key_0, rounds)
    (X_0, Y_0) ← PT
    for r ≤ rounds do
        (X_{r+1}, Y_{r+1}) ← ARXcipherEncRound_r(X_r, Y_r, Key_r).
        Key_{r+1} ← keyScheduling(Key_r)
    end for
    return (X_{r+1}, Y_{r+1})
end function
Initialize (PT, Key); CT ← ARXcipherEncAlg(PT, Key, rounds).
Declare PT = (X_0, Y_0), CT = (X_n, Y_n) and Key_r (0≤r≤n−1) = k^r_j |^n_{j=1}.
s = Solver()
s.add(keyscheduling) ▷ Perform key scheduling to obtain key equations.
for r ≤ rounds do
    (X_{r+1}, Y_{r+1}) ← Z3bvARXcipherOneRoundEqu(X_r, Y_r, Key_r, r)
    s.add(X_{r+1}, Y_{r+1}) ▷ Add every cipher round constraint to the SMT solver.
end for
if s.check() = UNSAT then
    (PT, CT) pairs are invalid.
end if
count = 0
while s.check() = SAT do
    if count > solution then
        break
    end if
    Generate the model, then create solution block as s.model().
    for d ≤ block do
        block.append(solution)
    end for
    s.add(Or(block))
    count = count + 1
end while
count the number of solutions.
```

and works with 64-bit data blocks. There are several rounds in the SPECK cipher [17], and each round has a round function and a key schedule. The key schedule generates the key, and the round function proceeds to execute a sequence of operations on the encrypted data block. In 2015, [3, 13, 17] proposed an SMT-based attack against SPECK that took advantage of an operational vulnerability. The attack successfully recovered the SPECK secret key with a success rate of 2^{16} by using only 2^{36} plaintexts. There are many variations, all represented as $2n/mn$, where m is the number of keywords, $2n$ is the block size, and mn is the key size, where $n \in \{16, 24, 32, 48, 64\}$ and $m \in \{2, 3, 4\}$.

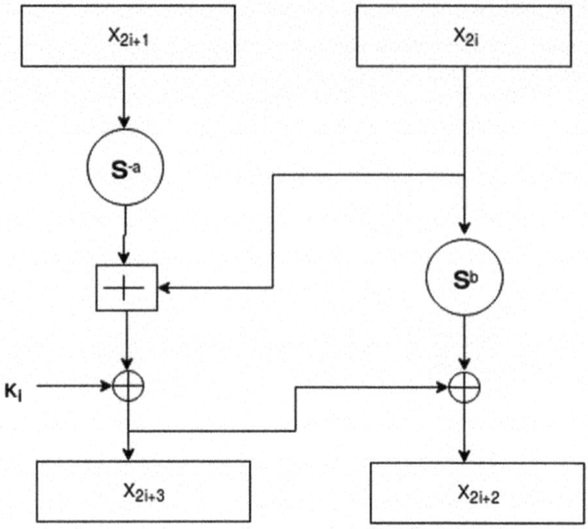

Fig. 4 SPECK block cipher round function

5.1 Encryption

Each round uses two n-bit parts, each of which is designed according to the classic Feistel design. The encryption algorithm of a single round of the SPECK block cipher [17] is depicted in Fig. 4. Each round of the cipher introduces a nonlinear, non-objective function.

$$F: GF(2)^n \rightarrow GF(2)^n$$

XOR, bitwise AND, and circular left shift (S), where S^{-b} implies circular right shift by b bits, as well as circular left shift (S) where S^a implies circular left shift by a bits, are the operations associated with the round function. Let x_i and y_i represent the input's left and right halves, respectively.

Then, R_k represents the round function as follows:

$$\begin{aligned} R_k(x_{i+1}, y_{i+1}) &= (f(x_i) \oplus k_i,\ f(y_i) \oplus f(x_i) \oplus k_i) \\ \text{where } f(x) &= (S^{-a}x + y) \oplus k \text{ and } f(y) = (S^b y) \end{aligned} \quad (1)$$

Key Scheduling: The key value is $K = (l_{m-2}, \ldots, l_0, k_0)$, where $l_i, k_i, \in GF(2)^n$, for a value of $m \in \{2, 3, 4\}$. k_i and l_i $\forall\, i \in [0, T)$ are sequences.

$$\begin{aligned} l_{i+m-1} &= (k_i + S^a l_i) \oplus i \\ k_{i+1} &= S^b k_i \oplus l_{i+m-1}. \end{aligned} \quad (2)$$

Table 2 Experiment results on PT-CT pairs for possible keys

#Rounds	PC pair (PL,PR) = (0x3b726574,0x7475432d)	10 Possible keys (in s)
1	(X_1, Y_1) = (0x3e901e99, 0x9d3a07f2)	1 m 25.715 s
2	(X_2, Y_2) = (0xbccd9352, 0x551dacc6)	2 m 2.494 s
3	(X_3, Y_3) = (0xf48f455a, 0x5c622368)	12 m 0.352 s
4	(X_4, Y_4) = (0x426fa5e2, 0xa17ebea0)	1214 m 42.129 s

5.2 *Attack Procedure*

To find keys to fixed plaintext and ciphertext pairs, follow these steps: Select a test case consisting of both plaintext and ciphertext for a predetermined number of rounds, utilizing the provided key. Subsequently, construct an interface for the given Z_3 solver to generate potential keys, and validate the resulting keys in order to identify the initial key. This experiment examines the execution of ten rounds of computation involving pairs of plaintext and ciphertext.

The following are two round constraints for recovering the key for a fixed plaintext/ciphertext pair in SMT format.

```
from z3 import *
x0,x1,x2,y0,y1,y2,k0,k1 = BitVecs('x0 x1 x2 y0 y1 y2 k0 k1',32)
s = Solver()
s.add(x0==0x3b726574, y0==0x7475432d)
s.add(x2==0x66da0218, y2==0x69258870)
s.add(x1==simplify((RotateRight(x0,7)+y0) XOR k0))
s.add(y1==simplify(x1 XOR RotateLeft(y0,2)))
s.add(x2==simplify((RotateRight(x1,7)+y1) XOR k1))
s.add(y2==simplify(x2 XOR RotateLeft(y1,2)))
print s.check() #check satisfiability
print s.model() #produce model of the result
```

Experiment Results: The SPECK cipher cryptanalytic technique is employed by the SMT framework to identify known (PT, CT) pairs for every round. The execution time is calculated in the table below according to the possible keys. In accordance with the information in Table 2, compute the execution time for ten possible keys using different ciphertexts and plaintexts (PL, PR) = (0x3b726574, 0x7475432d) for each round, which is required for twelve CPUs.

6 Conclusion

In this paper, we propose a framework for generating possible keys for given pairs of PT-CT pairs using SMT solvers. In order to generate potential keys for the SMT Solvers algorithm, we devise equations for an ARX-based cipher that employs fixed plaintext and ciphertext. A case study of the lightweight IoT cipher, that is, the SPECK block cipher, is utilized to assess the performance of the implemented algorithm. Test vectors are required to authenticate the probable keys derived from plaintext and ciphertext using the SMT tool. We observed that the SMT Solver performed better than alternative algebraic tools when we used ARX cipher algorithms. In the future, a hybrid strategy will be implemented to obtain the original key. The SMT solver-based cryptanalysis methodology will incorporate a statistical-based approach to constructing the maximum number of rounds to get the key.

References

1. Sorensson N, Een N (2005) Minisat v1. 13-a SAT solver with conflict-clause minimization. SAT 53:1–2
2. Kolbl S, Leander G, Tiessen T (2015) Observations on the SIMON block cipher family. In: Gennaro R, Robshaw M (eds) Advances in cryptology—CRYPTO 2015. CRYPTO 2015. LNCS, vol 9215. Springer, Berlin, Heidelberg
3. Huo Y, Zhang F, Feng X, Wang L-P (2015) Improved differential fault attack on the block cipher speck. In: 2015 workshop on fault diagnosis and tolerance in cryptography (FDTC). IEEE, pp 28–34
4. de Moura L, Bjorner N (2008) Z3: an efficient SMT solver. In: Ramakrishnan CR, Rehof J (eds) Tools and algorithms for the construction and analysis of systems. Springer Berlin Heidelberg, Berlin, Heidelberg, pp 337–340
5. Mironov I, Zhang L (2006) Applications of SAT solvers to cryptanalysis of hash functions. In: Biere A, Gomes CP (eds) Theory and applications of satisfiability testing-SAT 2006. LNCS, vol 4121. Springer, Berlin, Heidelberg
6. Brummayer R, Biere A (2009) Boolector: an efficient SMT solver for bit-vectors and arrays. In: Kowalewski S, Philippou A (eds) Tools and algorithms for the construction and analysis of systems. Springer Berlin Heidelberg, Berlin, Heidelberg, pp 174–177
7. Bjorner N, de Moura L, Nachmanson L, Wintersteiger CM (2019) Programming Z3. Springer International Publishing, Cham, pp 148–201
8. Sahu HK, Rajesh Pillai N, Gupta I, Sharma RK (2020) SMT solver-based cryptanalysis of block ciphers. SN Comput Sci 1(3):169
9. Chen H, Wang X (2016) Improved linear hull attack on round-reduced simon with dynamic key-guessing techniques. In: Proceedings of the international conference on fast software encryption, Bochum, Germany 20–23 Mar 2016, pp 428–449
10. Najim Alam M, Gundaram PK, Yerukala N (2023) Cryptanalysis of tiny encryption algorithm based on SMT solvers using HPC. In: Brahmananda Reddy A, Nagini S, Balas VE, Srujan Raju K (eds) Proceedings of third international conference on advances in computer engineering and communication systems. Springer Nature Singapore, Singapore, pp 59–68
11. Abed F, List E, Lucks S, Wenzel J Differential and linear cryptanalysis of reduced-round Simon. IACR Cryptology ePrint Archive, Report 2013/526. https://eprint.iacr.org/2013/526.pdf
12. Courtois N, Mourouzis T, Song G, Sepehrdad P, Susil P (2014) Combined algebraic and truncated differential cryptanalysis on reduced-round Simon. In: 2014 11th international conference on security and cryptography (SECRYPT). IEEE, pp 1–6

13. Beaulieu R, Shors D, Smith J, Treatman-Clark S, Weeks B, Wingers L (2015) The Simon and speck lightweight block ciphers. In: Proceedings of the 52nd annual design automation conference, pp 1–6
14. Gundaram PK, Tentu AN, Muppalaneni NB (2021) Performance of various SMT solvers in cryptanalysis. In: 2021 international conference on computing, communication, and intelligent systems (ICCCIS), pp 298–303
15. Beaulieu R, Shors D, Smith J, Treatman-Clark S, Weeks B, Wingers L (2013) The SIMON and SPECK families of lightweight block ciphers. Cryptology ePrint Archive, Paper 2013/404
16. Alkhzaimi HA, Lauridsen MM (2013) Cryptanalysis of the Simon family of block ciphers. Cryptology ePrint Archive, Paper 2013/543
17. Beaulieu R, Shors D, Smith J, Treatman-Clark S, Weeks B, Wingers L (2015) Block ciphers for the internet of things. Cryptology ePrint Archive, Simon and speck

Artificial Intelligence Integration in Programming Education: Implications for Pedagogy and Practice

Venushini Rajendran and R. Kanesaraj Ramasamy

Abstract In the swiftly evolving domain of programming education, Artificial Intelligence (AI) has emerged as a transformative force, prompting a reevaluation of pedagogical strategies and educational practices. This paper critically examines the integration of AI technologies in programming education and elucidates their pedagogical implications. With the accelerating development of programming languages and frameworks, traditional educational models struggle to maintain relevance. AI-driven intelligent search engines and recommendation systems are now pivotal in curating personalized educational content, thereby enhancing learning efficiency and reducing learner frustration. Furthermore, AI-facilitated platforms have redefined the conventional classroom, introducing virtual instructors and democratizing access to programming knowledge. Significantly, Natural Language Processing (NLP) technologies have reduced barriers to entry, enabling individuals with limited technical backgrounds to interact with programming interfaces in vernacular language. This has facilitated broader participation and has scaffolderd the learning process. Additionally, AI interventions have automated aspects of code generation and provided instantaneous feedback mechanisms, essential for real-time error correction and debugging, thus fostering a supportive learning environment. The collaborative aspects of AI are manifested through peer-to-peer learning ecosystems, bolstered by forums and code-sharing platforms. The synthesis presented in this paper highlights how AI not only enriches the learning experience but also promotes inclusivity and engagement in programming education. The implications of these AI-driven pedagogical innovations are vast, suggesting a paradigm shift towards more adaptive, personalized, and collaborative learning frameworks. This paper contributes to the discourse by not only charting the current landscape but also projecting the future trajectory of AI's role in programming education.

Keywords Artificial intelligence in education · Programming pedagogy · Personalized learning systems · Natural language processing · Collaborative learning environments

---

V. Rajendran · R. K. Ramasamy (✉)
Faculty of Computing Informatics, Multimedia University, Cyberjaya, Malaysia
e-mail: r.kanesaraj@mmu.edu.my

B. Annappa et al. (eds.), *High Performance Computing, Smart Devices and Networks*, Lecture Notes in Electrical Engineering 1262,
https://doi.org/10.1007/978-981-97-7794-5_16

1 Introduction

Programming has traditionally been seen as a complex and intimidating field, posing challenges for beginners [1]. However, the landscape of programming education has been transformed by the emergence of Artificial Intelligence (AI) tools, making programming more accessible to individuals from diverse backgrounds [2]. These AI tools have revolutionized the learning experience, providing valuable support and guidance to aspiring programmers. From code suggestion systems (CSS) to automated testing and error detection tools, these tools have enhanced productivity, reduced barriers, and fostered inclusivity in programming education.

As the integration of AI tools in programming education continues to advance, it is crucial to assess their advantages, impact, and limitations to ensure their effective incorporation into the curriculum [3]. This article delves into the advantages of AI tools in programming education, exploring their transformative potential and their impact on learners. Additionally, the evaluation of their effectiveness and future directions for their development will be examined. By understanding the implications of AI tools in programming education, educators and policymakers can make informed decisions about their integration and utilization, thereby fostering a more effective and inclusive learning environment.

According to Google Trends data, the interest in programming AI has been steadily increasing over the past few years as shown in Fig. 1. The trend indicates a growing curiosity and engagement in AI programming among individuals worldwide. The accessibility and advancements in AI tools have played a significant role in driving this upward trajectory.

The Google Trends data reveals that more people are searching for programming languages and frameworks specifically related to AI, such as Python, TensorFlow, PyTorch, and scikit-learn. This suggests a rising interest in acquiring the necessary skills to develop AI applications and work with AI technologies. The increasing

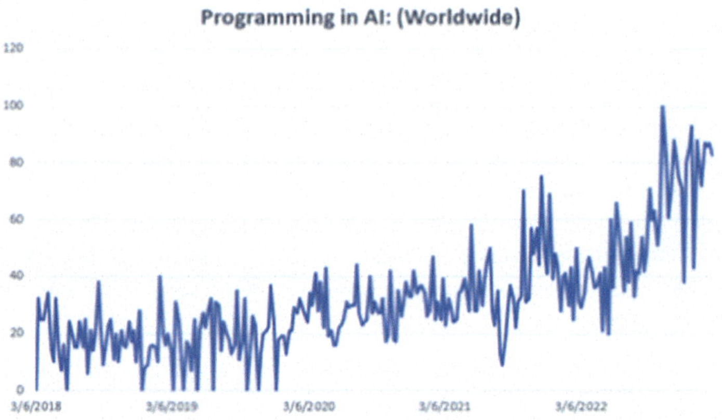

Fig. 1 Google trend for programming in AI

popularity of AI programming is also reflected in the rising demand for AI-related job roles. Employers are actively seeking professionals with AI programming skills, leading to a surge in enrollment in AI-related courses and certifications [4]. As a result, programming education platforms and institutions are incorporating AI tools into their curriculum to meet this growing demand.

Through this exploration, we aim to shed light on the transformative power of AI tools and their role in shaping the future of programming education. By embracing these advancements, we can overcome barriers and empower individuals from diverse backgrounds to engage with programming and acquire valuable skills in this rapidly evolving digital age.

2 Evolution of AI Tools

The evolution of AI tools and their contributions to programming education. It explores the development of CSS, automated debugging assistance, and the provision of interactive learning materials. This research emphasizes how these tools have significantly reduced the entry barriers and made programming more inclusive.

2.1 Code Suggestion System (CSS)

CSS are AI tools that aim to enhance developers' productivity by providing intelligent recommendations and autocompletion suggestions while writing code [5]. These systems leverage machine learning algorithms and natural language processing techniques to understand the context, syntax, and semantics of code, enabling them to offer relevant suggestions to developers.

2.1.1 Benefit of CSS

CSS increases efficiency by helping developers write code faster and with fewer errors [6, 7]. By offering context-aware suggestions, these tools reduce the time spent searching for code snippets or consulting documentation, enabling developers to focus more on implementing complex logic or solving critical problems.

The suggestions provided by CSS can help improve code quality, enforce coding best practices and standards [7]. By recommending optimized or more efficient code snippets, these tools contribute to writing cleaner, maintainable, and bug-free code. CSS serves as learning aid for developers, especially those who are new to a particular programming language or framework [7]. By suggesting relevant code patterns and examples, developers can quickly grasp the syntax and idiomatic usage, accelerating their learning curve.

Table 1 Example of CSS

CSS	Description	Provider
Tabnine [8]	AI-powered code completion tool	Codota
Kite [7]	AI-powered code completions and documentation	Kite
IntelliSense (VS Code) [9]	Intelligent code completion, suggestions, and tooltips	Microsoft
CodeIntel [10]	Intelligent code completion, jump-to-definition, and more	Sourcegraph
DeepCode [11]	AI-powered static code analysis and suggestions	DeepCode AG
Jedi [12]	Static analysis and autocompletion library for Python	David Halter
CoPilot (GitHub Copilot) [7, 8]	AI-powered code suggestions and completions in real-time	GitHub

CSS can facilitate collaboration among developers [7]. By suggesting code snippets or functions that have been previously used by other team members, these tools foster knowledge sharing and reduce duplicated efforts. CSS also promotes consistency across projects and development teams. By offering consistent suggestions for common tasks or code patterns, these tools help maintain a standardized codebase, reducing the risk of introducing errors or diverging coding styles. Table 1 includes an example of CSS and providers, but there are many more available in the market.

2.1.2 Challenges and Limitation

CSS systems face challenges in understanding the intricate context of code, particularly in complex scenarios or with dynamic typing. Although they have made significant progress, there are still instances where the suggestions might not align perfectly with developers' intentions. Certain programming domains or industries may require specialized knowledge that might not be readily available CSS. These tools may struggle to provide accurate suggestions for niche or emerging technologies, limiting their effectiveness in such cases.

CSS may have varying levels of support for different programming languages. While popular languages like Java, Python, or JavaScript typically receive extensive support, less mainstream or niche languages may have limited or no suggestion capabilities. CSS often operates based on large amounts of training data, including code repositories and examples. Ensuring the privacy and security of sensitive code or intellectual property is a crucial concern when using these tools, especially in enterprise or proprietary software development.

2.2 Automated Debugging Assistance

Automated Debugging Assistance (ADA) refers to the use of automated tools and techniques to assist software developers in the process of identifying and resolving bugs or errors in their code. ADA systems leverage various approaches, such as static analysis, dynamic analysis, and machine learning, to provide developers with relevant information and suggestions to aid in debugging. Table 2 shows the example of ADA.

The evolution of AI tools in programming education has significantly impacted the accessibility and inclusivity of the field. Code suggestion systems have empowered beginners by offering real-time suggestions and helping them learn correct syntax and coding practices. Automated debugging assistance tools have streamlined the debugging process, reducing the time and effort required to identify and fix errors. This enables programmers to focus more on problem-solving and development. Interactive learning materials have transformed the learning experience, providing adaptive and personalized content that caters to individual learning styles and skill levels. These tools collectively contribute to reducing entry barriers and making programming more approachable to a wider range of individuals.

Table 2 Example of ADA

Type of ADA	Description	ADA Tool
Static analysis	Static analysis techniques analyze the source code without executing it. These techniques can identify potential issues by examining the code structure, syntax, and data flow. They help detect common programming errors, such as null pointer dereferences, type mismatches, or unused variables	i. Pylint for Python [13] ii. FindBugs for Java [14]
Dynamic analysis	Dynamic analysis techniques involve observing the execution of code during runtime. These techniques can help identify issues related to program behavior, such as unexpected crashes, memory leaks, or performance bottlenecks	i. Valgrind, for C/C + + [15] ii. Apache JMeter for web application [16]
Machine learning-based approaches	Machine Learning-Based Approaches techniques use historical data and patterns to train models that can predict and identify potential bugs or suggest fixes	i. DeepBugs [17] ii. DeepFix [18]

Table 3 Example of AI tool type

AI tool type	Advantages	Analysis
Code suggestion systems	Enhance productivity by providing real-time code suggestions and autocomplete functionality	Code suggestion systems reduce manual effort, improve code quality, and help beginners especially students to learn proper coding practices by suggesting syntactically correct and contextually appropriate code
Automated testing tools	Aid in identifying and resolving issues efficiently by automating testing processes and detecting errors	Automated testing tools enhance software quality by detecting bugs and errors early in the development process, allowing programmers to fix issues promptly and reduce the time spent on manual testing
Error detection and analysis tools	Assist programmers in identifying and analyzing errors, offering insights and recommendations for resolution	Error detection and analysis tools leverage AI techniques to detect and diagnose errors, providing valuable insights and recommendations for resolving issues, and ultimately improving the debugging process

3 AI Tools in Programming Education

Table below discusses the advantages that AI tools offer to student programmers (Table 3).

AI tools in programming education offer numerous advantages to aspiring programmers, especially student learners. Code suggestion systems improve productivity and learning outcomes by providing real-time suggestions and autocomplete functionality. These tools assist beginners in understanding proper syntactic and semantic structures, fostering better coding practices. Automated testing tools automate the testing process, leading to more efficient bug identification and resolution, thereby enhancing software quality. Additionally, error detection and analysis tools leverage AI techniques to identify and analyze errors, providing programmers with valuable insights and recommendations for resolving issues effectively.

4 AI Tools in Programming Education

While AI tools provide valuable support, this section addresses their limitations. It emphasizes that these tools should be seen as aids rather than replacements for foundational programming knowledge. This research stresses the importance of developing critical thinking skills and problem-solving abilities alongside utilizing AI tools. It also acknowledges that not everyone may have an inclination or aptitude for programming, regardless of the tools available (Table 4).

AI tools have limitations in the context of programming education. While they provide valuable support, it is crucial to understand that they should complement,

Table 4 Example of AI tool type with its limitation

Limitations and considerations	Description	Analysis
AI tools as aids, not replacements [19]	AI tools should be viewed as aids rather than substitutes for foundational programming knowledge	While AI tools provide valuable support, they should not replace the need for developing a strong foundation in programming concepts and principles
Importance of critical thinking and problem-solving [20]	Emphasizes the significance of developing critical thinking skills and problem-solving abilities alongside utilizing AI tools	Critical thinking and problem-solving skills are essential for programming success and should be cultivated in conjunction with leveraging AI tools
Varied aptitudes and inclinations [21]	Recognizes that not everyone may have an inherent inclination or aptitude for programming, regardless of the availability of AI tools	AI tools can provide valuable support by offering real-time suggestions, aiding in learning syntax, and streamlining processes like debugging. However, these tools cannot necessarily create an inherent interest or aptitude for programming in individuals who may not naturally gravitate towards the field

rather than replace, foundational programming knowledge. AI tools can assist in code completion, error detection, and learning materials, but they cannot substitute the need for understanding programming concepts and principles.

5 Implications for Education

This section focuses on the implications of AI tools for programming education. It discusses how these tools can be integrated into curricula to enhance learning experiences. The article explores the role of educators in guiding students to effectively utilize AI tools while emphasizing the need for a balanced approach that encourages both independent thinking and reliance on AI assistance (Table 5).

The integration of AI tools into programming curriculum has profound implications for education. By incorporating these tools, students can gain hands-on experience and deepen their understanding of programming concepts. This integration enables students to leverage AI assistance effectively, enhancing their learning experiences and productivity.

Educators play a crucial role in guiding students' use of AI tools. They provide instruction and support, helping students understand when and how to use AI tools

Table 5 Example of AI tool type with implication

Implications for education	Description	Analysis
Integration into curriculum [4]	AI tools can be integrated into programming curricula to enhance learning experiences	Integrating AI tools into curricula allows students to gain hands-on experience with these tools, enhancing their understanding of programming concepts and improving their ability to leverage AI assistance effectively
Educators' role [22]	Educators play a vital role in guiding students to effectively utilize AI tools. They provide guidance and instruction, encouraging a balanced approach that fosters independent thinking alongside reliance on AI assistance	Educators serve as facilitators, helping students navigate AI tools and teaching them when and how to use them appropriately. They also foster critical thinking skills, enabling students to evaluate AI suggestions and make informed decisions in their programming tasks

appropriately. Educators foster a balanced approach that encourages independent thinking and problem-solving alongside reliance on AI assistance. This approach ensures that students develop critical thinking skills and make informed decisions while benefiting from the advantages of AI tools.

6 Future Directions

The research concludes with a discussion on the potential future developments in AI tools for programming education. It highlights the importance of continuous improvement and adaptation to cater to diverse learning needs. The article also explores the role of AI tools in fostering creativity and innovation within the programming community.

AI tools for programming education are expected to undergo continuous improvement and adaptation to meet the evolving needs of learners. This includes refining code suggestion systems, error detection algorithms, and interactive learning materials to provide more accurate and contextually relevant assistance. Future AI tools may focus on personalization, tailoring the learning experience to individual learners' needs and preferences. This could involve adaptive learning systems that analyze learners' strengths and weaknesses and provide targeted support and challenges accordingly.

AI tools can facilitate collaboration and social learning among programmers. Future developments may integrate features that enable collaborative coding, peer code review, and knowledge-sharing platforms, fostering a community of programmers who can learn from and support each other. Besides, AI tools have the potential to augment programmers' creativity and innovation. By offering novel code suggestions, identifying patterns, and analyzing code quality, AI tools can inspire programmers to explore new approaches and improve their problem-solving skills.

As AI tools become more prevalent in programming education, ethical considerations, and responsible AI practices become increasingly important. Future developments may include mechanisms to address biases in AI models, promote transparency, and ensure responsible use of AI tools in education.

7 Conclusion

In conclusion, this investigation delineates the significant strides AI tools have made in enhancing the inclusivity and reach of programming education. These tools have successfully demystified the art of coding, rendering it more accessible than ever before. Nonetheless, it is imperative to acknowledge that such technologies supplement rather than supplant the essential, foundational programming competencies and the rigorous commitment required to master them. As the trajectory of AI innovation presses forward, it is incumbent upon both educators and students to embrace and optimize these tools to unlock their full educational potential.

Peering into the horizon of educational technology, the evolution of AI within the sphere of programming instruction is poised to accelerate. There is an exigent call for perpetual innovation and adaptation to support an increasingly diverse array of learning paradigms. Anticipated future enhancements are likely to converge on heightened personalization, adaptive learning frameworks, and enriched collaborative coding ecosystems. Additionally, AI stands to catalyze creative and inventive impulses within the programmer community. Nevertheless, the ascendancy of AI in education necessitates a steadfast commitment to ethical practices and responsible implementation to ensure equitable and transparent educational opportunities.

References

1. Munawar K, Naveed MS (2022) The impact of language syntax on the complexity of programs: a case study of java and python. Int J Innov Sci Technol 4:683–695
2. Holmes W, Bialik M, Fadel C (2023) Artificial intelligence in education. Globethics Publ
3. Malinka K, Peresíni M, Firc A, Hujnak O, Janus F (2023). On the educational impact of ChatGPT: is artificial intelligence ready to obtain a university degree?. In: Proceedings of the 2023 conference on innovation and technology in computer science education Vol 1, pp 47–53
4. Southworth J, Migliaccio K, Glover J, Reed D, McCarty C, Brendemuhl J, Thomas A (2023) Developing a model for AI across the curriculum: transforming the higher education landscape via innovation in AI literacy. Comput Educ Artif Intell 4:100127
5. Vaithilingam P, Glassman EL, Groenwegen P, Gulwani S, Henley AZ, Malpani R, Yim A (2023) Towards more effective AI-assisted programming: a systematic design exploration to improve visual studio intelli-code's user experience. In: Proceedings of the IEEE/ACM international conference on software engineering: software engineering in practice (ICSE-SEIP'23). Association for computing machinery, New York, NY, USA

6. Fraiwan M, Khasawneh N (2023) A review of ChatGPT applications in education, marketing, Software Engineering, and Healthcare: Benefits, Drawbacks, and Research Directions. arXiv preprint arXiv:2305.00237
7. Wang R, Cheng R, Ford D, Zimmermann T (2023) Investigating and designing for trust in AI-powered code generation tools. arXiv preprint arXiv:2305.11248
8. Liang JT, Yang C, Myers BA (2023) Understanding the usability of AI programming assistants. arXiv preprint arXiv:2303.17125
9. Tan J, Chen Y, Jiao S (2023) Visual studio code in introductory computer science course: an experience report. arXiv preprint arXiv:2303.10174
10. Developing code navigation. Sourcegraph docs. (2023). https://docs.sourcegraph.com/dev/background-information/codeintel
11. Snyk powered by DeepCode AI. Snyk. (2023) https://snyk.io/platform/deepcode-ai/
12. Davidhalter (2023) Davidhalter/jedi: awesome Autocompletion, static analysis and refactoring library for python. GitHub. https://github.com/davidhalter/jedi
13. Pylint PyPI (2023) https://pypi.org/project/pylint/
14. FindBugsTM—find bugs in Java programs (2023) https://findbugs.sourceforge.net/
15. The VALGRIND Quick Start Guide (2023c) https://valgrind.org/docs/manual/quick-start.html
16. Apache JMeter, Apache JMeter TM (2023a) https://jmeter.apache.org/
17. Michaelpradel (2023) Michaelpradel/DeepBugs: DeepBugs is a framework for learning bug detectors from an existing code corpus. GitHub. https://github.com/michaelpradel/DeepBugs
18. Gupta R, Pal S, Kanade A, Shevade S (2017) DeepFix: fixing common C language errors by deep learning. Proceedings of the AAAI conference on artificial intelligence, pp 31(1). https://doi.org/10.1609/aaai.v31i1.10742
19. Deiana G, Dettori M, Arghittu A, Azara A, Gabutti G, Castiglia P (2023) Artificial intelligence and public health: evaluating ChatGPT responses to vaccination myths and misconceptions. Vaccines 11(7):1217
20. Almulla MA, Al-Rahmi WM (2023) Integrated social cognitive theory with learning input factors: the effects of problem-solving skills and critical thinking skills on learning performance sustainability. Sustainability 15(5):3978
21. Odeh M (2023) Games and AI: how did game engines become unbeatable and does AI dominance entail any risks
22. Kizilcec RF (2023) To advance AI use in education, focus on understanding educators. Int J Artif Intell Educ 34(1):12–19

Strategies for Enhancing Software Reliability: A Focus on Fault Avoidance Techniques

Venushini Rajendran and R. Kanesaraj Ramasamy

Abstract This paper provides an in-depth analysis of fault avoidance strategies in software applications, a critical aspect for enhancing software reliability and dependability. Fault avoidance encompasses various techniques aimed at preventing, eliminating, and predicting software faults to ensure the creation of fault-free software. The intertwined nature of software quality and reliability is a key focus of this study. The primary objective is to synthesize empirical and theoretical insights on the attributes of software fault avoidance. This is achieved through the analysis of diverse metrics, modules, and methodologies, leading to recommendations aimed at bolstering software dependability. Achieving high levels of software reliability is a challenging task due to the inherent nature of software faults, which are latent programming errors that manifest upon discovery. This review critically examines a range of strategies aimed at mitigating these issues, including Byzantine Fault Avoidance (BFA), formal and semi-formal methods, and verification and validation techniques involving fault detection. Additionally, the study delves into software reliability growth models and the concept of fault removal through relative correctness (NHCTMC), among others. Recommendations are formulated based on a comprehensive analysis of various research studies in the realm of software fault avoidance. Ultimately, this paper highlights that fault avoidance is a multifaceted approach, integral throughout the software development lifecycle, aimed at minimizing software faults and averting unnecessary failures.

Keywords Fault avoidance · Fault prevention · Fault removal

V. Rajendran · R. K. Ramasamy (✉)
Faculty of Computing Informatics, Multimedia University, Cyberjaya, Malaysia
e-mail: r.kanesaraj@mmu.edu.my

B. Annappa et al. (eds.), *High Performance Computing, Smart Devices and Networks*, Lecture Notes in Electrical Engineering 1262,
https://doi.org/10.1007/978-981-97-7794-5_17

1 Introduction

Software fault avoidance aims to keep faults from emerging in a working system, and the presentation of faults will be limited during system development. Validation and verification should be codified, and all of the specifications' codes should be thoroughly verified to eliminate any linked flaws [1]. The majority of software flaws should be avoided by utilizing simulation and modeling techniques during the specification requirements and software life cycle phases [2]. It entails fault prevention, fault removal, and fault forecasting.

Problem prevention aims to eliminate the possibility of a fault infiltrating the system before it starts to operate. Identifying and eliminating common errors is the goal of fault removal. Fault forecasting is the process of using models to predict probable software flaws. To eliminate software faults caused by software failure and environmental changes, fault avoidance strategies and solid principles are essential when building the software program. Software dependability is a crucial aspect of software quality. It becomes more difficult to ensure high dependability as software becomes more complicated.

Software flaws are one of the criteria that define software reliability; it is challenging to create flawless software. Software flaws are created by design flaws, which are difficult to predict since software flaws can be caused by errors in developing code, erroneous software specifications, and other unforeseen issues. There is no way to determine defect-free software due to a lack of understanding of software properties. There are several existing models, but none of them can be said to be correct for resolving the complexity issue. Software reliability is a vast cry from reality and cannot be measured directly [3].

The purpose of this study is to examine various software fault prevention modules and methodologies. To provide methods and modules for improving software reliability and investigate various coverage metrics in relation to software design for testability.

2 Literature Review

2.1 Fault-Avoidant Scheduler

According to [4], for task execution, a fault-avoidant scheduler was proposed, evolving scheduling techniques and incorporating them into the Tasklet system which is a method for distributing computing systems. The failure model was demonstrated where Tasklet execution may be terminated due to a variety of reasons because the system contains user-managed and persuasive edge resources Tasklet execution might fail due to the resources might always close or pass the system. Therefore, these faults must be considered while designing a fault avoidance scheduling algorithm.

The failure model in this paper defines the shortcomings that may occur while the Tasklet is executing. Heartbeat mechanism's fault awareness has been implemented in the Tasklet systems to identify these faults. When the provider performs the Tasklet on behalf of the customer, it will tell the consumer on a regular basis that the execution is ongoing. Consumers either passed a lack of heartbeat or provider's precise message indicating that the execution was canceled. Next, the Tasklet system incorporates fault tolerance by restarting Tasklet execution once discarding is identified. However, this strategy will lead to a delay for consumers because the scheduling procedure must be rebuilt, and the failed execution's progress will be misplaced. The ideal scheduling approach tries to prevent problems mentioned in the failure models.

2.2 Formal Method/Semi-formal Method

As said by Ioana and Martin's paper, formal methods are defined as mathematically rigorous methods and tools for specification, verification, and design. Meaning of the mathematically rigorous is which specification is built from well-formed statements based on mathematical logic and the formal verification is made up of rigorous deductions in the logic [5].

Almeida et al. proposed various classifications for formal techniques that may be utilized in different areas which is to implement wanted behavior in system specifications and to ensure implementation contains the same characteristics as the specification. Formal techniques are further classified into four types: specify and analyze, specify and prove, specify and derive, and finally specify and transform [6].

An internal state should be considered in specifying and analyzing, as well as the actions that alter the state. The requirements may be classified into several types, including model-based, state-based, and model-based specifications, which are used to create a unique model of the system. It is expressed in a language where internal state is the key focus like automata-based modeling and abstract state machines for model-based specification. Aside from that, an abstract state machine is utilized to describe the system using states and state transformations. One of the languages is, for example, B Method in conjunction with B specification language [5].

According to [7], author proposed a system of distributed reconfigurable control by integrating formal and semi-formal approaches during the development process. In this research, Unified Modeling Language (UML) is intended to offer a uniform way for visualizing system design and is distinguished because of its clear syntax and low semantics. Then, in the article, B is a method for formalizing software development that includes all stages of a software development cycle, beginning with early concepts and ending with coding. There are strong tools that allow the B method to be used in practice such as the Atelier B or B4free, which are based on mathematical formalism [8].

Formal techniques, often known as formal verification, can be used to verify properties about the specification when specifying and proving. According to Rushby's study, he divided the formal frameworks for verification into three levels, the first level makes use of natural languages and hands-on demonstrations rather than computers. To go on to the second level, demonstrations must be executed but with a very specific formulation, and the third level requires demonstrations and proofs to be carried out using computer-based tools.

Model checking is also one of the formal methods for verifying finite-state systems that are widely utilized in industry. The goal is to define the system using temporal logic, then explore the model completely and check the necessary characteristics. A counterexample is provided if a property is not valid. Additionally, Bounded Model Checking (BMC) is a kind of model monitoring that examines the state space in detail and is mostly used for semiconductor verification, however, it is equally relevant to software. Another approach is program annotations, which is used to test the behavioral characteristics of programming. Any program that needs to have its expressions validated must have some kind of annotation associated with it [5].

To get to an implementation with the required behavior, one must first provide a specification with the desired characteristics. Languages based on logic such as Prolog or functional such as Scheme and OCalm, may be used to generate a specification that is instantly executable. It's also possible to design an implementation that has all of the necessary features. There are many methods for demonstrating the correctness of a deviation. For example, one may focus on the derivation process and show that it's accurate or can establish a set of proof duties and show that they are correct as part of the process [5].

Ioana and Martin's research also emphasized the necessity to build a transformation of a specification in order to abstract features or enrich the definition with new details. Large issues are frequently split into smaller sections while building software because, in formal verification, transformations enable modularity. The job is eased by breaking the global model's behaviors into distinct perspectives. Transformations are also linked to the modeling language and the properties that will be demonstrated. As a result, only a few tools are available for making alterations [5].

Finally, the defining step of formal methods projects takes more time resulting in lesser problems in the latter stages of development. Formal techniques may be applied gradually throughout the development process and will have an immediate impact. As a result, it is evident that using formal techniques may be budgeted while reducing the number of potential faults to near zero [5].

2.3 Verification and Validation Through Fault Detection

Verification and validation are two approaches for removing possible faults while working on fault avoidance. The process of verifying that software fulfils its objectives without mistakes is known as verification. The method of evaluating is known as static testing whether the software developed fulfils the specified requirements. The process of assessing whether a software product fulfils high-level requirements is then followed by validation. It is the process of validating a product as well as the

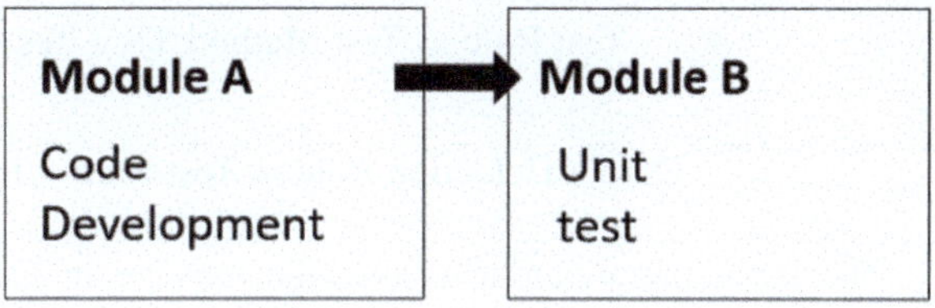

Fig. 1 Development phase and test phase in template model

validation of actual and expected software. Carina Andersson noted in the study that examinations and testing are a simplification of whether inspections are worthwhile putting work into when compared to other problem detection techniques [9].

In her study, she also found that simulation is a useful tool for better understanding the development methodology's effect aspects since it generates a detailed model of the actions involved. Simulations can then be utilized as a data-collecting technique and an analysis tool. It may be used to build awareness of the interdependence of various operations in the process or as a tool for organizations to assess process improvements. If a certain event occurs, only then does the state of the system change. If no specific event occurs, then the state remains the same throughout time. This is known as an infinite simulation [9]. In their article, [10] claimed that verification and validation operations in an organization provided a benchmark for the critical elements and intended action of the simulated system. To validate the validity of the simulation model, walkthroughs are done, with programmers and testers' comments mostly concerned with terms and impacts of in-parameters, as well as cost elements of identifying faults for different stages. For example, after receiving feedback from programmers and testers, test coverage is improved in the model.

A template model containing modules for code development and testing may be utilized when simulating since the model allows users to simulate the number of inserted flaws during coding stages, the number of discovered defects during testing, and the duration for development and testing (Fig. 1).

The amount of work received, the number of programmers, the average amount of faults injected per programmer per day, and the average amount of KLOC produced per programmer per day will be calculated throughout module A. The model's output will be the number of undefined faults in coding, which is sent to module B for additional testing. In module A, the top flow reflects the defect injection rate during coding, while the lower flow represents the coding rate.

$$\text{Coding Rate} = \text{Coding Method CR} * \text{Programming Resources} \tag{1}$$

$$\text{Defect Injection Rate During Coding} = \text{Coding Rate} * \text{Coding Method DI} \tag{2}$$

A defect detection rate is shown in module B, whereas a testing rate is determined by the number of testers and an average KLOC checked each day by each tester. Below is a graphic illustrating the formulas used in Module B of the course.

$$\text{Test Rate} = \text{Test Method TR} * \text{Tester Resource} \tag{3}$$

$$\text{Defect Detection Rate} = \text{Test Rate} * \text{Test Method DD} \tag{4}$$

The input parameter test coverage may be included in a model extension by estimating the equivalent number of KLOC tests based on the number of requirements tested. This presumption is used in industrial simulation.

As a result, it can be concluded that a template model can be used in process of verification and validation to assist programmers and testers in discovering undiscovered defects in their software because they may have numerous significant recommendations and corrections in their work with the created model.

2.4 Detection Software Fault Detection and Prediction Techniques

The software testing process may be improved, and the product testing effort and expenses can be reduced by focusing on defect-prone software modules. According to Sabia and Bushra, the techniques of "bagging and swarm optimization" are proposed as the new methodology to achieve fault prediction for software. Based on the results, this proposed methodology has the greatest improvement in improving the software quality than other techniques for fault prediction. With that, an examination of the continuous framework is determined, and a fault prediction program with Artificial Immune Systems (AIS) is characterized to evaluate the software's reliability. Besides that, the framework and modules' characterization are developed in the level of fault criticality in regard to the fault recurrence, where this model is motivated from multiple applications.

In order to observe the metrics value of predicting fault-prone classes, the evaluation of several design characteristics is performed. Various methods of software development are also examined with the use of an object-oriented approach. This can reduce the testing time and effort, reliability analysis of development, case analysis, as well as the prediction model. After all, software complexity and coupling are to be focused on and be responsible for the fault occurrence. Moreover, there are a lot of techniques suggested for software fault prediction. Nevertheless, none of them has been attested to achieve perfection and completeness. Thus, the proposed density-based spatial clustering of applications with noise (DBSCAN) is experimented and it indicates that it is effective against object-oriented software that is based on Java, to attain prediction for defect-prone software.

This proposed technique is proven to be the best technique in predicting the software fault after carrying out different measurements and matrices, in which it significantly increases the accuracy. With the aim of achieving a better module, the quality vector analysis and the prediction model are characterized for examination of performance as well as software fault. Besides that, the authors also proposed an analysis-based review so as to predict the fault in software. This helps to obtain the software acceptability from a level program and software applications. An examination for code-based and the investigation of attributes and nature are performed. With that, the direct-scaled estimation of code is developed so that the fault prediction and evaluation of unit change will be accomplished appropriately [11].

There are many techniques to handle software faults which consist of the techniques for fault avoidance, fault prediction, and fault tolerance. These techniques are important to minimize the faults when developing software which leads to reduced cost, time, and testing efforts.

The authors [11] proposed a Fuzzy Inferences System (FIS) model without project data while supporting a data-driven approach. The ANN and the Adaptive Neuro FIS (ANFIS) are also enlisted along with the approach to demonstrate its applicability. This significantly improves the fault prediction based on the evaluation of functioning features. The authors also proposed a technique with the methods of classification and selection, along with the approaches of data mining and machine learning, in order to accurately perform prediction for the framework. The result indicates that this hybrid approach is able to enhance the MCC (Mathew's Correlation Coefficient), prediction, and also accuracy [12].

In addition to the ANFIS technique, Artificial neural networks and Support-vector machines are used for analysis. This reduces occurring faults by predicting the vulnerabilities with the use of fault data and metrics. With that, the McCabe metrics and ROC-AUC are applied to achieve better performance and a comprehensive dedication to programming [13]. According to [14], they proposed a technique that uses Bayesian grid for the metrics and vulnerabilities. The authors presented a metric called Lack of Coding Quality (LOCQ) to predict software faults and parameter efficiency. For example, Weighted Methods per Class (WMC) metric and Coupling Between Object classes (CBO) metric. This proves that the LOCQ, LOC, have a great effect on vulnerabilities. Based on the methods, they examined several ways including algorithm ensembling and threshold adjusting, and they proposed the techniques of class imbalance learning which provides greater results in predicting software faults [15]. The parameters considered for different techniques applied are shown in Tables 1, 2 and 3.

Table 1 Parameters for the techniques applied

Technique applied	Parameters						
	Precisions	Recall	ROC_{AUC}	MCC	Accuracy	Balance	G-mean
FIS, ANN, ANFIS	✓	✓	✓				
Hybrid feature selection				✓	✓		
ANN, SVM, ANFIS			✓				
Bayesian networks					✓		
Naïve Bayes, random forest						✓	✓

Table 2 Software fault detection approaches

Approach	Description
Automated static analysis (ASA)	Able to analyze the code and locate 3% software faults
Graph mining	Discovers non-crashing software faults and simplifies the functions
Classifiers	A neural network approach is used to recognize abnormal occurrences. For example, this is applied to Bagging and Naïve Bayes
Pattern mining	To find relationships between the handling functions and the usual processes

Table 3 Machine learning techniques for software fault detection

Approach	Description
Log informally regarded data	Logs each single point and keeps track of the flow of data
Visual definition of logs	Connects the messages of different types since usual log files are uncoupled which causes longer time to draw relation
Support-vector machine (SVM)	An algorithm on a machine-learning basis that determines the software faults with the help of log files. This is commonly applied in the data mining approach
Random indexing	Determines the operation sequence from the log files by using SVM for classifications
Area-based fault localization	An automated technique used to discover the software faults found inside erroneous code in a segment manner

2.5 *Software Reliability Growth Models*

Software reliability is the degree to which the possibility of errors occurring in the software. Software growth models are the statistical functions for detection data and are commonly used to detect faults in order to predict the failure rates of the code. However, this is insufficient in detecting the software faults for software reliability since the testing time and coverage are also important to achieve reliability. Hence, the authors, Sabia and Bushra, discussed several techniques for fault detection and correction. Thereby, Maximum Likelihood Estimation performs greater than Least Square Estimation in terms of capability in predicting faults. The authors proposed a 2D method formed on the growth models for software reliability in which the testing resources are taken as inputs since the previous approach is either focused on testing coverage or testing time.

2.6 *Fault Removal Through Relative Correctness*

A concept using relative correctness for fault removal was introduced by [16]. The faults in iterative programs are considered removed if the static analysis of the software's source code proves that the newer program is more correct than the previous version. The line of code, statement, condition, and lexical token that may have faults are to be identified to locate the faults in a program thus determining the correctness of the program. At first, it is to be determined whether the loop in the program has faults. Given the condition and the while loop, all the invariant relations were generated and categorized as compatible relations or incompatible relations. If there are any incompatible relations, then the loop is incorrect thus proving that the program is faulty. If there are no incompatible relations and all compatible invariant relations satisfy the sufficient criterion of correctness, the loop is correct. The correctness of the loop is uncertain if none of the conditions above are satisfied, which means that there are improvements for the invariant relations generator. After the loop is identified, the localization of the fault is executed. The invariant relation with the fewest number of variables will be selected. If the number of variables are the same, the invariant relation with the variables that appear in the smallest number of statements in the loop is selected. These statements are selected as features that will be corrected. Then, the selected statement will be modified by replicating the current incompatible invariant relation with a compatible invariant relation without involving the compatible invariant relations. This is done by using the condition of compatibility preservation. Compatibility preservation is defined as:

$$\exists x_3 \ldots x_{n,}, x_3^1, \ldots x_n', \tag{5}$$

where x_n are the program variables with x_1, x_2 being assigned with the two variables that are found in the selected invariant relation and C is the intersection of the compatible invariant relations. After the modification, the invariant relations are recomputed, and the selected incompatible invariant relation is now taken over by a compatible invariant relation which increases the correctness of the program. Then, all the steps from above are repeated until the loop is identified as correct.

3 Analysis

Fault avoidance is crucial in preempting the occurrence of defects in software systems. Techniques such as Byzantine Fault Avoidance (BFT) [1], formal and semi-formal methods [2], and rigorous testing processes [3] are pivotal. Byzantine Fault Tolerance, for instance, is critical in systems where there is a need for high reliability despite malicious or arbitrary faults [1]. The method's resilience and the ability to function correctly even if some of the components fail maliciously or unpredictably set a high standard in fault avoidance. Conversely, formal and semi-formal methods provide a structured approach to software development, minimizing errors in the design phase [2]. These techniques are complemented by comprehensive verification and validation processes, integral in identifying and rectifying faults early in the software lifecycle [3].

Software reliability growth models like the Non-Homogeneous Continuous Time Markov Chain (NHCTMC) offer a mathematical approach to understanding and predicting software reliability growth over time [4]. These models are instrumental in quantifying the reliability of software systems and in forecasting future reliability based on observed failure data [4]. The NHCTMC model, in particular, provides a nuanced view of software reliability, taking into account various operational environments and usage conditions [4].

Relative correctness, a relatively new concept in fault removal, posits that software can be deemed 'relatively correct' if it meets certain predefined operational criteria, even if not entirely fault-free [8]. This approach contrasts with traditional fault removal strategies that aim for absolute correctness. Relative correctness offers a pragmatic approach in complex systems where achieving absolute correctness might be infeasible [8]. Based on our analysis, it is recommended that a combination of traditional and emerging fault avoidance techniques be employed to enhance software reliability future research could focus on the integration of artificial intelligence in fault prediction and avoidance, exploring its potential to revolutionize the field.

This chapter has presented a comprehensive analysis of various fault avoidance techniques and their impact on software reliability. The evidence suggests that a multifaceted approach, incorporating both traditional and emerging strategies, is essential for developing highly reliable software systems.

4 Recommendations

4.1 Adoption of Static Code Analysis and Relative Correctness in Fault Removal

In alignment with our comprehensive analysis, we strongly advocate the integration of static code analysis, particularly in its automated form, in tandem with the principles of relative correctness for enhanced fault removal. Automated static code analysis stands out in its ability to swiftly and accurately pinpoint defects within the software source code. This methodology significantly facilitates the early detection of faults, thereby substantially diminishing the associated costs of correction and expediting the remediation process. Furthermore, the concept of relative correctness, with its foundation in logical reasoning and its harmonious balance between formality and user accessibility, augments static code analysis. It provides a pragmatic and effective framework for fault evaluation and mitigation, especially in environments where automated solutions are implementable.

4.2 Role of Fault Prediction in Software Fault Avoidance

Our study underscores the critical role of fault prediction in the context of software fault avoidance. Among the array of techniques evaluated, the Classifier-based detection and prediction technique emerges as particularly noteworthy. This technique, rooted in a fuzzy-based classifier system that synergistically combines neural networks and decision tree algorithms, demonstrates exceptional proficiency in identifying aberrant behaviors within code. The robust predictive capacity of this approach, enhanced by sophisticated sampling methodologies and rigorous test data training, establishes it as an exemplary model for both fault detection and prediction. This technique not only elevates the quality of software but also plays a significant role in reducing developmental costs, aligning with the objectives of contemporary software engineering practices.

4.3 Critical Evaluation of Formal Methods in Software Development

While formal methods are integral to the precise documentation of software requirements, our analysis identifies critical limitations within these methodologies. Although effective in verifying the accuracy of systems in accordance with specific specifications, these methods do not inherently validate the correctness of the specifications themselves. Moreover, formal methods frequently lack a cohesive theoretical framework for the elicitation and documentation of user requirements in a formalized specification language. Consequently, it becomes imperative to exercise discernment in the application of formal methods within the software development lifecycle, particularly in the realm of fault avoidance. This cautious approach is essential to ensure the alignment of software specifications with actual user requirements and overall system objectives.

5 Conclusion

In conclusion, the goal of fault avoidance is to keep defects out of the software. It prevents errors from being introduced into the software as it is being built. It entails fault forecasting, fault prevention, and fault removal. Software fault avoidance approaches are a collection of methodologies and strategies aimed at reducing the presence of and preventing the introduction of defects. When building a reliable software, it is a must to address the reliability concerns from the start which is to include the mechanisms for fault tolerance into the software system design, as well as to use suitable techniques for fault avoidance throughout the design process.

According to Mili et al., fault tolerance, avoidance and removal are the defense lines against software faults. It is best to apply all these methods to gain the most efficient results.

Fault avoidance techniques attempt to decrease the likelihood of a fault occurring, whereas fault tolerance techniques attempt to keep the system functioning even when faults are present. Due to the total problem avoidance or removal is impossible, a critical system must constantly apply the attempts in fault tolerance to be remunerated and guard across the failure effects while the system is in operation. There is an example of software failure in the year of 2018 which is the data loss at GitLab. GitLab can be considered the world's most used software in the world, but they accidentally removed the data when trying to copy the data from the original server to another server and made this failure appear. Luckily the company was successful in recovering all the data. Although GitLab is a big company but still will appear in this kind of failure, as one can see that fault avoidance is very important in software applications.

References

1. Saha GK (2006) Software fault avoidance issues. Ubiquity 2006:1–15
2. Lambert N, Howard SM, Ian H, Stephen SD, Mauro R, Stefan B (1993) Reliability engineering for future telecommunication networks and services. IEEE
3. Okwong AE, Umoh EE (2011) Software reliability: failures, consequences and improvement. Glob J Pure Appl Sci 17(4):503–508
4. Edinger J, Schäfer D, Krupitzer C, Raychoudhury V, Becker C (2017) Fault-avoidance strategies for context-aware schedulers in pervasive computing systems. In: 2017 IEEE international conference on pervasive computing and communications (PerCom). IEEE, pp 79–88
5. Rodhe I, Karresand M (2015) Overview of formal methods. In: Software engineering, FOI-4-4156-SE, ISSN 1650-1942
6. Andersson C (2003) Exploring the software verification and validation process with focus on efficient fault detection
7. Berling T, Andersson C, Höst M, Nyberg C (2003) Adaptation of a simulation model template for testing to an industrial project
8. Oueslati R, Mosbahi O, Khalgui M, Li Z, Qu T (2018) Combining semi-formal and formal methods for the development of distributed reconfigurable control systems. IEEE Access 6:70426–70443. https://doi.org/10.1109/ACCESS.2018.2878896
9. Sabia S, Bushra N (2018) A review of software fault detection and correction process. JSEIS 3(1)
10. Jacob S, Geetha R (2017) Software defect prediction in large space systems through hybrid feature selection and classification. Int Arab J Inf Technol 14(2):208–214
11. Erturk E, Ebru A (2015) A comparison of some soft computing methods for software fault prediction. Expert Syst Appl 42(4):1872–1879
12. Okutan A, Olcay T (2014) Software defect prediction using Bayesian networks. Empir Softw Eng 19(1):154–181
13. Wang S, Xin Y (2013) Using class imbalance learning for software defect prediction. IEEE Trans Reliab 62(2):434–443
14. Ghardallou W, Diallo N, Mili A, Frias MF (2016) Debugging without testing. In: 2016 IEEE international conference on software testing, verification and validation (ICST). IEEE, pp 113–123

15. Gokhale SS, Lyu MR, Trivedi KS (2004) Analysis of software fault removal policies using a non-homogeneous continuous time Markov chain. Softw Qual J 12(3):211–230
16. Zhu M, Pham H (2018) A two-phase software reliability modeling involving with software fault dependency and imperfect fault removal. Comput Lang Syst Struct 53:27–42

Detection and Classification of Brain Tumor Using Deep Learning Model

Manoj Sitaram Routhu, Rohith Vigna Govindaraju, and P. Saranya

Abstract Accurate and efficient brain tumor classification is paramount for timely clinical diagnosis and effective treatment planning. In this groundbreaking research, we introduce an innovative Convolutional Neural Network (CNN) architecture intricately integrated with customized preprocessing techniques, resulting in an exceptional classification accuracy of 98.5% on the challenging br35H dataset. By harnessing the power of MRI scans and leveraging diverse datasets, our model expedites brain tumor assessments and sets the stage for advanced classification methodologies. With the global incidence of brain tumors rising, the need for technology-driven diagnosis becomes increasingly evident, and CNNs emerge as pivotal tools in enhancing diagnostic precision. This study not only underscores the profound significance of CNN models but also transcends geographical boundaries, reducing the frequency of misdiagnoses and ultimately empowering global healthcare. This paper comprehensively explores our methodology, delving into the intricate details of data collection processes, model development strategies, and experimental findings. Moreover, it sheds light on the implications of deploying CNN models in medical imaging. By contributing to the ongoing discourse on transformative healthcare technologies, this research aims to propel the adoption of CNN-based approaches, ushering in a new era of precise and efficient brain tumor classification for the benefit of healthcare professionals, patients, and society.

Keywords Brain tumor detection · CNN · MRI · Classification accuracy · Deep learning · Medical imaging

M. S. Routhu · R. V. Govindaraju · P. Saranya (✉)
Department of Computing Technologies, School of Computing, College of Engineering and Technology, SRM Institute of Science and Technology, Kattankulathur, Chennai 603203, India
e-mail: saranyap@srmist.edu.in

B. Annappa et al. (eds.), *High Performance Computing, Smart Devices and Networks*, Lecture Notes in Electrical Engineering 1262,
https://doi.org/10.1007/978-981-97-7794-5_18

1 Introduction

Brain tumors are a significant global health concern, with the World Health Organization reporting 308,102 new cases and 251,329 deaths annually [1]. Despite higher incidence rates in developed regions, the global landscape shows a steady rise in cases due to demographic shifts and improved diagnostic techniques.

Traditional brain tumor detection methods, relying on clinical evaluations and standard imaging, have limitations. However, advancements in medical imaging, especially Magnetic Resonance Imaging (MRI), have significantly improved accuracy. This progress has paved the way for integrating machine learning techniques, particularly Convolutional Neural Networks (CNNs), which have shown promising results in enhancing brain tumor detection accuracy [2].

This paper highlights the importance of utilizing CNN models for global brain tumor detection within MRI scans. The primary goal is to demonstrate how this technological approach, driven by CNNs, can not only improve diagnostic accuracy but also enhance treatment planning globally. By potentially reducing misdiagnoses and enabling timely interventions worldwide, this advancement aims to usher in a new era of global healthcare. The paper meticulously examines the methodology, data acquisition processes, model development intricacies, experimental findings, and broader implications of medical imaging, contributing significantly to the discourse on brain tumor detection and emphasizing the transformative impact of CNNs in global healthcare.

2 Related Work

In 2022, Vidhya J. and Rakshana R. developed a robust brain tumor detection and classification model [3]. Their approach included comprehensive preprocessing steps like grayscale conversion, OTSU binarization, and K-means clustering on the OASIS dataset. Feature extraction utilized DWT and PCA for dimensionality reduction. The core LSTM neural network achieved an impressive 90% accuracy, outperforming SVM at 74% [3]. This study underscores the potential of ML and DL algorithms in brain disease detection, emphasizing the need for larger datasets and algorithmic enhancements for precision.

Kibriya et al. introduced a successful multiclass brain tumor classification system using CNN models and SVM through deep learning techniques [4]. Vidhya and Rakshana [3] similarly transitioned to a different dataset, addressing accuracy challenges as in [4]. Kibriya employed CNNs like ResNet-18 and GoogLeNet, with SVMs, achieving an impressive 98% accuracy, emphasizing the efficacy of deep learning and machine learning techniques in brain tumor identification, echoing Vidhya J. and Rakshana R.'s work [3].

Hemanth et al. in [5] presented an innovative approach for this purpose, introducing two distinct deep learning-based methodologies. Their study achieved impressive outcomes by leveraging the YOLO (You Only Look Once) object detection framework and the FastAi deep learning library. Specifically, the YOLOv5 model exhibited an accuracy rate of 85.95%, while the FastAi classification model achieved a remarkable accuracy of 95.78%. Based on an analysis of a subset of the BRATS 2018 dataset, these findings offer substantial potential in automating brain tumor detection, thereby facilitating early diagnoses and advancements in healthcare outcomes.

In J. Mathew and N. Srinivasan's model [6], they compared five CNN architectures: VGG19, DenseNet169, AlexNet, InceptionV3, and ResNet101. With consistent hyperparameters and a shared dataset, ResNet101 stood out with an impressive 98.6% accuracy, while VGG19 achieved 97.2% [6]. This highlights ResNet101's superiority in brain tumor detection, emphasizing the importance of skip connections and hyperparameter tuning in CNN-based image analysis.

In Josmy Mathew and N. Srinivasan's model [7], they address early brain tumor detection using a dataset of 3064 brain MRI images, both healthy and tumor-afflicted. Their deep CNN model, AlexNet, achieved impressive results: 98.28% accuracy, 97.43% precision, and 97.51% recall [7]. This emphasizes the power of transfer learning, setting it apart from [4], which focuses on traditional CNN methods.

The studies discussed in this section have illuminated various facets of this critical area of medical diagnostics. Vidhya J. and Rakshana R.'s work [3] emphasized the significance of comprehensive preprocessing and LSTM neural networks, while Hareem Kibriya, Momina Masood, Marriam Nawaz, Rimsha Rafique, and Safia Rehman's study [4] introduced a robust multiclass classification system combining Convolutional Neural Networks (CNNs) and Support Vector Machines (SVMs). Furthermore, Çınar, Kaya, and Kaya's research [6] highlighted the effectiveness of the ResNet101 architecture. At the same time, Mathew and Srinivasan [7] showcased the power of transfer learning with the AlexNet model for early tumor detection.

3 Implementation

The architectural flow of our model, as depicted in Fig. 1, is a systematic process designed to enhance the accuracy of brain tumor detection in MRI images. It commences with preprocessing, encompassing essential steps such as grayscale conversion, thresholding, morphological operations (including contour detection, dilation, and erosion), and contrast enhancement, all aimed at refining and preparing the input image. Following preprocessing, the model proceeds to classification, where a Convolutional Neural Network (CNN) is employed to analyze and categorize the image, identifying potential tumors. To ensure the reliability of our results, a validation step rigorously assesses the outcomes, confirming their consistency and effectiveness in the critical task of brain tumor detection.

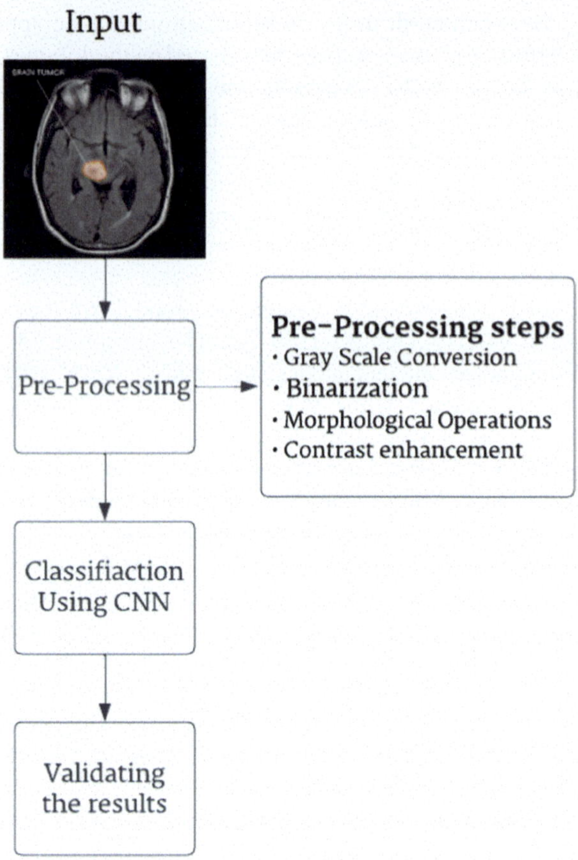

Fig. 1 Flow diagram

Sequential flow, from preprocessing through CNN-based classification to validation, constitutes a holistic approach that combines image enhancement, machine learning, and rigorous evaluation to achieve robust and precise brain tumor identification.

3.1 Preprocessing

The initial phase of our brain tumor detection methodology unfolds a meticulous sequence of preprocessing steps and contour extraction techniques, each playing a pivotal role in the journey toward precise detection. This section elucidates the processes to enhance the input brain MRI images, transforming them into polished canvases that facilitate detection.

In the brain tumor detection methodology, preprocessing plays a pivotal role, transforming raw brain MRI images into polished canvases for precise detection.

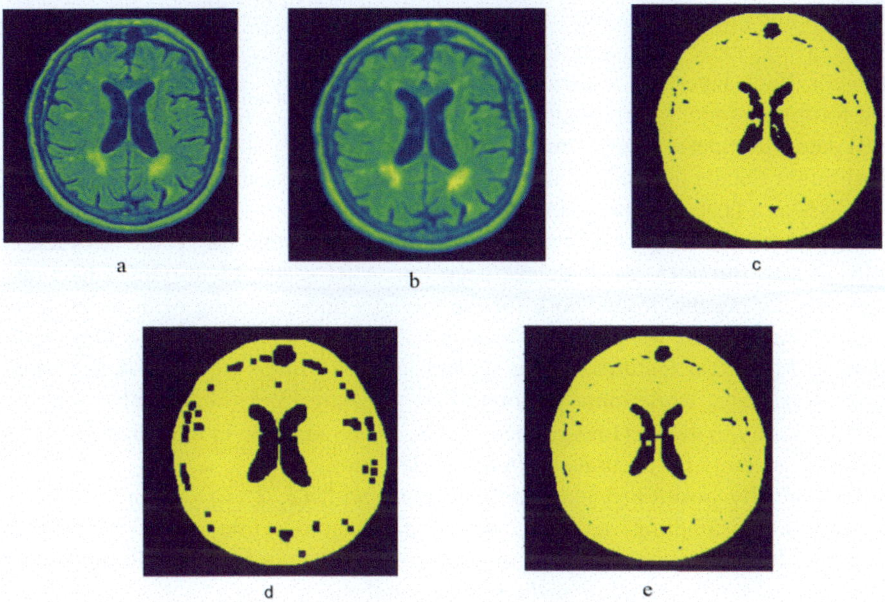

Fig. 2 **a** Grayscale, **b** Gaussian blur, **c** threshold, **d** erosion, **e** dilation

This orchestrated sequence includes grayscale conversion, where the RGB image becomes monochrome, preserving essential structural details. Gaussian blur then elegantly reduces noise and refines fine details. Thresholding simplifies the image into a binary form, revealing regions of interest. Erosion and Dilation delicately separate connected regions and smooth contours. Contour Extraction is pivotal, unveiling significant contours, with OpenCV's cv2.findContours() function as the star performer. The dominant contour represents the primary tumor region. Extreme Points Identification adds vital spatial context, identifying cardinal guides within the contour. This meticulous journey enhances image clarity and sets the stage for precise brain tumor detection, with preprocessing as the prima ballerina.

The outcomes of our preprocessing techniques and various stages of image enhancement are visually represented in Fig. 2. Figure 2a presents the initial grayscale conversion result, followed by Fig. 2b showcasing the image after the Gaussian blur application. Figure 2c provides insight into the binary thresholding process, while Fig. 2d, e demonstrate the effects of erosion and dilation, respectively.

3.2 Classification Using CNN

Our CNN architecture is a beacon of promise in brain tumor detection. Crafted for precise feature extraction from MRI scans, it is a marvel designed to elevate classification precision. Comprising 16 layers, including convolutional, pooling, and dense layers, these are not just structural elements but the pillars of the model's strength. Each layer has a distinct purpose, meticulously extracting vital features that form the model's core predictive capabilities. This layered approach processes MRI data with finesse, transforming and refining it at every stage, capturing even the most minor details for brain tumor detection. This orchestration of layers positions our CNN as a beacon of precision in medical imaging.

At the outset, the first Conv2D layer is the cornerstone for feature extraction with 32 size filters (5, 5). It delves deep into MRI scans, identifying edges and gradients, laying the groundwork for the model's understanding of these intricate medical images. As we progress through subsequent Conv2D layers, filter sizes adapt to MRI scan features. In the second layer, 64 filters of (5, 5) detect more nuanced patterns, while more profound layers feature (3, 3) filters, enhancing the model's ability to discern fine-grained details. Each reduction in filter size enhances the model's acuity for accurate brain tumor detection.

Strategically positioned max-pooling layers with (2, 2) size ensure precision through down-sampling, reducing spatial dimensions while preserving critical features. This balance optimally captures essential features while streamlining model efficiency, which is crucial in medical image analysis.

Densely connected layers are integral to the classification process in the proposed architecture. These layers are woven into our CNN, serving as intellectual hubs that capture high-level features and shape the model's predictions. Choosing the number of units in each dense layer is a precise task, controlling model complexity to optimize performance. These layers guard the model's cognitive abilities, ensuring it discerns intricate patterns, extracts relevant information, and provides precise classifications. Each unit acts as a neuron in a neural orchestra, working harmoniously to decode brain tumor images. This orchestration of dense layers achieves a delicate balance of accuracy, efficiency, and adaptability, advancing brain tumor detection in medical imaging.

The number of units within each dense layer is pivotal in our model's capacity to capture complex patterns. Our architecture adopts a well-balanced configuration, encompassing units ranging from 256 to 4096. This intentional diversity ensures that the model learns both low and high-level features. Smaller units, set at 256, serve as meticulous observers, detecting subtle nuances and low-level details. In contrast, larger units, towering at 4096, act as grand synthesizers, comprehending overarching complexities and weaving high-level features into a coherent narrative. This calibrated range equips our architecture to navigate the entire spectrum of features within brain tumor MRI scans, enhancing accuracy and enabling the model to decipher intricate facets of detection effectively.

The choice and configuration of pooling and dense layers are pivotal cornerstones in constructing our CNN architecture, profoundly influencing its performance. By judiciously employing max-pooling layers, we ensure effective down-sampling, which not only optimizes computational efficiency but also acts as a safeguard against the perils of overfitting. These pooling layers gracefully waltz with the spatial data, preserving the most salient features while condensing information. The architecture

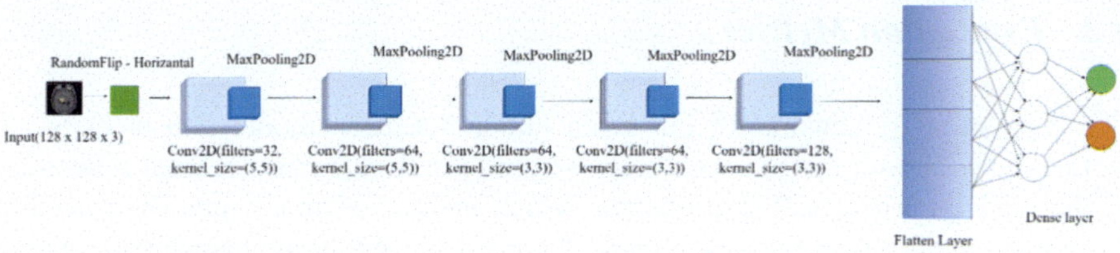

Fig. 3 Model architecture

diagram (Fig. 3) represents the structure of our CNN model. It visually illustrates the sequential flow of layers, including convolutional, pooling, and dense layers, providing a comprehensive view of our model's architecture. The presented CNN architecture stands as a testament to precision engineering in medical imaging, where meticulous design converges with modern computational prowess.

4 Results and Discussion

This section delves into a comprehensive analysis of the outcomes obtained from the brain tumor detection model we developed. The experimentation was carried out on a GPU T100 with Python 3.10 and TensorFlow 2.13.3.

4.1 Dataset Used

The foundation of our research is firmly rooted in the "Br35H" dataset, a meticulously curated collection that showcases a vast spectrum of brain MRI images. Boasting a comprehensive compilation, the dataset presents 1500 images marked with the presence of tumors, complemented by an equivalent 1500 images without any tumor manifestations. Such a harmonious symmetry in the distribution of tumorous and non-tumorous images within the repository is a testament to its robustness. This deliberate equilibrium ensures that during the critical training and evaluation phases, our model receives a fair and unbiased exposure to both categories. As a result, it significantly enhances the model's capability, honing its precision in distinguishing between the intricate nuances of tumorous and non-tumorous MRI representations, ensuring a robust and reliable diagnostic tool.

4.2 Evaluation Metrics

The evaluation framework for our brain tumor detection model is a tapestry of meticulously chosen metrics intricately woven to deliver a multi-faceted reflection of the model's prowess. These metrics, which include accuracy, precision, recall, the $F1$ score, and the critical area under the receiver operating characteristic curve (ROC AUC), are not merely numbers but pillars that uphold the model's credibility. Together, they paint a panoramic view of the model's performance, highlighting its strengths and areas for refinement. By leaning on this comprehensive suite of metrics, we delve deep into the model's capabilities, ensuring that its brain tumor detection efficiency is scrutinized. This rigorous evaluation approach guarantees that our assessment is both holistic and robust, setting a gold standard for performance analysis in medical imaging.

4.3 Model Performance and Training Results

The training spanned 200 epochs with a batch size of 32. Training accuracy reached an impressive 100%, while test accuracy achieved around 98.33%, showing good generalization. The training loss decreased to approximately 0.0021, while the test loss was higher at around 0.1339. The divergence between training and test loss suggests potential overfitting on training data, which regularization techniques could address. Nonetheless, the model efficiently minimizes training and test data classification errors (Table 1).

The training progress log presented herein illustrates the developmental trajectory of a machine learning model over multiple training epochs. An "epoch" signifies a complete iteration through the entire training dataset, during which the model

Table 1 Training log

Epoch	Loss%	Accuracy%
1	0.6925	0.5288
11	0.5593	0.7292
21	0.4597	0.79
31	0.2117	0.9179
41	0.0735	0.9742
51	0.018	0.9921
61	0.0038	0.9996
71	0.0058	0.9967
81	0.0068	0.9987
91	0.0056	0.9983
101	0.0017	0.9996

endeavors to minimize a loss function that quantifies the disparity between its predictions and actual target values. During the inaugural epochs, such as Epoch 1, the model commences its learning journey with an initial accuracy of approximately 52.88% and a relatively elevated loss of 0.6925. These early stages of training are characterized by a tentative model performance, as is to be expected. However, as training unfolds, training accuracy exhibits discernible enhancement. This progressive augmentation in accuracy indicates that the model is progressively acquiring the capacity to generate more precise predictions. This conveys a sense of stability in the model's learning process.

4.4 Comparison of Classification Results with Existing Methods

The proposed model's learning dynamics and generalization ability are insightful. In the first epoch (Epoch 1), the model starts with a loss of 0.6925 and an accuracy of 0.5288, indicating initial misclassification challenges. Validation metrics show similar performance. The time per epoch is 11 s, leaving room for efficiency improvements. However, as training progresses, improvements are evident. By Epoch 11, the loss drops to 0.5593, and accuracy rises to 0.7292, with validation metrics also improving. The time per epoch reduces to 3 s, indicating increased efficiency. Continuing to Epoch 21, both training and validation losses decrease further. Training accuracy surges to 0.79, and validation accuracy reaches 0.8083 (Fig. 4).

In subsequent epochs, performance continues to improve. By Epoch 51, training accuracy is 0.9921, and validation accuracy is 0.9683, showing intense learning and generalization. Validation loss remains low at 0.1208. Beyond Epoch 61, training

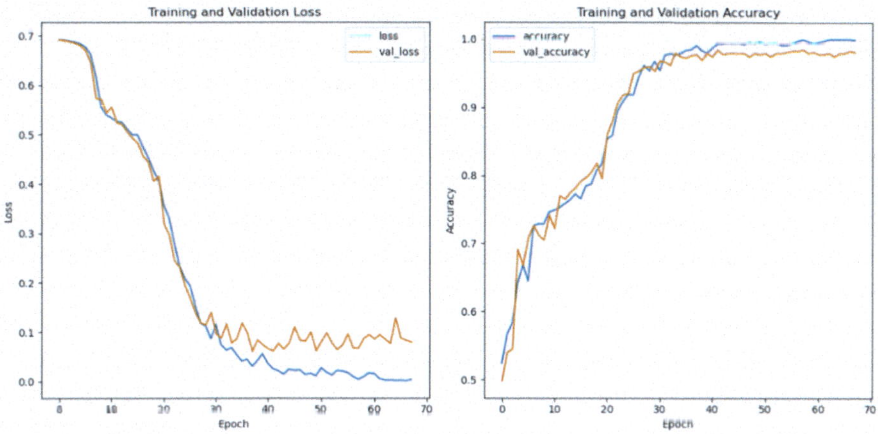

Fig. 4 Learning curves

Table 2 Comparison of the performance metrics

Author name	Accuracy	Precision	Recall	$F1$ score	ROC AUC
Vidhya and Rakshana [3]	90	–	–	–	–
VGG19 [8]	97.2	97.2	97.1	97.1	–
Mathew and Srinivasan [7]	98.38	97.43	97.51	–	–
Kolla and Mishra [9]	99.23	97.12	95.73	–	–
Dipu et al. [5]	95.78	96.70	95.65	96.17	
Proposed model	98.5	98.36	98.68	98.52	98.49

accuracy nears perfection at 0.9996, and validation accuracy remains robust at 0.9833, with a stable validation loss of around 0.1237. Epochs 71 to 101 show consistent performance with minor fluctuations, reaching a minimal training loss of 0.0017 by Epoch 101. Validation loss remains steady at around 0.1339, and validation accuracy stays robust at 0.985, showcasing the model's strong generalization to new MRI scans.

Table 2 compares performance metrics for our proposed model, labeled "Proposed Model," alongside state-of-the-art brain tumor detection models from the literature. The key metrics include accuracy, precision, recall, $F1$ score, and ROC AUC, all vital indicators of a model's effectiveness in brain tumor detection.

Our proposed model excels with an outstanding 98.5% accuracy, demonstrating its proficiency in accurate classification. It achieves an impressive precision of 98.36, minimizing false positives, and exhibits an outstanding recall of 98.68, effectively identifying genuine tumor cases. This balance is reflected in an impressive $F1$ score of 98.52, indicating robustness. The ROC AUC score of 98.49 reaffirms its exceptional discriminatory power. These metrics underscore the model's effectiveness in brain tumor detection, promising contributions to medical diagnosis and treatment planning. Compared to models like VGG19 with 97.2% accuracy, our model outperforms in accuracy and precision, enhancing diagnostic accuracy and reducing misdiagnoses, making it a promising advancement in brain tumor detection.

In summary, our proposed model showcases exceptional accuracy, precision, recall, F1 score, and ROC AUC performance. It outperforms existing state-of-the-art models, contributing significantly to medical imaging and improving brain tumor detection methodologies, benefiting healthcare professionals and patients.

5 Conclusion

Our proposed model represents a groundbreaking advancement in medical imaging and diagnostic accuracy. Through state-of-the-art Convolutional Neural Networks (CNNs) and advanced preprocessing techniques, we achieved an extraordinary 98.5% classification accuracy. This remarkable accuracy and our comprehensive model set a new standard in brain tumor diagnosis. Our model's robustness, evident in its balanced precision-recall equilibrium and high $F1$ score, ensures precise distinctions between tumor and non-tumor cases, reducing the risk of misdiagnosis.

Our research has laid a strong foundation for ongoing exploration of brain tumor diagnosis. While we initially focused on binary classification, identifying tumor presence or absence, our future trajectory aims for multiclass classification to categorize specific brain tumor types. This transition addresses the diversity of brain tumors, each requiring unique treatment strategies. We will refine our model's precision through extensive medical datasets and collaborate with global medical institutions to enhance our research. By employing advanced algorithms and preprocessing techniques, our goal remains to create a versatile tool that bridges AI with medical diagnostics, ultimately improving patient care.

References

1. Malzkorn B, Reifenberger G (2016) Practical implications of integrated glioma classification according to the World Health Organization classification of tumors of the central nervous system 2016. Curr Opin Oncol 28(6):494–501
2. Ahmed SF, Rahman FS, Tabassum T, Bhuiyan MTI (2019) 3D U-Net: fully convolutional neural network for automatic brain tumor segmentation. In: 22nd international conference on computer and information technology (ICCIT), 1–6
3. Vidhya J, Rakshana R (2022) Advanced detection of brain disease using ML and DL algorithm. In: International conference on edge computing and applications (ICECAA), 1170–1174
4. Kibriya H, Masood M, Nawaz M, Rafique R, Rehman S (2021) Multiclass brain tumor classification using convolutional neural network and support vector machine. In: Mohammad Ali Jinnah University international conference on computing (MAJICC), 1–4
5. Dipu NM, Shohan SA, Salam KMA (2021) Deep learning based brain tumor detection and classification. In: International conference on intelligent technologies (CONIT), 1–6
6. Çınar N, Kaya B, Kaya M (2022) Comparison of deep learning models for brain tumour classification using MRI images. In: International conference on decision aid sciences and applications (DASA), 1382–1385
7. Mathew J, Srinivasan N (2022) Deep convolutional neural network with transfer learning for automatic brain tumor detection from MRI. In: International conference on computing, communication, security and intelligent systems (IC3SIS), 1–6
8. Deepa PL, Ponraj N, Sreena VG (2021) A comparative analysis of deep neural networks for brain tumor detection. In: 3rd international conference on signal processing and communication (ICPSC), 36–40
9. Kolla M, Mishra RK, Zahoor Ul Huq S, Vijayalata Y, Gopalachari MV, Siddiquee KE (2022) CNN-based brain tumor detection model using local binary pattern and multilayered SVM classifier. Comput Intell Neurosci

Improving Safety and Air Quality in the Mining Industry with IoT-Enabled Monitoring and Air Purification Solutions

Mohan Dholvan, Krishna Samalla, and G. Narsimhulu

Abstract Ensuring industrial safety remains a critical concern within the mining sector, where the welfare and efficiency of employees depend on effective healthcare and communication protocols. Dependable communication is indispensable for promptly addressing potential threats, while access to medical personal protective equipment (PPE) and regular health screenings is imperative. Acknowledging the limitations of current safety systems, there is a pressing need for enhancing safety measures within the mining industry. Existing safety protocols, such as gas monitoring, emergency response plans, and ventilation systems, play a vital role in mitigating risks and safeguarding workers. Emergency procedures mitigate accidents, gas monitoring identifies hazardous gases, and ventilation ensures a steady supply of fresh air. However, these systems exhibit shortcomings, including their inability to fully prevent accidents, detect all types of gases, and maintain optimal airflow. Consequently, there is a demand for system improvements. Our approach to enhancing safety in the mining industry involves leveraging advanced technology to bolster communication and implement Internet of Things (IoT) solutions for real-time monitoring of air quality, toxicity levels, and workers' vital signs. Through the integration of Blynk software, ESP32 boards, and sensors, our system enables continuous monitoring and immediate reporting of potential threats. The overarching objective is to prioritize worker well-being by closely tracking vital signs and promptly addressing any emerging risks. Through the application of innovative concepts and cutting-edge technology, our proposed system represents a significant advancement in bolstering safety within the mining sector.

M. Dholvan (✉) · K. Samalla
Sreenidhi Institute of Science and Technology Ghatkesar, Medchal, Telangana, India
e-mail: mohan.aryan19@sreenidhi.edu.in

K. Samalla
e-mail: krishnas@sreenidhi.edu.in

G. Narsimhulu
Vignana Bharathi Institute of Technology, Hyderabad, India
e-mail: gorre.narsimhulu@vbithyd.ac.in

B. Annappa et al. (eds.), *High Performance Computing, Smart Devices and Networks*,
Lecture Notes in Electrical Engineering 1262,
https://doi.org/10.1007/978-981-97-7794-5_19

Keywords ESP32 · Air quality detection · Air purifier · Blynk · IoT

1 Introduction

Mechanical security concerns loom large over the mining industry, underscoring the critical need for robust communication and healthcare protocols to ensure worker safety and productivity. Reliable communication is pivotal for swiftly identifying and addressing potential threats, while regular health examinations and access to protective equipment are indispensable. Recognizing the inadequacies in current safety standards, there is a pressing call for advancements in security measures within the mining sector.

Ventilation, emergency response plans, and gas monitoring constitute primary security measures employed to mitigate risks and protect workers in mining operations. However, shortcomings such as insufficient ventilation and the inability to detect all gases necessitate urgent improvements in these systems. Our proposed solution aims to enhance safety in the mining industry by leveraging IoT technology to monitor air quality, toxicity levels, and workers' vital signs in real-time.

By integrating Blynk software, ESP32 boards, and sensors, our system enables proactive threat detection and immediate reporting, thereby safeguarding worker well-being and increasing productivity. Through innovative approaches and state-of-the-art technology, our solution represents a significant leap forward in bolstering security within the mining sector.

Despite the pivotal role the mining sector plays in global economic growth, it faces inherent risks to worker safety and well-being. Our solution addresses these risks by harnessing IoT innovation to improve communication and monitoring in mining operations, thereby ensuring real-time risk assessment and enhancing safety protocols.

Furthermore, our solution offers additional benefits beyond security and productivity enhancements. It facilitates data collection on equipment performance, environmental conditions, and worker health, enabling the creation of predictive models to pre-empt potential issues. Additionally, through data analysis and information sharing, mining companies can proactively enhance safety cultures and mitigate the recurrence of accidents.

The adaptability of our solution allows it to be tailored to mining operations of any scale, from large-scale mines to small-scale endeavors. By fostering collaboration between mining companies, technology vendors, and safety experts, our solution can be seamlessly integrated into existing safety frameworks and routines.

While initial implementation costs may pose a challenge, the long-term benefits of our solution outweigh these concerns. By reducing expenses associated with accidents, downtime, and equipment failures, our innovation ultimately contributes to cost savings and operational efficiency for mining companies.

To ensure the successful adoption of our solution, we advocate for collaborative partnerships between mining companies and technology vendors. By working together to tailor the solution to industry-specific needs, companies can maximize its effectiveness and seamlessly integrate it into their existing safety protocols.

2 Literature Review

Huang Mengtao and Feng Zunxiang presented an electrostatic adsorption-based depleting gas framework in 2016. The framework aims to reduce the amount of discuss contamination through the use of electrostatic adsorption to capture harmful chemicals and subsequently cleanse the exhaust gas [1].

A clever framework for discussing contamination checking was developed by Kennedy Okokpujie and Etinosa Noma-Osaghae in 2018. It continuously monitors talk quality within a range and displays the measured discuss quality on an LCD screen. This system provides real-time data on discussion quality levels and encourages discussion of contamination [2].

Ms. S. Menaga and Ms. J. Paruvathavardhini presented a quality-checking system based on IoT innovation for automobiles in 2019. By installing a module in a moving vehicle, the framework can screen and discuss contamination of any necessary place. This approach aims to provide a comprehensive view of air pollution levels in various regions [3].

Asrudin Container Tangle Ali and Farih Container Deraman proposed in 2020 to construct an IoT-enabled discussion quality detector for a passenger vehicle. The framework has a control window structure that can be adjusted as a result. When the framework detects high CO_2 levels in an automobile, the control window automatically rolls down. The window automatically closes when the car's CO_2 levels return to a safe level. The goal of this framework is to provide a safe and enjoyable driving experience [4].

Xue Dong presented a concept for a sifting automotive air filter in 2021 that can effectively advance harmful chemicals like formaldehyde in an automobile. This system aims to provide a more comfortable ride for both drivers and passengers [5].

The discuss quality observing and location framework in vehicle cabin based on Web of Things was proposed by Jacquline Waworundeng and Priana Sari Adrian in 2022. It is a framework for discussing quality observations and discoveries that is implemented inside the vehicle cabin to promote mindfulness and assist people in screening the discuss quality [6]. In general, these suggestions show the efforts being done to address discuss pollution and the various strategies being used to address this global problem.

3 Proposed System

The sensors can detect several gas concentrations, including methane, propane, hexane, ammonia, and carbon monoxide, as well as particulate matter, which includes smoke and fine dust, around the miner. A threshold value is assigned to the ESP-32 microcontroller, and an alert is set off when this value is exceeded. The provided mask should be worn by the mine worker as a result of this alarm. The air filtration system starts the process of cleaning the ambient air and forces the cleaned air into the respirator mask. The liquid crystal display, or LCD (Fig. 1).

The detected gas values for each gas will be displayed once integrated into the system. The Blynk application will likewise provide access to the same values. The worker will only take off the mask after the number of contaminants in the air has decreased. The backpack contains an oxygen tank that supplies the miner with oxygen in case they need it.

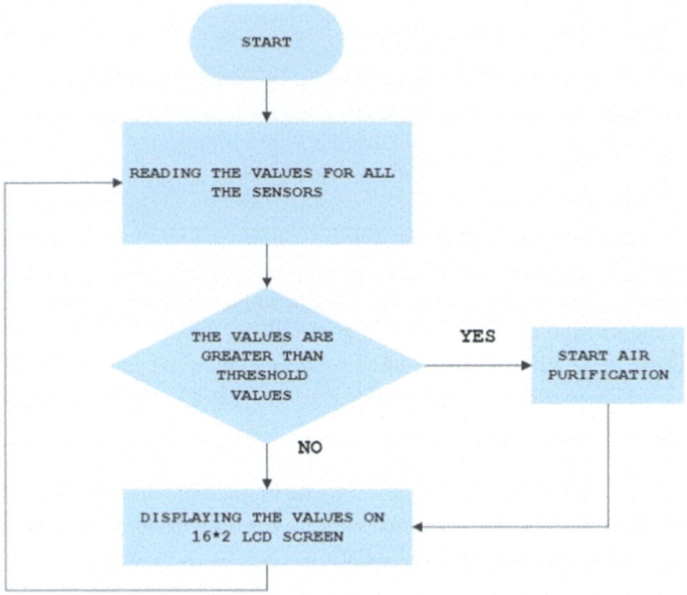

Fig. 1 Workflow

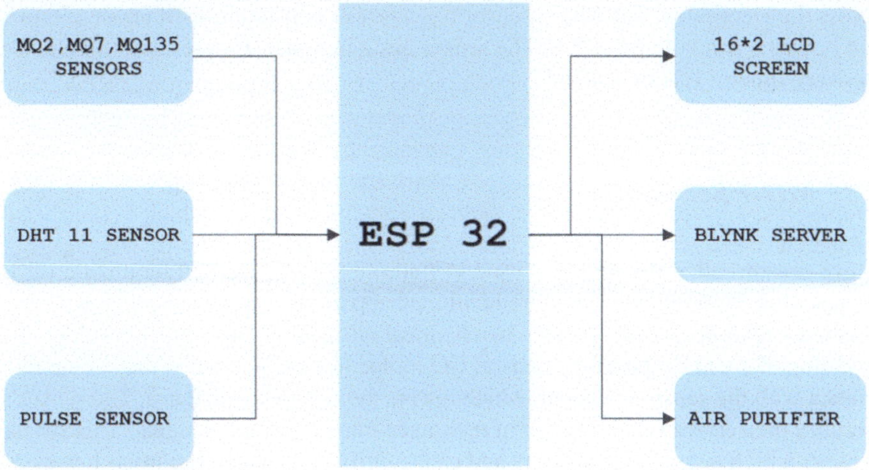

Fig. 2 Block diagram

4 Framework

4.1 MQ-2 Sensor

A common gas sensor module used in fire detection and prevention systems is the MQ-2 smoke sensor. The instrument can identify a number of gases, such as alcohol, carbon dioxide, propane, methane, smoke, and liquefied petroleum gas (LPG), by measuring changes in resistance that result from the contact of the gas with the sensing element. The sensor operates in the DC voltage range of 5–24 V and can detect various gases in the range of 100–10,000 ppm and smoke in the range of 300–10,000 ppm. The response time and recovery times of the MQ-2 sensor module are around 10 and 30 s, respectively. It is easily incorporated with microcontrollers to create Internet of Things (IoT) smoke detection systems [7] (Fig. 2).

4.2 MQ-3 Sensor

Hexane, benzene, and methane are just a few of the volatile organic compounds that can be found using the MQ-3 gas sensor. Tin oxide (SnO_2), a semiconductor material, makes up the detecting element of the MQ-3 sensor. When exposed to the gasses of interest, SnO_2 changes electrical resistance. The detecting element is located inside a small, electrode-equipped container. The resistance of the sensing component changes in response to the gases being measured when a voltage is applied across the electrodes. The MQ-3 sensor has a power consumption of about 150 mA and operates at a voltage of 5 V. The sensor has a less than 10-s reaction time and a recovery time of less than 30 s [7].

4.3 MQ-7 Sensor

A gas sensor called the MQ-7 is used to find out whether carbon monoxide (CO) is present in the ambient. The aforementioned apparatus is a space-saving and cost-effective technology that operates by chemical adsorption. A catalyst envelops the detecting element of the MQ-7 sensor. CO molecules oxidize when they come into contact with the sensing element, which lowers the sensor's resistance. The sensor's circuitry then converts this change in resistance into an electrical signal. This sensor exhibits a high degree of sensitivity and can identify CO concentrations as low as 20 parts per million. Because of its small size and low power consumption, the MQ-7 sensor is a great option to be integrated into a variety of safety and surveillance systems [7].

4.4 MQ-135 Sensor

A gas sensor called the MQ-135 is used to detect a number of gases, including CO_2, ammonia, nitrogen oxides, benzene, and smoke. The device's working mechanism is based on how a sensing component changes in resistance when exposed to a specific gas. The sensing device consists of a small heating element to raise the temperature of the sensing element and an electrical circuit to measure the resistance changes. For nitrogen dioxide, 10–300 ppm, benzene, and carbon dioxide, 100–10,000 ppm are the reported sensitivity ranges of the MQ-135 sensor. The gadget has a power consumption of 800 mW and runs at a voltage of 5 V. The sensor shows a recovery time of less than 60 s and a reaction time of less than 10 s. This entity's small size makes it easier for it to be integrated into a variety of systems and devices [7].

4.5 DHT11 Sensor

This digital instrument was created especially to measure the humidity and temperature in the immediate surroundings. The device's components, a thermistor, and a capacitive humidity sensor are used to detect changes in temperature and relative humidity, respectively. This sensor measures humidity levels from 20 to 90% with an accuracy of $\pm$ 5% and operates in the voltage range of 3.5–5.5 V. The instrument can measure temperature with an accuracy of $\pm$ 2 °C and a range of 0–50 °C.

Through a single wire, the sensor's data is transferred in the form of a digital signal. A microcontroller or other device then interprets and analyzes this signal [8].

4.6 Pulse Oxy Sensor

A patient's finger, earlobe, or another appropriate site serves as the home of a pulse oxygen sensor. The two light wavelengths that the sensor emits—red and infrared—are absorbed by the blood's oxygenated and deoxygenated hemoglobin. Hemoglobin-free light passes through the tissue and is detected by the sensor. The sensor's microcontroller looks over the light absorption data and calculates the pulse rate and SpO_2 measurements. A light source, photo detectors, and an electronic circuit that amplifies and filters the signal are often included in the sensor's circuitry. It is a medical gadget that is frequently used to monitor patients suffering from respiratory and cardiovascular conditions in a variety of healthcare settings, such as clinics, hospitals, and ambulatory care centers. Moreover, pulse oximetry sensors are integrated into modern wearable technologies to monitor factors linked to sleep and physical fitness [9].

4.7 HEPA Filter

One kind of air filtration system that is intended to successfully remove pollutants and airborne particles from the surrounding area is the HEPA filter. HEPA filters are widely used in a variety of environments, including homes, offices, and medical facilities. In homes, they are usually found in air purifiers, vacuum cleaners, and HVAC systems. They are used in clean rooms, operating rooms, and isolation rooms in hospitals to reduce the spread of infectious diseases. The effectiveness with which HEPA filters remove airborne particles is the basis for its evaluation. To satisfy the requirements for a true HEPA filter, the filter must be able to remove 99.97% of particles with a size of 0.3 μm or larger. Smaller particles, such as ultrafine particles measuring less than 0.1 μm, can be removed by some high-efficiency particulate air (HEPA) filters [10].

4.8 Activated Carbon Filter

These filters are frequently used to remove dangerous pollutants and impurities from the air. The material that makes up the filters is activated carbon, which is carbon that has been exposed to oxygen to create a highly porous substance with a large surface area. It is possible to customize the filter to target specific types of contaminants, such as volatile organic compounds (VOCs), chlorine, and other chemicals. In order to protect against dangerous pollutants, these filters are widely used in water treatment systems, air purifiers, and face masks. These substances demonstrate a great level of effectiveness in getting rid of smells in the air, such as those that come from cooking or cigarette smoke [11].

4.9 ESP32 Board

Because of its durability, the ESP32 board is a microcontroller development board that is widely used for Internet of Things (IoT) applications. The ESP32 microcontroller, which has two central processing units and supports Bluetooth and Wi-Fi communication protocols, forms the foundation of the device. Many functions, such as GPIO (General Purpose Input/Output) pins, ADC (Analog to Digital Convertor), DAC (Digital to Analog Convertor), SPI (Serial Peripheral Interface), I2C (Inter integrated Controller), and UART (Universal Asynchronous Receiver/Transmitter), are available on the ESP32 board, making it an excellent option for a variety of tasks. The device's small size, energy efficiency, and ability to run on battery power make it suitable for Internet of Things applications [12].

4.10 Blynk Software

Blynk is a software program that works with online and mobile platforms to provide remote control of Internet of Things (IoT) devices. Without the need for programming knowledge, users can design dynamic, individualized graphical user interfaces (GUIs) with the software. Blynk offers a wide range of widgets, including graphs, buttons, sliders, and displays that can be easily dropped using a drag-and-drop interface onto an electronic canvas. Blynk libraries are available for widely used platforms like ESP32, Arduino, Raspberry Pi, and Node MCU, allowing users to connect their Internet of Things devices to Blynk. To protect user data, Blynk provides secure communication protocols and cloud connectivity [13].

4.11 16 * 2 LCD Display

A sum of 16 columns and 2 rows of characters make up the commonly used 162 LCD display, an electronic display module that allows a maximum of 32 characters to be displayed at once. Because alphanumeric displays can display both numeric and alphabetical characters, they are frequently used in a variety of applications, including as digital clocks, thermometers, and home automation systems. A microcontroller

Fig. 3 3D model

allows the exchange of data input and output by sending signals to the display module, which in turn manages the display. The minimal interface and low power consumption of the 162 LCD display make it a popular choice for electronic projects requiring basic display capabilities [14].

5 Model Design

Here, the intended configuration for the air quality detection and purification system is shown in the CAD model. The main circuit will be installed in a backpack that mine workers can comfortably wear, together with an oxygen cylinder, breathing mask, and air purifier. Within the backpack, an essential communication and alarm system will also be included (Figs. 3 and 4).

6 Results

The good performance of the suggested system demonstrates the effectiveness of the chosen sensors, in particular the MQ-series gas sensors, in identifying air pollutants. The sensors exhibit high levels of sensitivity and precision, making them highly suitable for applications requiring rigorous monitoring of air quality. The combination of the Blynk software and the ESP32 microcontroller provides an efficient and intuitive platform for controlling the system from a distance and gathering data in real time.

Fig. 4 **a** Blynk output. **b** Output

Furthermore, the 16 × 2 liquid crystal display (LCD) serves as a stable and cost-effective method of information presentation, making understanding and analysis easier. Because of the display's simple and intuitive layout, users may quickly and easily determine the quantities of pollutants present and take the necessary corrective action. The system's effective functioning demonstrates the promise of using IoT platforms and advanced sensor technology to create effective air quality monitoring systems. Concerns about air pollution and its effects on human health and the environment are growing, which highlights the importance of these systems in promoting safer and cleaner air.

7 Conclusion

A dependable and affordable method of determining the concentrations of various air pollutants is to use the ESP32 board and MQ sensors in conjunction with Blynk software for air quality monitoring. In the end, this approach provides a workable resolution to the current issue. The system uses a range of sensors to detect several types of pollutants, including nitrogen dioxide, carbon monoxide, and so on. The sensors' data is shown on an LCD screen and in the Blynk application. The system may be remotely monitored and controlled through the use of Blynk software, making it an appropriate solution for indoor air quality surveillance in residential, commercial, and other indoor environments.

This project, in general, demonstrates how mine workers and other industries can benefit from the availability of reasonably priced and easily accessible air quality monitoring systems to help them make informed decisions about their surrounding environment. Future research efforts can focus on expanding the range of detected pollutants by adding more sensors or combining machine learning algorithms to improve accuracy and real-time analysis. On the other hand, this research provides a solid foundation for creating effective air quality monitoring systems with readily available parts and software.

References

1. Menaga (2019) Air quality monitoring system using vehicles based on the IoT. https://www.irjet.net/volume-6-issue03
2. Dong X (2021) Design of a filtering car air purifier. www.iopscience.iop.org, https://doi.org/10.1088/1755-1315/632/5/052095
3. Mengtao H, Zunxiang F (2016) Automobile exhaust purification system research based on ARM. In: 2016 international symposium on computer, consumer and control (IS3C), Xi'an, China, pp 728–731.https://doi.org/10.1109/IS3C.2016.186
4. Gowda MP, Harshitha GY, Jyothi KN, Srushti PR (2021) Air quality monitoring system. Int J Eng Res Technol (IJERT) NCCDS 09(12)
5. Deraman FB, Ali ABM, Muhamad NB (2020) Innovation of air quality detector in passenger car using IoT. 1(1):16–20
6. https://robu.in/mq-series-gas-sensor/
7. https://www.electronicwings.com/
8. https://robocraze.com/blogs/post/what-is-pulse-sensor
9. https://cowaymega.com/blogs/blog/what-is-a-hepa-filter
10. https://www.indiancrusaders.com/blog/importance-of-activated-carbon-filter-in-an-air-purifier/
11. https://www.electronicshub.org/getting-started-with-esp32/
12. https://techexplorations.com/guides/blynk/1-what-is-%20blynk
13. https://www.elprocus.com/lcd-16x2-pin-configuration-and-its-working/

Deep Anomaly Detection Approaches for Enhancing Pedestrian Safety

Surekha Samsani and Sabbina SuryaSaranya

Abstract Anomaly detection in pedestrian walkways is a critical research area to enhance pedestrian safety. The overspeeding of vehicles, drunk driving, bad weather, and failures in autonomous vehicles pose a threat to pedestrians. The imperative for a smart surveillance system capable of autonomously detecting abnormal activities is paramount. In this scenario, this paper presents a deep learning-based anomaly detection system for enhancing pedestrian safety. The proposed approach analyzes the data captured by the IP cameras to identify the movement of vehicles on pedestrian walkways and sends text alert messages. The deep anomaly detection approaches presented in this research study integrate Convolutional Autoencoder and Fuzzy Logic with Data Encryption techniques for developing a context-sensitive smart surveillance model. The model has been trained with 1200 images consisting of pedestrian paths which are taken from the University of California San Diego (UCSD) dataset. The proposed anomaly detection systems' effectiveness is assessed using metrics such as accuracy, precision, recall, F1-score, and area under curve (AUC).

Keywords Anomaly detection · Pedestrian walkways · Deep learning · Convolutional autoencoder · Data encryption

1 Introduction

In recent times, there has been a concerning rise in incidents where vehicles deviate from designated paths, encroaching into pedestrian zones which leads to pedestrian accidents. The different types of pedestrian accidents that may happen in pedestrian zones are crosswalk accidents, intersection accidents, collisions with bicycles or scooters, jaywalking, ignoring traffic signals, skateboarding, or rollerblading. Among

S. Samsani (✉) · S. SuryaSaranya
University College of Engineering Kakinada, JNTUK, Kakinada, India
e-mail: surekha.samsani@jntucek.ac.in

B. Annappa et al. (eds.), *High Performance Computing, Smart Devices and Networks*, Lecture Notes in Electrical Engineering 1262,
https://doi.org/10.1007/978-981-97-7794-5_20

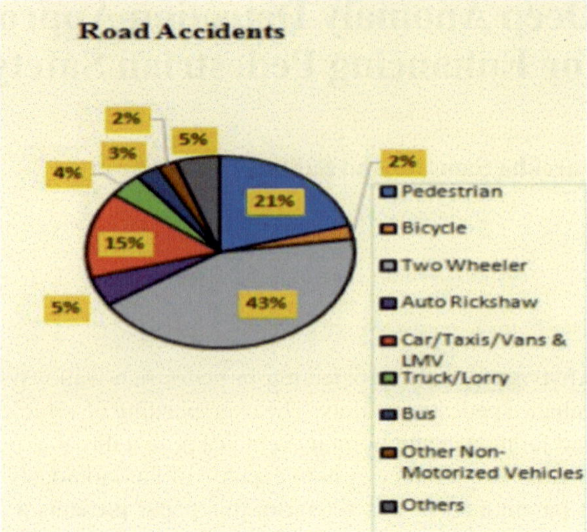

Fig. 1 Road accidents by type of impacting vehicles [1]

the presented anomalies, there has been a noticeable surge in incidents involving vehicles straying from their designated routes and encroaching into pedestrian zones, resulting in a concerning uptick in pedestrian accidents.

As per the statistics of the Indian road accidents survey, the pedestrian accident rate is up to 21% due to vehicle deviation. Figure 1 shows the percentage of road accidents that occurred by impacting vehicles [1].

Nowadays, smart surveillance systems are prevalent everywhere, and they can be developed to detect anomalous activities thus minimizing pedestrian accidents by taking necessary actions and sending alerts. As shown in Fig. 2, the general framework of a video surveillance system consists of various phases such as motion and object detection, object classification, behavior understanding, and activity analysis [2].

Motion and Object Detection: This stage involves identifying regions in a video frame where there is a significant change in pixel values compared to the previous frame. These changes can indicate the presence of motion.

Object Classification: After motion is detected, the system proceeds to classify the objects within the regions of interest. This step involves determining what type of object is present, such as a person, vehicle, animal, or other relevant categories.

Behavior Understanding and Activity Analysis: This stage focuses on interpreting the interactions and movements of objects over time. It aims to understand the behavior and activities within the surveillance area.

Alert Notification: When suspicious or unusual activities are detected, the system generates alerts such as text alerts or voice alerts. These alerts can be seen by human operators, security personnel, or automated systems for further action.

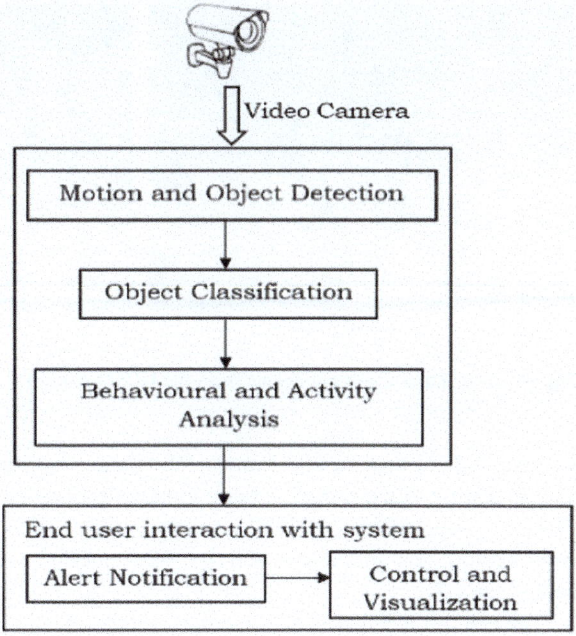

Fig. 2 Framework of automated visual surveillance system [2]

Pedestrian accidents can be minimized through proper surveillance and decision policies. This paper proposes an intelligent monitoring system for detecting vehicle intrusion on pedestrian walkways to ensure pedestrian safety. The proposed system is affordable as it uses the data captured by existing CCTV cameras strategically positioned at key intersections, crosswalks, and areas prone to vehicle intrusion posing a direct threat to the lives and well-being of individuals on foot. Over the past decade, the performance of Deep Learning approaches has been improved for solving various real-time problems. Therefore, in this paper, the Deep Learning-based Convolutional Autoencoder is used to learn the patterns for detecting vehicles in pedestrian zones. The paper introduces privacy-preserving anomaly detection to enhance privacy for individual data by encrypting the video. Encryption ensures that sensitive information remains confidential and inaccessible to unauthorized parties, which is especially important when dealing with personal or sensitive data. Accurate and reliable data is crucial for the success of any Deep Learning-based anomaly detection model. The De-facto UCSD dataset is used to train the Deep Learning-based anomaly detection system presented in this paper. The description of the dataset is given below.

UCSD Dataset

The UCSD [3] anomaly detection dataset is taken from the UCSD Library Research Collections dataset and consists of 8-bit grayscale images. These images are the clips of groups of people walking toward and away from the camera, and some amount of perspective distortion. The video footage recorded from each scene was split into

Fig. 3 Normal and abnormal events from the UCSD dataset

various clips of around 200 frames. The dataset contains 34 training video samples and 36 testing video samples. Each video clip is saved as a set of .png files. Figure 3 shows a snapshot of the UCSD training videos. Figure 3a represents a pedestrian zone without any anomalous activity, whereas Fig. 3b shows the bicycle entering in pedestrian zone.

The rest of the paper is organized as follows: Sect. 2 summarizes the related work; Sect. 3 presents the proposed model. The experiment results and observations are discussed in Sects. 4, and 5 concludes the paper.

2 Related Work

Li et al. [4] introduced a spatiotemporal representation learning model, which involves extracting spatial–temporal features from videos from the *UCSDSped and Subway* datasets using a multi-scale 3D Convolutional Neural Network. The extracted information is utilized to model the scene background through a high-dimensional mixed Gaussian model for anomaly detection. On the *Subway dataset*, the frame-level Equal Error Rate (EER) stands at *15.9%* at the exit and *10.1%* at the entrance. Similarly, on the *UCSDSped2 dataset*, the EER is *10.5%* at the frame level and *13.8%* at the pixel level.

Wu et al. [5] proposed three pre-trained models to extract both appearance and spatiotemporal features on the *UCSD Ped1 and Ped2*dataset: *Panoptic Feature Pyramid Network (PFPN)*, *Joint Detection and Embedding (JDE)*, and *ResNet101 (R101)*. The proposed approaches achieve an impressive area under curve (AUC) of *84.1%* with an equal error rate (EER) of *23.8%* on the Ped1 dataset. On the Ped2 dataset, it attains an AUC of *92.4%* with an EER of *14.9%*.

Nayak et al. [6] proposed approach in this paper which is the use of a *Convolutional Spatiotemporal Autoencoder* on the *UCSD Ped1 and Ped2* datasets. The architecture of the proposed model consists of three key sections: Spatial Encoder employs three convolutional layers for spatial feature extraction. The approach achieves a remarkable *area under curve (AUC) of 90.1%* and an *equal error rate (EER) of 11.9%* on the *UCSD Ped1 dataset*. On the *UCSD Ped2 dataset*, it attains an *AUC of 88.3%* and an *EER of 11.3%*.

Mangai et al. [7] proposed a Deep Learning-based anomaly detection from the UCF crime dataset that includes the keyframe extraction process, temporal feature extraction, and classification. The process of keyframe extraction involves utilizing the HSV histogram in conjunction with the k-means clustering technique. Subsequently, temporal features are extracted utilizing a Convolutional Neural Network. Finally, a ConvLSTM model is deployed to comprehend activity sequences and identify anomalies within the keyframes. This comprehensive approach enables the effective detection of irregularities in the analyzed video data. The proposed approach gave a *recall value of 90%*, *precision of 80%*, and *F1-score of 84.7%*.

Nguyen et al. [8] proposed an anomaly detection model as a combination of two streams, each addressing a specific aspect on *CUHK Avenue, UCSD Ped2, Subway Entrance Gate and Exit Gate, Traffic-Belleview, and Traffic-Train* datasets: The appearance stream incorporates a *Convolutional Autoencoder (Conv-AE)* and the motion stream employs a *U-Net architecture*. The proposed model achieves an area under the curve (AUC) score of *0.869* on the *Avenue dataset*, on the *UCSD Ped2 dataset*, the proposed model attains an AUC score of *0.962*, and also provides frame-level anomaly detection results, with an average *precision* of *0.751* on the *Belleview dataset,* and *0.490* on the *Train dataset*.

Nithesh et al. [9] introduced a method that includes the system equipped with a *GSM module*. The methodology comprises two essential components. The first component involves feature extraction and learning from the video frames. The second component is dedicated to anomaly identification. The study focuses on *two pedestrian datasets from UCSD*. The proposed model achieves the highest *accuracy rate* of *92%*.

Hasib et al. [10] introduced spatiotemporal models for video action recognition, employing methods like *Pseudo-3D Residual Network, 3D ResNet, and 3D Convolutional Neural Network (C3D)* on the *UCF-Crime* dataset. Among the methods, C3D emerges as the frontrunner, boasting an AUC of *97.4%* and an accuracy of *89.3%*.

3 Methodology

The proposed work consists of two phases, namely

i. Fuzzy Logic-based deep training for anomaly detection
ii. Privacy-preserving anomaly detection.

3.1 Fuzzy Logic-Based Deep Training for Anomaly Detection

The model is built for both localization and identification of anomalous events occurring due to vehicle intrusion using a Convolutional Autoencoder. Convolutional Autoencoder (CAE) is a deep pre-trained model that integrates the concepts from Convolutional Neural Networks (CNN) and Autoencoders to learn a compressed representation of input data and then use this representation to reconstruct the input. It consists of an Encoder and a Decoder. The Encoder part of the CAE consists of convolutional layers followed by pooling layers. The decoder part of the CAE consists of transposed convolutional layers and up-sampling layers. This part of the network aims to reconstruct the input frames from the latent representation. The CAE is trained to minimize the reconstruction loss, which measures the difference between the original input frames and the frames reconstructed by the decoder. The architecture of the Convolutional Autoencoder used in this paper is shown in Fig. 4.

The first convolutional layer takes an image with a height of 238 and a width of 158 and applies 32-size filters. To extract features, the ReLU activation function is

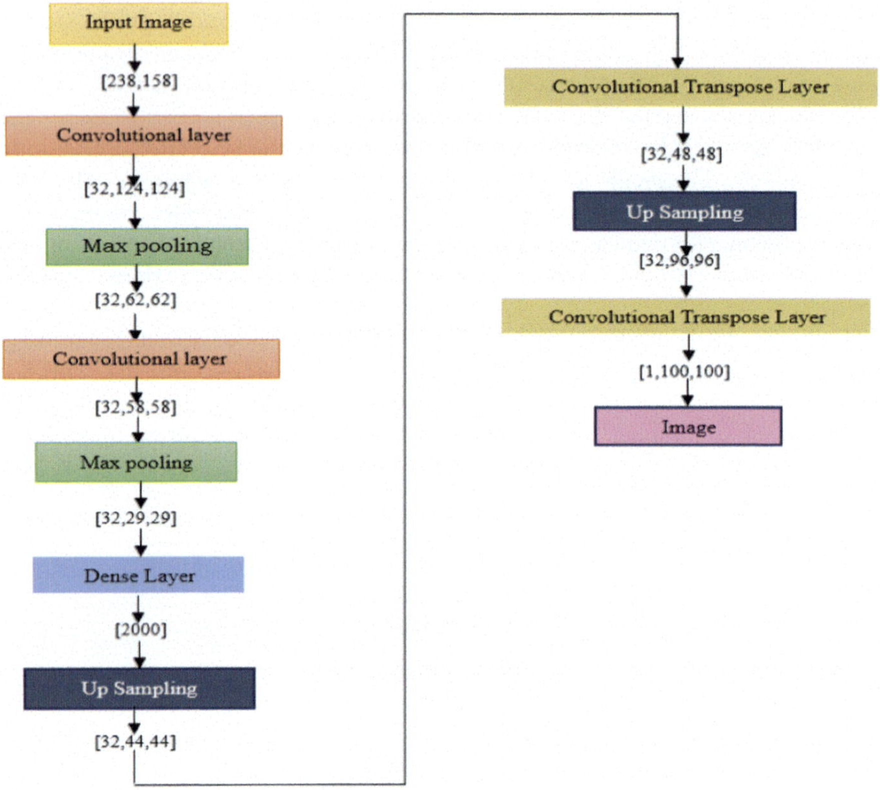

Fig. 4 Architecture of convolutional autoencoder

applied. Next, the Max Pooling is performed to reduce the spatial dimensions of the feature maps by half. The output of Max Pooling is passed to the fully connected layer. Fully connected (Dense) layer flattens the feature maps into a 1D vector and connects every element from the flattened vector to every neuron in the dense layer to produce a 2000-dimensional feature vector. The first fully connected layer receives the 2000-dimensional feature vector from the encoder and applies ReLU activation. The up-sampling layer increases the spatial dimensions. The deconvolutional (transposed convolutional) layer applies 32 filters to up-sample the feature maps. The final transposed convolutional layer applies a single filter and sigmoid activation function, which squashes the values between 0 and 1, suitable for image reconstruction.

To deal with uncertainties in detecting anomalous activities, CAE is hybridized with Fuzzy Logic. The fuzzy triangular membership function is used to enhance the power of CAE by adjusting the threshold to classify the membership scores into binary labels.

3.2 Privacy-Preserving Anomaly Detection

The proposed model is further enhanced to ensure privacy by encrypting video. The image encryption algorithm is used for encrypting the video. For image encryption, the original image is to be processed by shuffling the colors of the image. This means the pixel values for different color channels are rearranged. After shuffling, XOR operation is performed between the color channels and filtering is performed to replace each pixel value with the average value of its neighboring pixels. A blurring operation is performed to replace each pixel's value with the median value of its neighboring pixels. After performing all the processing steps, the encrypted or encoded image is formed.

The step-by-step process of the image encryption algorithm used in this paper is shown below.

Algorithm for image encryption

Input: An Image of size 238 × 158

Output: Encrypted image.

1. *R*_int ← 100, color_shuffle ← 0.
2. Resize the image to a fixed size of 200 x 200 pixels.
3. Convert the colour format of image from BGR to RGB.
4. Get the *h* ← height, *w* ← width of the image.
5. Loop through the pixels in the image:
6. For each *y* in the range of 0 to *h*-1
7. For each *x* in the range of 0 to *w*-1
8. Get the red (*r*), green (*g*), and blue (*b*) components.
9. Increment *R*_int by 1.
10. Check the value of color_shuffle
11. if color_shuffle is 0

    Apply XOR operation between *r*, *g*, *b*, and *R*_int, set color_shuffle to 1.
12. Otherwise

    Apply XOR operation, shuffling channels differently.

    Update color_shuffle to 2 if it was 0 or to 1 if it was 1.
13. Apply an average filter with a kernel size of 8 × 8.
14. Apply a median blur with a kernel size of 3 × 3 on the result of the average filter.
15. Apply a max filter to the result of the median filter.
16. Apply a min filter to the result of the maximum filter.

4 Experimental Results

The proposed model was implemented in Python on a system with an Intel iRISXei5 processor and 8 GB RAM. The data from the IP cameras are analyzed continuously to detect the anomalies in the video. Anomalies can be detected in two ways: one way is detecting anomalies from the clear video, and the other is from the encrypted video which is used for data privacy.

a) Normal Video b) Anomaly Detection c) Anomaly Detection for Privacy Preserving

Fig. 5 Real-time anomaly detection

4.1 Anomaly Detection and Localization

By using the pre-trained Convolutional Autoencoder model, the video is read as frames. It detects bicycles passing on the pedestrian path in the video by employing a rectangular red bounding box in the area, accompanied by a text alert indicating where the vehicle is passing, signifying anomalous behavior. Figure 5a shows the video without any anomalies in the pedestrian zone. Whenever anomalous activity is detected on pedestrian walkways, the type of the anomaly is indicated with red bounding boxes along with text alerts, as shown in Fig. 5b.

4.2 Privacy-Preserving Anomaly Detection

When the anomaly is detected using Convolutional Autoencoder, the video is encrypted using image encryption. The encrypted video utilizes the proposed model to detect bicycles passing on a pedestrian path, using rectangular red bounding boxes, and displaying the message, "Vehicle passing on Pedestrian path is detected." Figure 5c illustrates anomaly detection on encrypted video for data privacy.

The performance of the proposed anomaly detection system is compared against Faster RCNN. The Faster RCNN exhibited 96.9% accuracy, whereas the proposed model attained 98.5%. The performance comparison of the proposed model and Faster RCNN is shown in Fig. 6. Figure 6 depicts that the proposed model outperformed Faster RCNN and achieved a precision of 98.7%, recall of 98.2%, and $F1$-score of 98.4%.

Figure 6 depicts that the proposed model outperformed Faster RCNN and achieved a precision of 98.7%, recall of 98.2%, and $F1$ score of 98.4%.

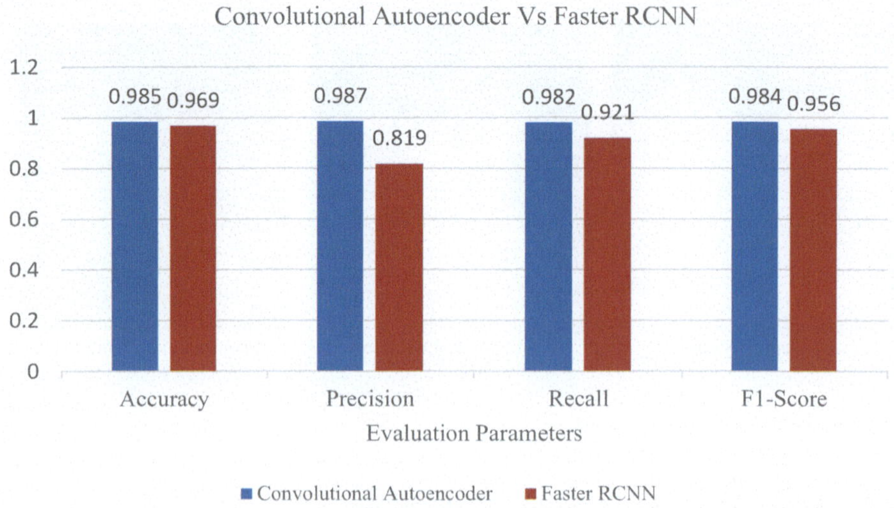

Fig. 6 Comparison between the proposed model and faster RCNN

5 Conclusion

This paper introduces a novel approach for anomaly detection on pedestrian paths. The presented model employs a combination of Convolutional Autoencoder and Fuzzy Logic to construct a robust detection model with red bounding boxes and sends text alerts. This model has been further enhanced by integrating data encryption for developing a context-sensitive smart surveillance model. The experimental study unveils that the proposed approach attained a 98.5% accuracy, which is significant compared to the 96.9% accuracy attained by Faster RCNN. The proposed model outperformed the existing approaches with a precision of 98.7%, recall of 98.2%, $F1$-score of 98.46%, and area under curve of 0.819. The methodologies developed in this paper collectively demonstrate promising results in enhancing the safety and privacy aspects of pedestrian pathways.

The Deep Learning approaches presented in this work addresses anomaly detection in pedestrian walkways. Most of the pedestrian accidents occur due to the violation of traffic rules. Hence, this work can be further extended in the following ways to minimize the risk of pedestrian accidents and injuries.

(1) Anomaly detection in critical pedestrian zones, specifically at zebra crossings, intersections, and signal violations.
(2) Detecting the age of the individual performing the anomalous activity may be further helpful to take necessary actions as follows.

(a) If a child violates the rule, voice alerts can be sent to guide them in the proper route.
(b) If an adult is violating the rule, then fines can be imposed based on the IPC law for pedestrian safety.

References

1. Road Accidents in India 2021. https://morth.nic.in/road-accidents-india-2021
2. Gawande U, Hajari K, Golhar Y. Pedestrian detection and tracking in video surveillance system: issues, comprehensive review, and challenges [online]. Available www.intechopen.com
3. http://www.svcl.ucsd.edu/projects/anomaly/dataset.htm. UCSD dataset available online on 1st Dec 2023
4. Li Z, Li Y, Gao Z (2020) Spatiotemporal representation learning for video anomaly detection. IEEE Access 8:25531–25542. https://doi.org/10.1109/ACCESS.2020.2970497
5. Wu C, Shao S, Tunc C, Hariri S (2020) Video anomaly detection using pre-trained deep convolutional neural nets and context mining. In: Proceedings of IEEE/ACS international conference on computer systems and applications, AICCSA. IEEE Computer Society, Nov 2020. https://doi.org/10.1109/AICCSA50499.2020.9316538
6. Nayak R, Pati UC, Das SK (2020) Video anomaly detection using convolutional spatiotemporal autoencoder. In: 2020 international conference on contemporary computing and applications, IC3A 2020. Institute of Electrical and Electronics Engineers Inc., Feb 2020, pp 175–180. https://doi.org/10.1109/IC3A48958.2020.233292
7. Mangai P, Geetha MK, Kumaravelan G (2022) Temporal features-based anomaly detection from surveillance videos using deep learning techniques. In: Proceedings of the 2nd international conference on artificial intelligence and smart energy, ICAIS 2022. Institute of Electrical and Electronics Engineers Inc., pp 490–497. https://doi.org/10.1109/ICAIS53314.2022.9742960
8. Nguyen TN, Meunier J (2019) Anomaly detection in video sequence with appearance-motion correspondence. In: Proceedings of the IEEE international conference on computer vision. Institute of Electrical and Electronics Engineers Inc., Oct 2019, pp 1273–1283. https://doi.org/10.1109/ICCV.2019.00136
9. Nithesh K, Tabassum N, Geetha DD, Kumari RDA (2023) Anomaly detection in surveillance videos using deep learning. Institute of Electrical and Electronics Engineers (IEEE), Mar 2023, pp 1–6. https://doi.org/10.1109/ickecs56523.2022.10059844
10. Hasib R, Jan A, Khan GM (2022) Real-time anomaly detection for smart and safe city using spatiotemporal deep learning. In: 2nd IEEE international conference on artificial intelligence, ICAI 2022. Institute of Electrical and Electronics Engineers Inc., pp 79–83. https://doi.org/10.1109/ICAI55435.2022.9773464

CASe_UNet: Multi-level Multi-scale UNet for Medical Image Segmentation

Arrun Sivasubramanian, Jayanth Mohan, and V. Sowmya

Abstract Encoder-decoder architectures have been extensively used for semantic segmentation tasks, including biomedical image segmentation. Most architectures utilize UNet as the base deep learning encoder-decoder network for biomedical image segmentation. To extract global and local features, multi-scale feature extraction using convolution involving different kernel sizes is very important for an effective analysis. UNet3+ can extract multi-level features from different layers of the encoder-decoder, which could be appropriately concatenated with decoder blocks. This paper leverages the combination of multi-scale features from convolution to extract higher-level semantics and multi-level hierarchical features from different blocks of the encoder-decoder as performed with UNet3+. We also incorporate multi-head attention in the bottleneck to further enhance extracted features from convolution, thus making a UNet architecture combining Convolution and Attention for Segmentation (CASe_UNet), combined with a hybrid loss for reducing precision and recall in pixel classification. The proposed architecture also tries to encounter class imbalance by assigning class weights inversely proportional to the probability distribution of each class. The approach is trained and validated on LiTS 2017 to identify tumors in the Liver and EndoVis 2018 Medical Instrument Segmentation data. Experimental results, along with ablation experiments, show that the proposed architecture achieves good results with an IOU and Dice coefficient of 0.9438 and 0.8684, respectively, on the EndoVis 2018 dataset and the benchmark IoU and Dice coefficient of 0.9166 and 0.9482 on LiTS 2017 data.

Keywords Semantic segmentation · Multi-level skip connection · Multi-scale feature extraction · Self attention · Hybrid loss function

---

A. Sivasubramanian · J. Mohan · V. Sowmya (✉)
Amrita School of Artificial Intelligence, Amrita Vishwa Vidyapeetham, Coimbatore, India
e-mail: v_sowmya@cb.amrita.edu

B. Annappa et al. (eds.), *High Performance Computing, Smart Devices and Networks*, Lecture Notes in Electrical Engineering 1262,
https://doi.org/10.1007/978-981-97-7794-5_21

1 Introduction

Semantic segmentation, a foundational computer vision task, involves assigning meaningful class labels to individual pixels within an image. This process enables deciphering complex visual scenes and plays a crucial role in various applications, such as autonomous driving [1], augmented reality, and image editing. In the medical domain, the significance of semantic segmentation analysis is elevated, as it entails partitioning images into distinct regions corresponding to anatomical structures, lesions, or other relevant regions of interest and facilitating the identification and quantification of abnormalities [2]. In turn, this assists medical professionals in devising effective treatment strategies tailored to individual patients' needs. It also enhances the monitoring of disease progression over time, helping to evaluate the efficacy of interventions and predicting patient outcomes. Together, these attributes underscore the role of accurate medical image segmentation as a cornerstone in modern medical diagnostics and patient care.

However, challenges persist within medical image analysis despite severe advancements in semantic segmentation methodologies. Variability in imaging modalities, anatomical variations among patients, noise, artifacts, and indistinct boundaries between structures pose substantial hurdles. The current methods [3] often grapple with achieving consistent and precise segmentations, especially when faced with these challenges. Inaccurate segmentations can lead to misdiagnosis, inappropriate treatment planning, and suboptimal disease monitoring, potentially compromising patient outcomes. In light of these challenges, this paper seeks to address limitations within existing methods, specifically focusing on the popular UNet architecture for medical image segmentation. While UNet has demonstrated its capacity to capture low-level and high-level features through its encoder-decoder structure, it faces difficulty effectively integrating global multi-scale context and handling semantic gaps. These limitations hinder the model's ability to produce comprehensive and contextually informed segmentations.

Thus, we propose a novel modification to the existing UNet architecture called CASe_UNet to bridge this gap, introducing attention mechanisms that facilitate robust channel-wise information fusion across multiple scales. This augmentation aims to enhance the encoder-decoder connection, effectively addressing the challenges associated with global context integration and semantic gaps. By enriching the UNet architecture with attention mechanisms, we aim to facilitate the creation of more accurate and context-aware segmentation in medical images. Our proposed approach can potentially improve diagnostic accuracy, treatment planning, and disease monitoring, ultimately advancing the field of medical image analysis and contributing to improved patient care. Through comprehensive experimentation and analysis of medical image datasets, we showcase the efficacy of our method and the potential impact it can have on the accuracy and reliability of medical image segmentation.

2 Related Works

Research on image segmentation can be classified into traditional and deep learning techniques. Traditional image processing techniques involve Otsu+Kmeans [4], Gaussian and Gabor pre-filter operators [5] for segmentation and output generation. Genetic algorithms [6] were introduced in image segmentation due to challenges in determining global thresholds. However, not all these techniques are data-driven and incapable of capturing the distribution of data and the contextual information that needs to be understood by a model. With increased data, they were gradually replaced by deep learning techniques that perform better on medical data. Fully convolutional networks (FCNs) [7] for pixel-level segmentation inspired and transformed subsequent advancements. The paper innovates by preserving spatial data for pixel-wise classification using convolutional layers instead of discarding details through pooling. PSPNet [8], a deep learning architecture for scene parsing and segmentation, has a pyramid structure with dilated convolutional layers that can analyze context from multiple scales, enhancing scene understanding accuracy. A global pooling module efficiently captures global context, contributing to robust segmentation. UNet [9] addresses the challenge of maintaining spatial details while capturing high-level context. The U-shaped structure comprises a contracting path for feature extraction, an expansive path for precise localization, and skip connections to preserve spatial information. Zhou et al. introduced UNet++ [10], an enhanced version of the UNet architecture, significantly improving the accuracy of medical image segmentation. It employs nested skip connections to enrich the UNet structure, enhancing information capture across various scales. This hierarchical approach in contracting and expansive paths refines segmentation progressively, leading to better delineation of object boundaries and details in medical images. Several subsequent works use UNet as the base model [11].

Oktay et al. [12] introduced attention mechanisms to the UNet framework, altering how the network focuses on relevant information during segmentation. Differentiating from the traditional UNet architecture, the Attention UNet incorporates attention modules that enable the network to emphasize crucial regions for accurate segmentation. They were introduced within the expansive path of the UNet, specifically within the skip connections. This allows the network to dynamically learn where to focus its processing resources, concentrating on the most relevant features and effectively reducing the risk of misclassification or missed boundaries. Huang et al. [13] refine UNet for accurate medical image segmentation by introducing a Full-Scale Connected (FSC) structure, connecting deep layers for better context incorporation and feature extraction. UNet 3+ retains UNet's architecture while adding FSC and Multi-Scale Feature Fusion modules, handling object size variations more efficiently than UNet++. Recent architectures in the literature have introduced lightweight networks, such as LFU-Net [14] and CeLNet [15], to increase the efficiency of segmentation. However, no mechanism introduced in the literature uses the combined prowess of multi-level skip connections and multi-scale feature extraction with attention mechanism, which has proven to capture highly precise global and local contextual information. This paper aims to leverage all of this and try to analyze the use of each component while comparing the performance with other works elaborated above.

3 Methodology

3.1 Datasets Used

To analyze the model's efficacy, we trained and tested the proposed model on two publicly available datasets. The Liver Tumor Segmentation (LITS) dataset is a well-known medical imaging dataset used for liver and liver tumor segmentation. It was first introduced in the Liver Tumor Segmentation Challenge in 2017. The dataset, a collection of 131 3D volume contrast-enhanced CT scans from the LITS dataset, was assembled to support the creation and evaluation of algorithms for precise tumor and liver segmentation in computed tomography (CT) scans. Each scan includes multiple annotated slices showing the liver and any tumors that may be present. It includes a variety of liver shapes, sizes, and tumor characteristics. Due to detailed annotations, this open-source data set has become a valuable resource for researchers and practitioners in medical image analysis, aiding advancements in liver and liver tumor segmentation techniques [16]. Figure 1 contains a few sample images of the splices of the CT scan, with the ground truth segmentation maps.

The other dataset introduced as a part of the Endoscopic Vision Challenge at the MICCAI 2018 conference is the EndoVis2018 Surgical Equipment Segmentation dataset. With the help of this dataset, algorithms for precisely segmenting surgical instruments in endoscopic images will be developed and evaluated. The dataset consists of 1679 annotated training and 569 validation endoscopic images and depicts actual circumstances encountered during minimally invasive procedures. These images present difficulties like obstructions, varying lighting, and intricate

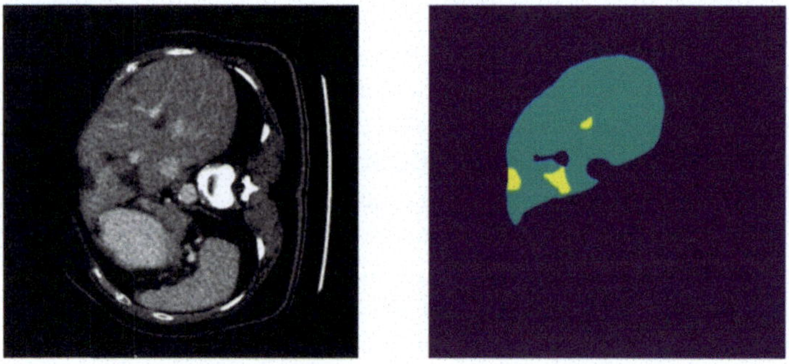

Fig. 1 Sample image and segmentation map of a splice of LiTS2017 dataset

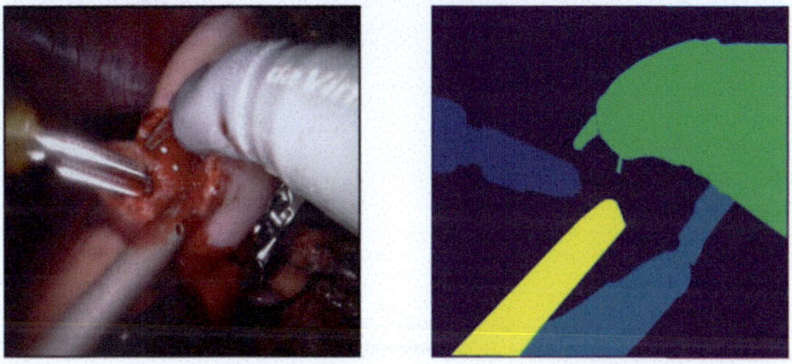

Fig. 2 Sample image and segmentation map of the EndoVis2018 dataset

instrument shapes that can help test the robustness of the segmentation models. The boundaries of surgical instruments are annotated, allowing for the training and benchmarking of segmentation algorithms. Figure 2 contains a few sample images of the surgical video frame and ground truth segmentation maps.

3.2 Architecture Proposed: CASE_UNet

The UNet architecture is a convolutional neural network (CNN) design widely used in medical image segmentation and other tasks requiring pixel-level predictions. However, extracting features using convolution heads of the same size will not help the model retrieve all important semantic information from the input image. Moreover, it will extract only local features by anchoring them on a particular pixel and obtain feature maps. Thus, in the proposed CASe_UNet, the encoder block consists of multiple convolution heads of size 3, 5, and 7, respectively, to extract higher-order spatial information from an anchor pixel (starting from an image of dimension $512 * 512$ and downsamples to a mere $16 * 16$ feature map). These feature maps of the same depth are then vertically stacked to enable the model to further extract features on relative scaling information (from 32 filters up to 1024 filters), thus retrieving both high-level global and local semantic contexts. Figure 3 contains the encoder of CASe_UNet, and Eq. (1) depicts the output given by the ith encoder block.

$$X^i_{\text{En}} = \text{Pool}_{2D}(V(C_{3*3}, C_{5*5}), V(C_{5*5}, C_{7*7}), V(C_{5*5}, C_{7*7}), n_f = 2^{5+i}) \quad (1)$$

Multiple computer vision tasks like image segmentation, object detection, and classification have benefited greatly from multi-head attention mechanisms. They were inspired by the Transformer architecture, which allows models to concentrate on various aspects of the input data simultaneously, capturing both local and global dependencies. Multi-head attention has proven useful in learning complex spatial relationships and hierarchies within visual data by focusing on various image regions

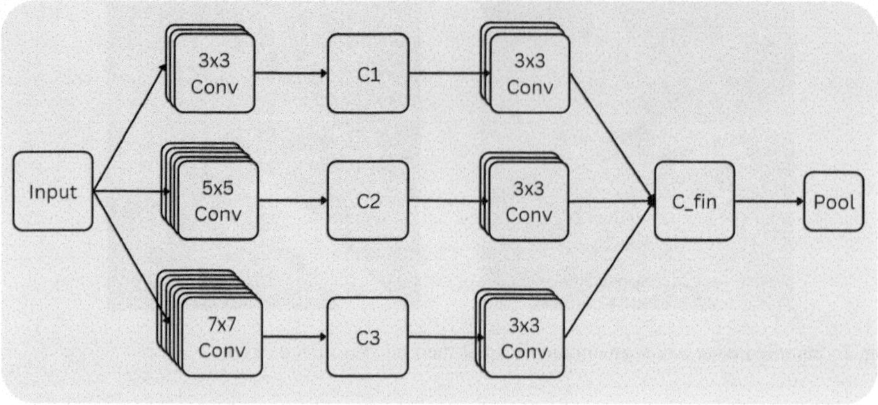

Fig. 3 Multi-scale encoder block used in CASe_UNet

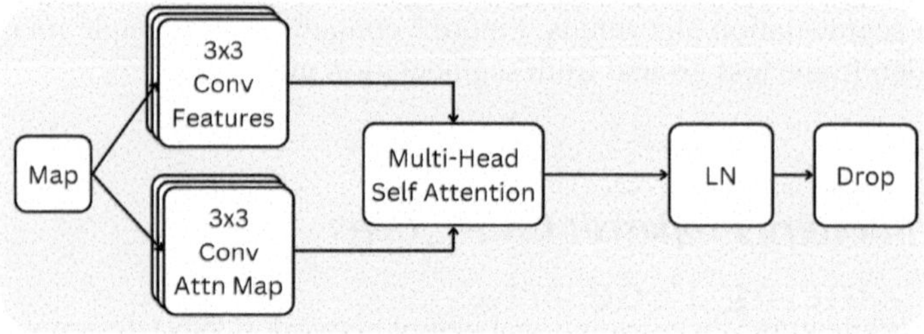

Fig. 4 Bottleck containing MHSA in CASe_UNet

at various scales and resolutions. They can then handle occlusions, recognize complex patterns, and enhance contextual awareness, enhancing computer vision applications' accuracy and robustness. However, since the operation is computationally expensive, the mechanism was introduced solely in the bottleneck of the architecture since it will retrieve maximum information from the highest number of feature maps obtained from the encoder. This is followed by applying Layer Normalization to help the model normalize the features across different feature maps and a dropout layer (set to 0.2) to avoid overfitting on the training dataset. Figure 4 contains the bottleneck of CASe_UNet. Since UNet3+ proves that multi-level skip connections in place of simple same-level encoder-decoder residual connections play a pivotal role in enhancing the UNet architecture's ability to capture intricate features and spatial information for image segmentation tasks, CASe_UNet is also inspired to introduce multi-level skip connections to process low-level, fine-grained details and high-level semantic context to handle various object sizes, shapes, and contextual relationships, making it extremely effective for tasks like medical image segmentation.

Moreover, the architecture processes multi-scale and multi-level features within all encoder and decoder layers, coupled with attention in the bottleneck. The decoder block contains stacked transpose convolutions, which are upsampled to the required

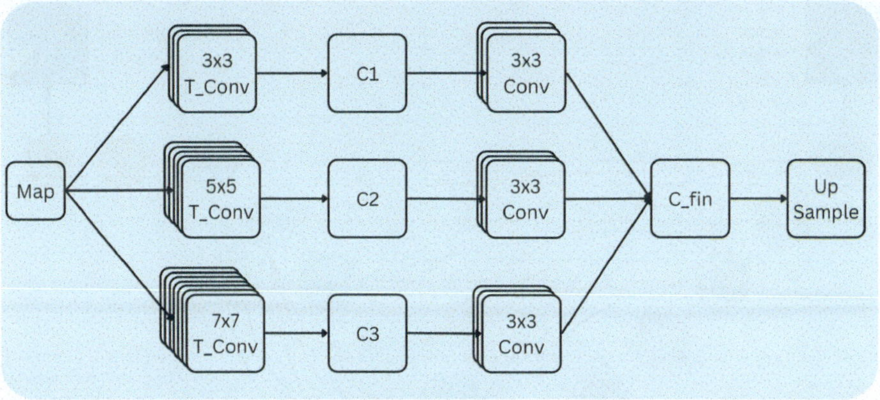

Fig. 5 Multi-scale decoder block used in CASe_UNet

dimension. The operation, which is exactly the reverse of the operations performed in the encoder, will help the model retrieve accurate pixel-wise predictions. Figure 5 contains the decoder of CASe_UNet, while Eqs. (2) and (3) depicts the input and output given by the ith decoder block, respectively. Combining all these figures, the overall architecture of the proposed CASe_UNet is shown in Fig. 6. The final two feature smoothening layers contain simple convolutions containing 16 filters to prevent the input image from directly influencing the segmentation output via the skip connection and prevent the model from predicting sharp edges during segmentation.

$$X^i_{\text{De}} = \begin{cases} X^i_{\text{En}} & \text{if } i = N \\ \mathcal{H}\left(\left[\underbrace{\mathrm{C}\left(\mathcal{D}\left(X^k_{En}\right)\right)_{k=1}^{i-1}, \mathrm{C}\left(X^i_{En}\right)}_{\text{Scales: 1th}\sim i\text{th}}, \underbrace{\mathrm{C}\left(\mathcal{U}\left(X^k_{De}\right)\right)_{k=i+1}^{N}}_{\text{Scales: }(i+1)\text{th}\sim N\text{th}}\right]\right), & \text{if } i = 1, \ldots, N-1 \end{cases} \tag{2}$$

$$X^i_{\text{De}} = \text{Pool}_{2D}\left(V(C_{T_{3*3}}, C_{T_{5*5}}), V(C_{T_{5*5}}, C_{T_{7*7}}), V(C_{T_{3*3}}, C_{T_{7*7}}), n_f = 2^{-5+i}\right) \tag{3}$$

3.3 Experimental Setup

The images of both datasets were resized to a dimension of $512 * 512 * 3$. The architectures were trained on an Nvidia A6000 GPU with 32GB RAM and 48GB vRAM. The cost function minimized during backpropagation of the model weights is a combination of Weighted Categorical Cross Entropy (CCE) for multiclass segmentation (class weights inversely proportional to the pixel probability distribution), Structural Similarity Index (SSIM) for evaluating the structural similarity between the predicted

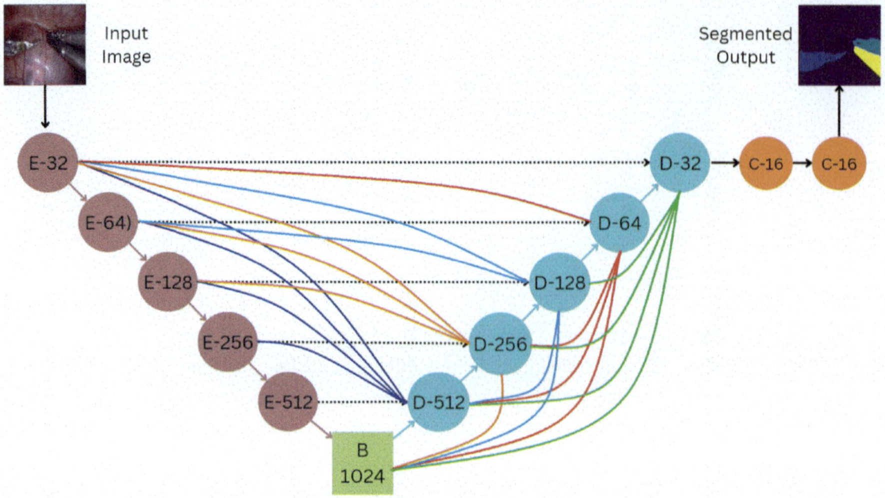

Fig. 6 Block diagram of the proposed CASe_UNet architecture

map and the ground truth and dice and dice loss to reduce false positive and false negative predictions. The optimizer used is Adamax to perform adaptive gradient descent. The combined cost function equations are shown in Eqs. (4)–(7).

$$\ell_{\text{wcce}} = -\sum_{i=1}^{N} w_i \sum_{j=1}^{C} y_{ij} \cdot \log(p_{ij}) \tag{4}$$

$$\ell_{\text{ssim}} = 1 - \prod_{m=1}^{M} \left(\frac{2\mu_p \mu_g + C_1}{\mu_p^2 + \mu_g^2 + C_1} \right)^{\beta_m} \left(\frac{2\sigma_{pg} + C_2}{\sigma_p^2 + \sigma_g^2 + C_2} \right)^{\gamma_m} \tag{5}$$

$$\ell_{\text{dice}} = 1 - \frac{2\sum_{i=1}^{N} p_i \cdot g_i}{\sum_{i=1}^{N} p_i^2 + \sum_{i=1}^{N} g_i^2} \tag{6}$$

$$\ell_{\text{seg}} = \ell_{\text{dice}} + \ell_{\text{wcce}} + \ell_{\text{ssim}} \tag{7}$$

4 Results and Discussion

Two benchmark medical image datasets were used to compare the results. The data was split into the train-to-test ratio 90:10 and trained with a batch size 32 for 100 epochs. The results obtained by the proposed architecture for the datasets, including

Table 1 Performance metrics of various architectures on LiTS 2017 data

Architecture	Conference	IOU	MIOU	Dice
PSPNet [8]	CVPR	–	–	0.9242
DeepLabV2 [17]	TPAMI	–	–	0.9021
DeepLabV3 [18]	arXiv	–	–	0.9217
DeepLabV3+ [19]	ECCV	–	–	0.9186
Attention UNet [12]	MIDL	–	–	0.9341
UNet 3+ (Hybrid loss + CGM) [13]	ICASSP	–	–	**0.9675**
CASe_UNet (CCE+Simple Skip)	-	0.8992	0.6988	0.9390
CASe_UNet (Hybrid loss+ Simple Skip)	–	0.9061	0.7056	0.9425
CASe_UNet (Hybrid loss+ Multilevel Skip)	–	**0.9166**	**0.7154**	0.9482

Table 2 Performance metrics of various architectures on EndoVis 2018 data

Architecture	Conference	IOU	MIOU	Dice
Mask2Former [20]	CVPR22	0.6547	0.2767	–
TernausNet-11 [21]	ICMLA18	0.4622	0.1419	–
MF-TAPNET [22]	MICCAI19	0.6787	0.2468	–
ISINET [23]	MICCAI20	0.7303	0.4021	–
TraSeTR [24]	ICRA22	0.7620	**0.4777**	–
S3Net (+MaskRCNN) [25]	WACV	0.7581	0.4258	–
CASe_UNet (CCE+Simple Skip)	–	0.9136	0.4065	0.8312
CASe_UNet (Hybrid loss+ Simple Skip)	–	0.9427	0.4224	0.8527
CASe_UNet (Hybrid loss+ Multilevel Skip)	–	**0.9438**	0.4166	**0.8684**

the ablation experiments, are compared with other state-of-the-art metrics obtained in the literature. Tables 1 and 2 contain the results obtained on the LiTS2017 and EndoVis 2018 datasets, respectively. It is evident from the results that the proposed CASe_UNet architecture performs significantly better on the LiTS than most UNet Architecture variants proposed in the literature. Since UNet contains only single-size kernel convolutions, it cannot extract the global-level semantic information. Although PSPNet contains multi-scale convolutions, making it perform better than the base UNet architecture, the performance is not as good as UNet with attention, which can capture much global and local contextual information. However, the performance of our architecture that combines multi-scale feature extraction and multi-level skip connections procures better results, proving the importance of multi-level skip connections and multi-head attention. However, the results of UNet3+ are slightly better than the results of the proposed architecture because they used input

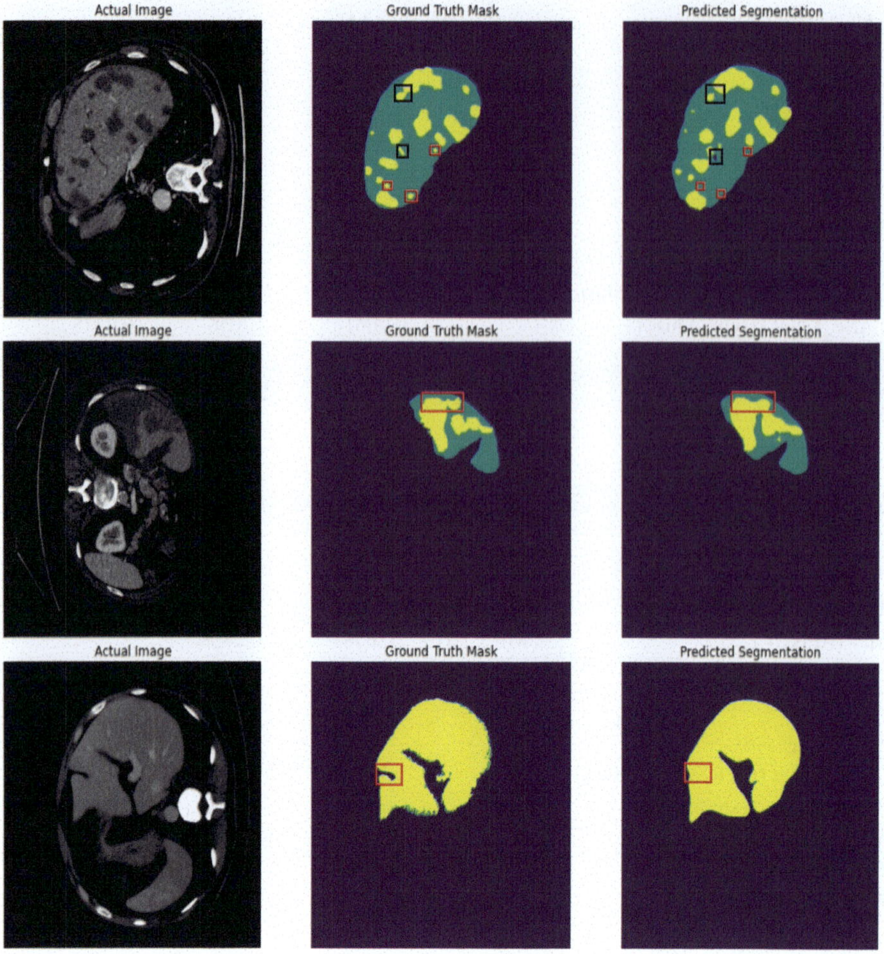

Fig. 7 Sample predictions made by CASe_UNet on the LITS 2017 data

data that contains the splice as well as the splice present behind and ahead of the actual splice, which assists the network in obtaining more supervision at every step of the decoder. To improve the results, they also used a classification-guided module (CGM) at the end of the architecture that helped to refine the class predictions.

On closely inspecting the results of the proposed architecture on LiTS 2017 data, one can notice from Fig. 7 that the segmentation made by the model is very smooth and has accurate boundaries of the organs. However, there are a few regions where this model fails to detect a very small tumor, as seen in the first example of the figure. Nevertheless, even if the feature of the tumor is not very distinct compared to the surroundings, the model can segment the regions very accurately but with slightly smoothened output edges, as seen in the last example. The model could

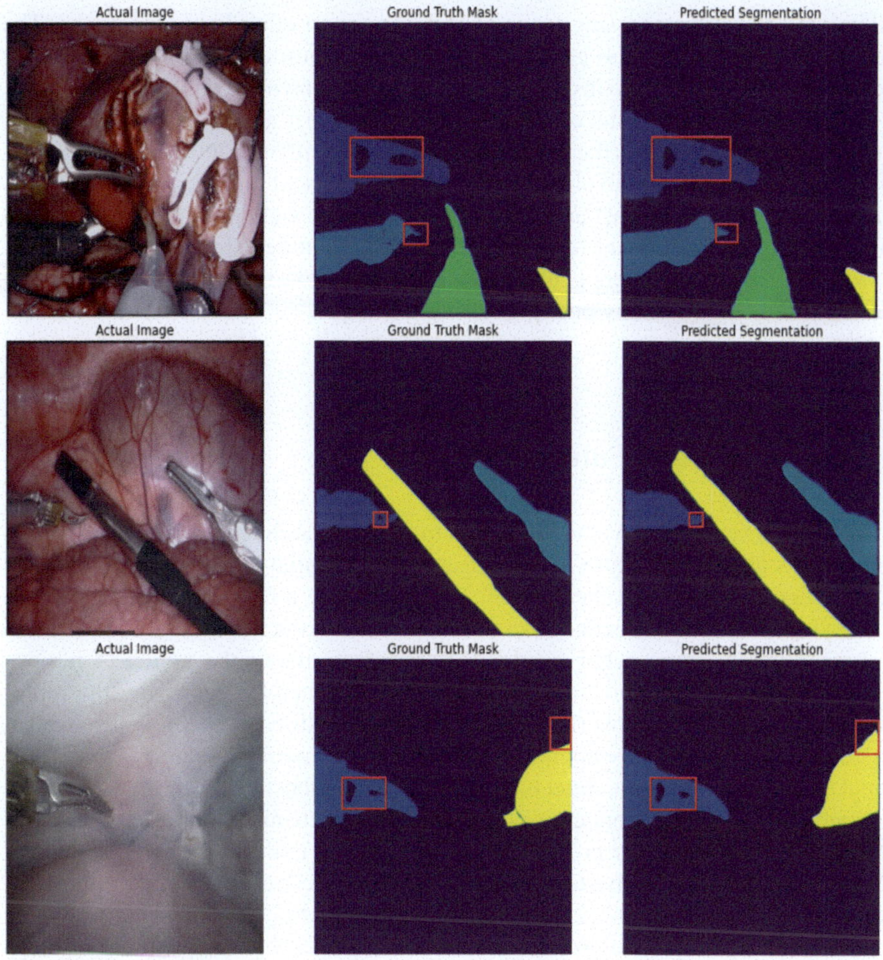

Fig. 8 Sample predictions made by CASe_UNet on the EndoVis 2018 data

also capture the semantic features from the input image more accurately than the ground truth annotation (highlighted in black boxes). Similarly, the edges obtained on the EndoVis2018 data (see Fig. 8) are very accurate, supported by the quantitative results. The segmentation masks are too good for edges due to capturing good global and local contextual semantics. Even when the instrument is buried under the tissue, the segmentation algorithm can identify and find the instrument's mask. However, the segmentation algorithm misses a few sharp edges, presumably due to several convolution-based features stacked together and the model's inability to extract the most appropriate feature maps, yielding a tradeoff between smoothening and redundant features in multi-scale feature extraction. The number of total trainable parameters is more than many other existing works, but this helps the algorithm eliminate any post-processing techniques necessary in other literature works. The segmentation masks are too good for edges due to capturing good global and local contextual semantics. Even in cases where the surgical instrument is buried below the translucent tissue, the model can predict the correct class and near-appropriate segmentation mask. However, the segmentation algorithm misses a few sharp edges, presumably due to several convolution-based features stacked together and the model's inability to extract the most appropriate feature maps, yielding a tradeoff between smoothening and redundant features in multi-scale feature extraction. The number of total trainable parameters is more than many other existing works, but this helps the algorithm eliminate any post-processing techniques necessary in other literature works.

5 Conclusion and Future Works

This paper presented a novel modification to the UNet architecture, CASe_UNet, to enhance its ability to capture intricate features and spatial information for medical image segmentation and related tasks. By introducing multiple convolution heads of varying sizes in the encoder block, we enable the extraction of higher-order spatial information from anchor pixels. This approach allows the model to retrieve both high-level global and local semantic contexts, thereby improving the representation power of the network. We incorporated a multi-head attention mechanism in the bottleneck to capture complex spatial relationships and hierarchies within the input data, enhancing the architecture's accuracy and robustness. Additionally, introducing multi-level skip connections and attention mechanisms at various stages of the architecture allows for processing multi-scale and multi-level features, making CASe_UNet highly effective for handling diverse object sizes, shapes, and contextual relationships in medical image segmentation. While CASe_UNet presents promising enhancements to the UNet architecture, several avenues remain for future research and improvement. While our proposed architecture demonstrates improved performance, it's important to explore methods to reduce the overall number of parameters without compromising accuracy such as model pruning, and knowledge distillation. Although CASe_UNet provides advanced features for image segmentation, integrating post-processing techniques could further refine the segmentation outputs. Techniques like conditional random fields (CRFs) or graph-based methods could be explored to enhance the segmentation quality. Given the complexity of the architecture, developing a hierarchical supervision strategy could potentially enhance the training process. By providing guidance and supervision at various levels of the network, we could achieve better convergence and more stable training. Investigating adaptive scaling strategies for the convolution heads in the encoder block could lead to more flexible feature extraction. This would allow the model to dynamically adjust the kernel sizes based on the input context and improve the extraction of relevant features.

References

1. Deepika N, Sajith Variyar V (2017) Obstacle classification and detection for vision based navigation for autonomous driving. In: 2017 international conference on advances in computing, communications and informatics (ICACCI), pp 2092–2097
2. Ramasamy G, Singh T, Yuan X (2023) Multi-modal semantic segmentation model using encoder based Link-Net architecture for BraTS 2020 challenge. Procedia Comput Sci 218:732–740
3. Singh T, Karanchery S (2019) Universal image segmentation technique for cancer detection in medical images. In: 2019 10th international conference on computing, communication and networking technologies (ICCCNT), pp 1–7
4. Liu D, Yu J (2009) Otsu method and K-means. In: 2009 ninth international conference on hybrid intelligent systems, vol 1. IEEE
5. Dhanalakshmi S, Ravichandran T (2012) A new method for image segmentation. Int J Adv Res Comput Sci Softw Eng 2(9):293–299
6. Kanungo P, Nanda PK, Samal UC (2006) Image segmentation using thresholding and genetic algorithm
7. Long J, Shelhamer E, Darrell T (2015) Fully convolutional networks for semantic segmentation. In: Proceedings of the IEEE conference on computer vision and pattern recognition, pp 3431–3440
8. Zhao H, Shi J, Qi X, Wang X, Jia J (2017) Pyramid scene parsing network. In: Proceedings of the IEEE conference on computer vision and pattern recognition, pp 2881–2890
9. Ronneberger O, Fischer P, Brox T (2015) U-net: convolutional networks for biomedical image segmentation. In: Medical image computing and computer-assisted intervention—MICCAI 2015: 18th international conference, Munich, Germany, 5–9 Oct 2015, proceedings, Part III 18, pp 234–241
10. Zhou Z, Rahman Siddiquee M, Tajbakhsh N, Liang J (2018) UNet++: a nested u-Net architecture for medical image segmentation. In: Deep learning in medical image analysis and multimodal learning for clinical decision support: 4th international workshop, DLMIA 2018, and 8th international workshop, ML-CDS 2018, held in conjunction with MICCAI 2018, Granada, Spain, 20 Sept 2018, proceedings 4, pp 3–11
11. Sachin T, Sowmya V, Soman K (2021) Performance analysis of deep learning models for biomedical image segmentation. Deep learning for biomedical applications, pp 83–100
12. Oktay O, Schlemper J, Folgoc L, Lee M, Heinrich M, Misawa K, Mori K, McDonagh S, Hammerla N, Kainz B, et al (2018) Attention U-Net: learning where to look for the pancreas. arXiv Preprint arXiv:1804.03999
13. Huang H, Lin L, Tong R, Hu H, Zhang Q, Iwamoto Y, Han X, Chen Y, Wu J (2020) UNet 3+: a full-scale connected UNet for medical image segmentation. In: ICASSP 2020—2020 IEEE international conference on acoustics, speech and signal processing (ICASSP), pp 1055–1059
14. Deng Y, Wang H, Hou Y, Liang S, Zeng D (2023) LFU-Net: a lightweight U-Net with full skip connections for medical image segmentation. Curr Med Imaging 19:347–360
15. Zhang B, Wang X, Liu L, Zhang D, Huang X, Xia M, Jiang W, Huang X (2023) CeLNet: a correlation-enhanced lightweight network for medical image segmentation. Phys Med Biol 68:115012
16. Bilic P, Christ P, Li H, Vorontsov E, Ben-Cohen A, Kaissis G, Szeskin A, Jacobs C, Mamani G, Chartrand G et al (2023) The liver tumor segmentation benchmark (LiTS). Med Image Anal 84:102680
17. Chen L, Papandreou G, Kokkinos I, Murphy K, Yuille A (2017) DeepLab: semantic image segmentation with deep convolutional nets, atrous convolution, and fully connected CRFs. IEEE Trans Pattern Anal Mach Intell 40:834–848
18. Chen L, Papandreou G, Schroff F, Adam H (2017) Rethinking atrous convolution for semantic image segmentation. arXiv Preprint arXiv:1706.05587

19. Chen L, Zhu Y, Papandreou G, Schroff F, Adam H (2018) Encoder-decoder with atrous separable convolution for semantic image segmentation. In: Proceedings of the European conference on computer vision (ECCV), pp 801–818
20. Cheng B, Misra I, Schwing A, Kirillov A, Girdhar R (2022) Masked-attention mask transformer for universal image segmentation. In: Proceedings of the IEEE/CVF conference on computer vision and pattern recognition, pp 1290–1299
21. Iglovikov V, Shvets A (2018) TernausNet: U-Net with vgg11 encoder pre-trained on ImageNet for image segmentation. arXiv Preprint arXiv:1801.05746
22. Jin Y, Cheng K, Dou Q, Heng P (2019) Incorporating temporal prior from motion flow for instrument segmentation in minimally invasive surgery video. In: Medical image computing and computer assisted intervention—MICCAI 2019: 22nd international conference, Shenzhen, China, 13–17 Oct 2019, proceedings, Part V 22, pp 440–448
23. González C, Bravo-Sánchez L, Arbelaez P (2020) IsiNet: an instance-based approach for surgical instrument segmentation. In: International conference on medical image computing and computer-assisted intervention, pp 595–605
24. Zhao Z, Jin Y, Heng P (2022) TraSeTR: track-to-segment transformer with contrastive query for instance-level instrument segmentation in robotic surgery. In: 2022 international conference on robotics and automation (ICRA), pp 11186–11193
25. Baby B, Thapar D, Chasmai M, Banerjee T, Dargan K, Suri A, Banerjee S, Arora C (2023) From forks to forceps: a new framework for instance segmentation of surgical instruments. In: Proceedings of the IEEE/CVF winter conference on applications of computer vision, pp 6191–6201

EffiSANet: EfficientNet Integration with Self-attention for Colorectal Cancer Classification

Shashank Girepunje and Pradeep Singh

Abstract The highly fatal disease of colorectal cancer still has a substantial effect on world public health. Early detection of this malignancy is essential for extending human life and stopping its spread. Recently, image classification tasks have been highly influenced by CNN's architecture because of its effective learning in hierarchical visual feature representation. However, CNNs are unable to recognize intricate patterns and variations in smaller datasets because they lack the diversity and volume of samples. Attention mechanisms in conjunction with CNNs can be utilized to alleviate these constraints. In this article, an efficient CNN-based model that is a combination of attention and an efficient net, denoted as the EfficientNet Self-Attention network (EffiSA-net), is proposed to classify colorectal cancer. Our experiment shows better results with the proposed model, achieving 95.49% test accuracy on the colorectal histology dataset. Our study offers a promising method for enhancing the effectiveness and precision of colorectal cancer categorization, which will eventually help patients by enabling earlier identification and intervention.

Keywords Colorectal cancer · Self-attention · CNN

1 Introduction

Colorectal cancer is responsible for 10% of cancer-related fatalities globally. According to statistics from the American Cancer Society, 1.5 million new cases are expected at the end of 2023. Among these cases, 21,000 are found in individuals under the age of 50 [1]. As per the World Health Organization's data, it is estimated that this disease is going to affect 75 million people and leading to diagnosis of 27 million new cases [2]. This cancer affects the large intestine or rectum

S. Girepunje (✉) · P. Singh
National Institute of Technology, Raipur, India
e-mail: sgirepunje.phd2022.cs@nitrr.ac.in

P. Singh
e-mail: psingh.cs@nitrr.ac.in

B. Annappa et al. (eds.), *High Performance Computing, Smart Devices and Networks*, Lecture Notes in Electrical Engineering 1262,
https://doi.org/10.1007/978-981-97-7794-5_22

causing severe harm or death. The development of colorectal cancer is influenced by various lifestyle factors, including elder age, unhealthy diets, obesity, smoking, and alcohol. The diagnosis of colorectal cancer frequently occurs in its later stages, when treatment options are severely restricted. In the domain of medical diagnostics, fine-needle aspiration and needle biopsy are widely utilized procedures for the identification of colon cancers [3]. These techniques involve the microscopic analysis of cellular or tissue samples on dedicated glass slides. As a result, histopathology imaging (HI) has emerged as a reliable and widely accepted method for evaluating colorectal cancer (CRC). While the use of HI analysis proves valuable in the clinical assessment of colorectal cancer (CRC), it is imperative to accurately demarcate tumor locations, assess their level of aggressiveness, and identify different carcinoma prototypes by employing whole slide images. Histopathological diagnosis is dependent on the knowledge and experience of clinical specialists, requires a large amount of time for inspection, and frequently relies on the personal judgment of medical professionals. There is a possibility that the use of artificial intelligence to the process of automated CRC detection may lead to an improvement in diagnostic accuracy and will offer significant assistance to physicians as they make decisions [4]. The use of efficient algorithms and models for computer-aided diagnosis can refine process to detection of colorectal cancer. However, a significant obstacle in this scenario is figuring out how to extract useful features from the histopathological images produced from colon tissue.

Many studies have utilized manually designed features to tackle issues such as poor classification accuracy, high computational demands, and low-quality health imaging. Alqudah and Alqudah proposed detection model using machine learning which extract features from GLCM matrices in different color spaces [5]. They performed the experiment with five different machine learning algorithms surpassing the previous approaches with 97% accuracy. These machine learning solutions have two primary shortcomings. Firstly, the process of extracting important features from the dataset is tedious task because we need to go through trial-and-error method to find the suitable combination for the models or algorithms. Secondly, it was difficult to choose a classifier that worked well because past studies had utilized a variety of classifiers with several parameters. Recently, deep learning algorithms have become a transformational force, ushering in a new era of diagnostic accuracy and efficacy [6]. Deep learning algorithms have a remarkable ability to recognize complex patterns and characteristics on their own in medical images when working with intricate and nuanced datasets, like those produced by X-rays, MRIs, CT scans, and histopathology slides [7]. This characteristic makes it more suitable for image processing as compared with machine learning algorithms. CNN models have been widely utilized to extensive and effective application in achieving different histology image analysis and processing such as nuclei detection and prediction [8], metastasis detection [9], tumor segmentation [10], and cancer grading [11].

The primary challenge for the histology image prediction is to recognize the cancer's grade, in the distinct subgroups of CRC that are quantitatively examined having apparent similarities in texturing between images [3]. Although, deep learning models are exceptionally efficient in classification tasks, but its training needs large amount of data to learn the representation. In the context of smaller medical image datasets, CNN models frequently encounter challenges in extracting meaningful information [7]. Attention models offer a potential solution to the constraints of training with little datasets. Insufficient training examples can provide a challenge for traditional models, as they may have difficulty capturing complex patterns and subtle details, resulting in less than optimal performance [12]. Attention methods enable models to selectively concentrate on the most pertinent aspects of the input data, enabling them to efficiently allocate attention to crucial features. The focused methodology improves the model's ability to identify significant patterns, even in scenarios with low data [13, 14]. In this work, a modified EfficientNet model is proposed to perform the disease classification for colorectal cancer. In the following sections, we will examine our methodology, dataset, experimental design, and findings. Additionally, we will also discuss the consequences of our findings for clinical practice, highlighting the importance of a precise and effective cancer diagnosis in the fight against this fatal condition.

2 Related Works

In this section, we will discuss the previous architectures based on machine learning (ML) and deep learning (DL) that was used for detecting and categorizing histology images in colorectal tissue. Masud et al. [15] created a classification system in their study that includes two benign and three malignant categories and is intended to distinguish between five different types of lung and colon tissues. Their study's conclusions show that the framework they designed was remarkably accurate in identifying malignant tissues in histological images, reaching a high accuracy of 96.33%. Ghosh et al. employed an ensemble of deep learning networks for tumor detection [16]. The model demonstrates strong performance through the utilization of the multi-class tissue features thorough different models. While the ensemble technique enhances the model, it also results in an increase in size and parameters. Kumar et al. [17] developed a lightweight CNN for automated colorectal cancer classification using histopathological images from two public datasets. Their model outperforms other models in terms of computational efficiency, achieving 93.50% and 96.26% accuracy on separate datasets and 99.21% on a combined dataset. In another recent work, Wang proposed method utilizes binary convolutional neural networks (BCNN) for the classification of histopathological images [18]. A novel aspect involves decomposing images into H and E components to address co-localization issues, enhancing tissue image accuracy. The BCNN model is then applied to both components for effective feature representation and better classification results. Ghosh et al. employed an ensemble of deep learning networks for tumor detection [16]. The model demonstrates strong performance through the utilization of the multi-class tissue features thorough different models. Zeid et al. performed tests on colorectal cancer (CRC) using a vision transformer and compact convolutional transformer (CCT). Both models exhibited robust performance in classifying CRC, attaining an accuracy rate of 93% [19].

Analysis of histopathological images (HI) is difficult due to a lack of labeled images and the inherent heterogeneity introduced by staining methods. Histopathological image classification and interpretation are further complicated, because HI capturing and characterizing textures of tissues rather than focusing on specific objects as opposed to many other image processing jobs. This study used an innovative strategy to tackle these problems using lightweight deep learning algorithms. The main goal of this study was to close current gaps in the histopathological imaging classification of colorectal cancer (CRC).

3 Proposed Work

CNNs have been modified through trial and error to enhance accuracy by adding layers (depth), neurons or filters (width), or higher-resolution input images. This manual method is time consuming and unorganized. EfficientNet [20] used compound scaling with coefficients to uniformly scale width, depth, and resolution for data-driven solutions. Scaling a model's depth enables it to extract intricate input information, but the vanishing gradient problem can make training difficult. On the other hand, expanding the network's size facilitates training and aids in the acquisition of finer data elements. The proposed architectural layout of our model is shown graphically in Fig. 1.

In this experiment, we assessed pretrained EfficientNet models on a dataset related to colorectal cancer. We improved the performance of the architecture by adding custom layers to the pretrained models. This procedure enhanced image classification. Our proposed design leverages the feature representations learned by an already effective architecture to harness the potential of transfer learning. However, we optimize the model's performance by doing some modification in existing architecture. The proposed architectural layout is shown graphically in Fig. 1.

3.1 Self-attention Layer

The self-attention layer is an essential component that plays an important role in analyzing the links between feature maps. It does this by computing attention scores, which are then used to highlight relevant places in these maps. By capturing spatial linkages, this approach improves the model's ability to anticipate. We employ queries, keys, and values when paying attention to ourselves. The feature representation (X) from the final layer of the EfficientNet is used as the input for both queries and keys in this situation. Effective spatial analysis is made possible by applying weight matrices to X in order to produce Q, K, and V.

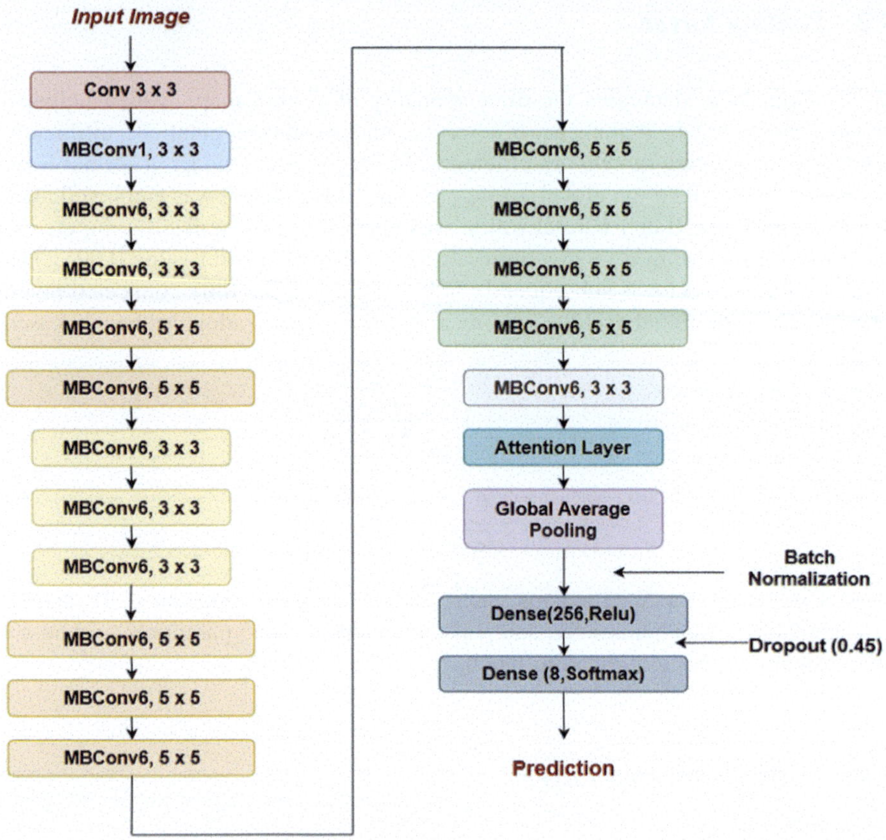

Fig. 1 Architecture of modified EfficientNetB0

$$Q = W^q X;\ K = W^k X;\ V = W^v X$$

W^q, W^k, and W^v are query, key, and value weight matrices; the value of one's own self-attention is computed using the following equation where characteristics b and d_k is the square root of the key vector dimension.

$$b = \text{softmax}\left(\frac{QK^T}{\sqrt{d_k}}\right)V$$

3.2 Pooling Layer

The pooling layer decreases the dimensionality of feature maps, which helps to manage complexity by making use of fewer parameters and computations, ultimately leading to an increase in network efficiency. In the proposed model, a special form of pooling layer called the global average pooling layer (GAP) was employed after the attention layer. Batch normalization was applied to standardize the inputs for each mini-batch, using a momentum value of 0.99 and epsilon value 0.001. The resulting feature vector is subsequently fed as input into the fully connected layer. The equations for global average pooling and batch normalization can be expressed as follows:

$$\mathrm{GAP}_C = \frac{1}{H \times W} \sum_{i=1}^{H} \sum_{j=1}^{W} X_{i,j,c}$$

$$X_{BN} = \frac{X - \mu}{\sqrt{\sigma^2 + \in}} * \alpha + \beta$$

Here, X represents the feature maps, each of which has the dimensions H, W, and C. α is learnable scaling parameter, and β is a learnable shifting parameter. μ And σ^2 are mean and variance of the mini-batch.

3.3 Fully Connected Layer

The model's fully connected (FC) layer is made up of a number of hidden layers that interact with higher-level representations. These FC layers are in charge of identifying whole objects, regardless of their placements or shapes. Our model's design included dense layer which were given L2 Kernel regularizer with a factor of 0.016. The model's capacity to properly generalize is improved by this regularization technique. Finally, one output dense layer is added to the model, which is followed by a dropout layer with a 0.45 value to avoid overfitting. These dense layers, combined with regularization, helped the model learn essential patterns from the data and performed the classification efficiently.

4 Experimental Setup and Results

4.1 Dataset Used

The visual properties of cell nuclei, the ratio of nucleus to cytoplasm, the geometry of cysts, and the general tissue look are important elements to consider when diagnosing colorectal cancer (CRC) by histological imaging. In this work, we have experimented on some models in addition to the proposed model for colorectal tissue histology. We have utilized public dataset for colorectal histology dataset. The study's data come from the German UMM (University Medical Center Mannheim) archives. Ten private histology slides of tissue samples from colorectal cancer (CRC) are included in this dataset. The hematoxylin and eosin (H&E) staining technique was used to emphasize the cellular features on these slides. This dataset covers a wide spectrum of CRC patients, including both high-grade and low-grade tumors. This variety in tumor grade offers a thorough illustration of the disease's pathological variability. Figure 2 provides detailed information about various classes within our dataset.

This public dataset contains 5000 colorectal histology images, each measuring 150 $\times$ 150 pixel size. These collections are classified into eight separate, non-overlapping categories. The dataset offers a valuable resource for researchers seeking to investigate colorectal cancer at tissue level. It enables the in-depth exploration of CRC features and characteristics in advancement of early prediction and diagnosis approaches. For our experiment, we have utilized 80% training set, 8% validation set, and 12% test set, expanding on the prior investigation [16, 17].

4.2 Performance Evaluation Metrics

We utilized different assessment metrics that were applied for the purpose of measuring the performance of proposed architecture. These metrics contain precision, recall, accuracy, $F1$ score, specificity, and confusion matrices. These metrics reveal how successfully a model distinguishes dataset classes. Precision, which measures the ratio of properly predicted positive examples to the total expected positives, and recall, which indicates the model's ability to identify all positive cases, are common metrics. Fundamentally, accuracy measures the percentage of right predictions, including genuine positives and negatives. The $F1$ score balances precision and recall with a single score that accounts for false positives and negatives. Specificity, also known as true negative rate, measures the model's ability to recognize negative events. Confusion matrices show the model's performance by breaking down true positives, true negatives, false positives, and false negatives. These assessment criteria allow academics and practitioners to evaluate and improve image classification models for medical diagnosis, object recognition, and other applications. The subsequent mathematical formula illustrates the metrics as follows:

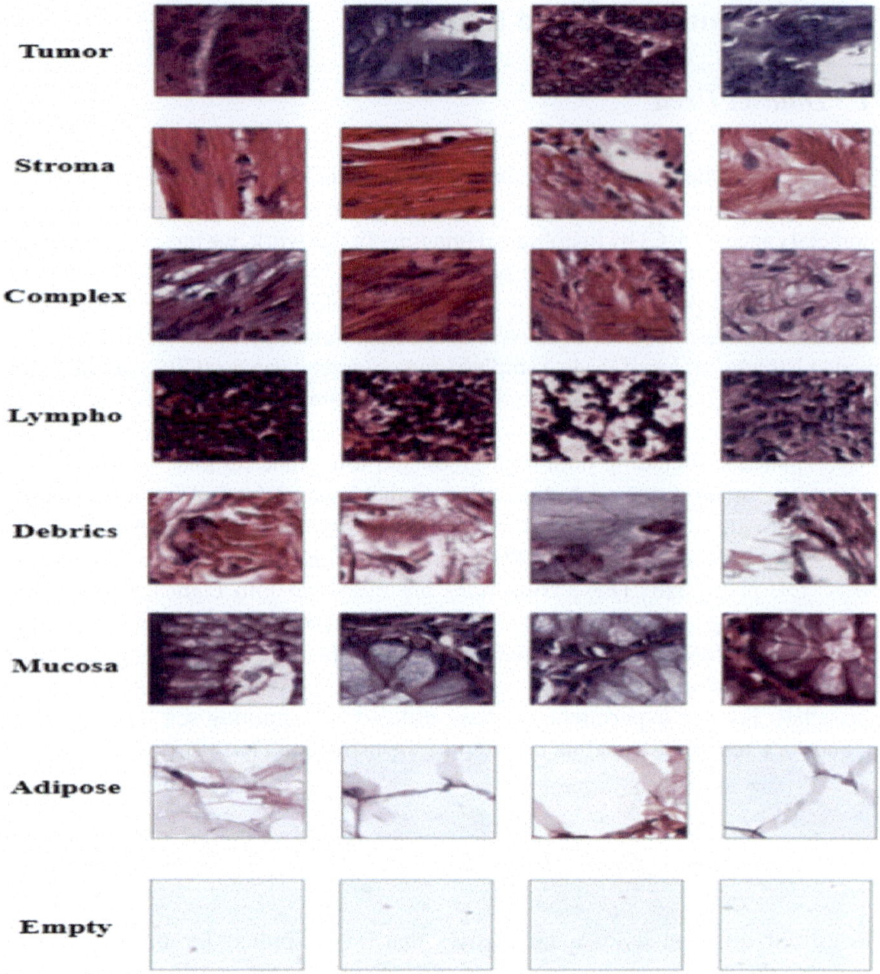

Fig. 2 Description of an eight-class colorectal histology dataset

$$\text{Accuracy} = \frac{\text{TP}}{\text{TP} + \text{TN} + \text{FP} + \text{FN}}$$

$$\text{Recall} = \frac{\text{TP}}{\text{TP} + \text{FN}}$$

$$\text{Precision} = \frac{\text{TP}}{\text{TP} + \text{FP}}$$

$$F1\,\text{Score} = \frac{\text{TP}}{\text{TP} + \text{FP}}$$

Table 1 Modified layer structure for EfficientNet B0

Layer type	Dimension	Activation function/regularizer
Attention layer	$5 \times 5 \times 1280$	N/A
Global average pooling	$1 \times 1 \times 1280$	N/A
Batch normalization	$1 \times 1 \times 1280$	Momentum = 0.99, epsilon = 0.001
Dense layer (256)	256	Relu, Kernel regularizer ($l2 = 0.016$)
Dropout (0.45)	256	N/A
Dense output (8)	8	Softmax

4.3 Implementation Details

The initial hyperparameters were all consistently maintained throughout the training of the proposed model. The batch size was set to 32. The network was trained with a learning rate of 0.001 with the Adam optimizer. It can adjust to shifting gradients more successfully, which may result in a faster and more stable convergence in these circumstances. A kernel regularizer is used in the added dense layer to prevent overfitting. The model was adjusted by training it over several epochs. After 20 epochs of training, the final results of all models were compared to find the maximum accuracy using these settings. Table 1 provides a description of the parameters for the added layers in the proposed architecture.

All models were set up to make use of GPU assistance for accelerated computation. The operating system used for this work is Ubuntu 20.04 LTS. The system used in the experiment has a central processing unit (Intel (R) i7) as well as two graphics processing units (Quadro RTX 5000, each with 32 GB of memory). The Tensorflow 2.13 framework, an open-source deep learning toolkit created in the Python programming language, was used to implement all of the code.

4.4 Results and Discussion

In this section, we will delve into the findings and their implications to provide a comprehensive understanding of the image classification task performed in this work.

Figures 3 and 4 exhibit the accuracy and training and validation losses for the proposed model's corresponding training and validation phases. The confusion matrix that was obtained for colorectal histology may be found in Fig. 5.

According to our findings, the proposed architecture performs better than other recent models with tiny modification, providing more accurate colorectal tissue classification using histopathological images (HIs). Convolutional neural networks (CNNs) frequently employ local receptive fields, which restricts their capacity to detect distant relationships in the image. Self-attention mechanisms are particularly adapted to handle complex relationships in the data because they get around this

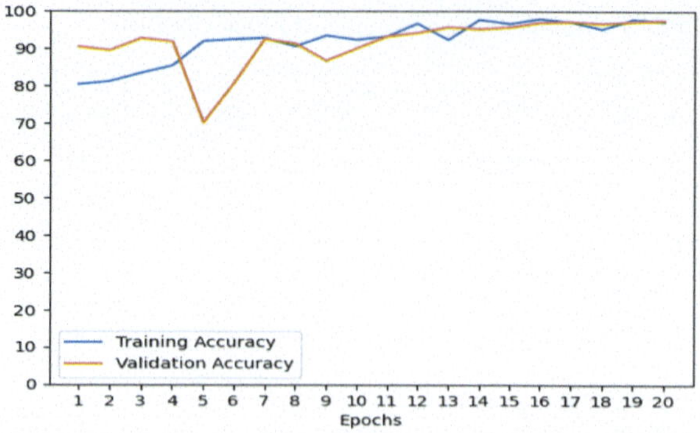

Fig. 3 Training and validation accuracy over epochs

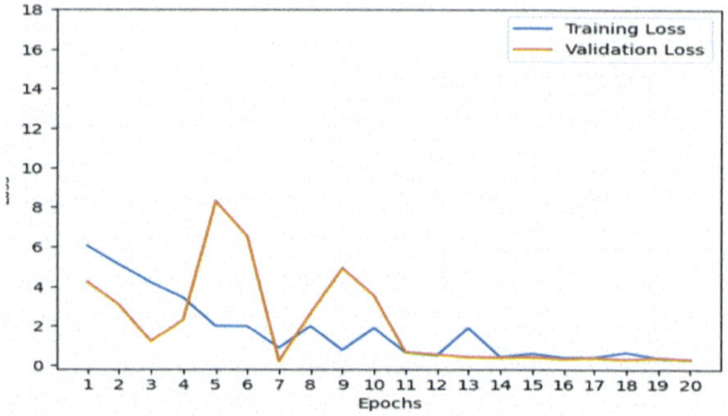

Fig. 4 Training and validation accuracy over epochs

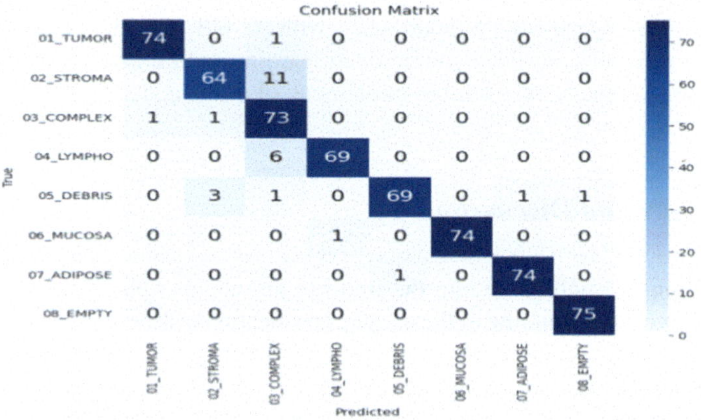

Fig. 5 Confusion matrices for testing data

restriction by taking into account interactions between all pixels or regions. As given in Table 2, our results show that this modification strategy is effective when compared to modern techniques.

Table 2 Performance comparison of proposed architecture with the existing state of art on the colorectal histology dataset

Author	Methods	Accuracy	Precision	Sensitivity	$F1$ score
Wang et al. [18]	Bilinear CNN	92.60	–	92.80	–
Ghosh et al. [16]	Ensemble deep learning models	90.00	89.00	86.00	88.00
Ohata et al. [21]	CNN + machine learning algorithms	92.083	–	–	92.12
Zeid et al. [19]	Vision transformers	94.75	93.44	93.33	93.3
CRCCN Net [17]	CNN	93.50	94.12	93.62	93.86
EfficientNet (fine-tuned)	CNN	93.20	93.55	92.33	93.22
Proposed model	Modified EfficientNet (EffiSA-net)	95.49	95.61	95.50	95.52

5 Conclusion

In this paper, a modified EfficientNet model for the colorectal cancer classification is proposed. The proposed modified EfficientNet attention mechanism improves the representation power by enabling feature maps to combine information from different channels or positions effectively. Also, it captures dependencies between distant locations in the input model to understand relationships between non-adjacent elements which enhance the overall colorectal cancer classification performance. The model's overall accuracy significantly increased after the addition of an attention layer and a dense layer. The model presented in this work demonstrated exceptional performance on the testing dataset with 95.49%, with a minimal number of false negatives. Experiments were conducted on the EfficientNet to assess the effectiveness of proposed model. The resulting model outperformed the majority of recently proposed solutions for colorectal cancer. In the future, we will investigate the various integration techniques of attention blocks into deep neural networks for image classification. Additionally, we will also explore the application of lightweight models and generative adversarial networks (GANs) for effective and efficient colorectal cancer-classification.

References

1. Siegel RL, Wagle NS, Cercek A, Smith RA, Jemal A (2023) Colorectal cancer statistics. CA Cancer J Clin 73:233–254. https://doi.org/10.3322/caac.21772
2. Cohen R, Platell CF (2023) Metachronous colorectal cancer metastasis: who, what, when and what to do about it. J Surg Oncol: 1–7. https://doi.org/10.1002/jso.27400

3. Davri A, Birbas E, Kanavos T, Ntritsos G, Giannakeas N, Tzallas AT, Batistatou A (2022) Deep learning on histopathological images for colorectal cancer diagnosis: a systematic review. Diagnostics 12:1–32. https://doi.org/10.3390/diagnostics12040837
4. Taghiakbari M, Mori Y, von Renteln D (2021) Artificial intelligence-assisted colonoscopy: a review of current state of practice and research. World J Gastroenterol 27:8103–8122. https://doi.org/10.3748/wjg.v27.i47.8103
5. Alqudah AM, Alqudah A (2022) Improving machine learning recognition of colorectal cancer using 3D GLCM applied to different color spaces. Multimed Tools Appl 81:10839–10860. https://doi.org/10.1007/s11042-022-11946-9
6. Kim SH, Koh HM, Lee BD (2021) Classification of colorectal cancer in histological images using deep neural networks: an investigation. Multimed Tools Appl 80:35941–35953. https://doi.org/10.1007/s11042-021-10551-6
7. Suganyadevi S, Seethalakshmi V, Balasamy K (2022) A review on deep learning in medical image analysis. Int J Multimed Inf Retr 11:19–38. https://doi.org/10.1007/s13735-021-00218-1
8. Song TH, Sanchez V, Eidaly H, Rajpoot NM (2019) Simultaneous cell detection and classification in bone marrow histology images. IEEE J Biomed Heal Inform 23:1469–1476. https://doi.org/10.1109/JBHI.2018.2878945
9. Lin H, Chen H, Graham S, Dou Q, Rajpoot N, Heng PA (2019) Fast ScanNet: fast and dense analysis of multi-gigapixel whole-slide images for cancer metastasis detection. IEEE Trans Med Imaging 38:1948–1958. https://doi.org/10.1109/TMI.2019.2891305
10. Qaiser T, Sirinukunwattana K, Nakane K, Tsang YW, Epstein D, Rajpoot N (2016) Persistent homology for fast tumor segmentation in whole slide histology images. Procedia Comput Sci 90:119–124. https://doi.org/10.1016/j.procs.2016.07.033
11. Arvaniti E, Fricker KS, Moret M, Rupp N, Hermanns T, Fankhauser C, Wey N, Wild PJ, Rüschoff JH, Claassen M (2018) Automated Gleason grading of prostate cancer tissue microarrays via deep learning. Sci Rep 8:1–11. https://doi.org/10.1038/s41598-018-30535-1
12. Hassanin M, Anwar S, Radwan I, Khan FS, Mian A (2022) Visual attention methods in deep learning: an in-depth survey. 1–20
13. Zhou P, Cao Y, Li M, Ma Y, Chen C, Gan X, Wu J, Lv X, Chen C (2022) HCCANet: histopathological image grading of colorectal cancer using CNN based on multichannel fusion attention mechanism. Sci Rep 12:1–12. https://doi.org/10.1038/s41598-022-18879-1
14. Xiao H, Liu Q, Li L (2023) MFMANet: multi-feature multi-attention network for efficient subtype classification on non-small cell lung cancer CT images. Biomed Signal Process Control 84. https://doi.org/10.1016/j.bspc.2023.104768
15. Masud M, Sikder N, Nahid AA, Bairagi AK, Alzain MA (2021) A machine learning approach to diagnosing lung and colon cancer using a deep learning-based classification framework. Sensors (Switzerland) 21:1–21. https://doi.org/10.3390/s21030748
16. Ghosh S, Bandyopadhyay A, Sahay S, Ghosh R, Kundu I, Santosh KC (2021) Colorectal histology tumor detection using ensemble deep neural network. Eng Appl Artif Intell 100:104202. https://doi.org/10.1016/j.engappai.2021.104202
17. Kumar A, Vishwakarma A, Bajaj V (2023) CRCCN-Net: automated framework for classification of colorectal tissue using histopathological images. Biomed Signal Process Control 79:104172. https://doi.org/10.1016/j.bspc.2022.104172
18. Wang C, Shi J, Zhang Q, Ying S (2017) Histopathological image classification with bilinear convolutional neural networks. In: Proceedings of annual international conference on IEEE engineering medical biology society. EMBS, 4050–4053. https://doi.org/10.1109/EMBC.2017.8037745
19. Zeid MAE, El-Bahnasy K, Abo-Youssef SE (2021) Multiclass colorectal cancer histology images classification using vision transformers. In: Proceedings of 2021 IEEE 10th international conference on intelligence computing information system. ICICIS 2021, 224–230. https://doi.org/10.1109/ICICIS52592.2021.9694125
20. Huang G, Liu Z, Van Der Maaten L, Weinberger KQ, Tan M, Le QV, He K, Arge FORL, Mage CI (2017) EfficientNet: rethinking model scaling for convolutional neural networks. In:

Proceedings of 30th IEEE conference on computing vision pattern recognition, CVPR 2017. 2019 June, 10691–10700. https://doi.org/10.1109/CVPR.2017.243

21. Ohata EF, Chagas JVS, Bezerra GM, Hassan MM, de Albuquerque VHC, Filho PPR (2021) A novel transfer learning approach for the classification of histological images of colorectal cancer. J Supercomput 77:9494–9519. https://doi.org/10.1007/s11227-020-03575-6

Smart Agriculture System

Vishal Kumar Kanaujia and Chitranjan Prasad Sah

Abstract Agriculture stands as a linchpin for national prosperity, especially in regions blessed with abundant water and fertile soil, capable of yielding a diverse range of food products crucial for human sustenance. However, escalating global population and diminishing freshwater reserves pose significant challenges to sustainable agriculture. Approximately, 96% of Earth's water is saline, leaving a mere 3% as freshwater resources, intensifying the need for advanced techniques to conserve water in farming practices. This research addresses the pressing issue by proposing a solution rooted in smart agriculture, leveraging modern technologies like the Internet of Things (IoT) to revolutionize conventional farming methods. The integration of automation and IoT technologies, such as the ESP8266 IoT automatic irrigation system, is presented as a means to enhance crop productivity. This system utilizes cutting-edge technologies like Arduino and wireless sensor networks, offering a comprehensive solution for monitoring environmental conditions crucial for maximizing crop yield. The paper underscores the paramount importance of connecting agriculture to the Internet, emphasizing the role of IoT in facilitating applications like crop monitoring and automated irrigation decisions. The proposed smart agriculture system, featuring energy autonomy and cost-effectiveness, aims to address persistent agricultural challenges that impede national development. The research introduces a novel approach to agriculture, offering a tangible pathway toward sustainable farming practices in the face of water scarcity and environmental uncertainties. Through the integration of sensors and live data feeds, accessible via mobile devices, this system represents a promising step toward a technologically advanced and efficient agricultural landscape, especially beneficial for water-scarce and geographically isolated regions.

V. K. Kanaujia (✉)
Computer Science and Engineering-Data Science Department, ABES Engineering College, Ghaziabad, Uttar Pradesh, India
e-mail: vishalkanaujia.cs@gmail.com

C. P. Sah
School of Engineering and Technology, Noida International University, Greater Noida, Uttar Pradesh, India
e-mail: chitranjan.prasad@niu.edu.in

B. Annappa et al. (eds.), *High Performance Computing, Smart Devices and Networks*, Lecture Notes in Electrical Engineering 1262,
https://doi.org/10.1007/978-981-97-7794-5_23

Keywords Agriculture · Farming · IoT · Smart agriculture · Sensors

1 Introduction

The global population is projected to reach 9.6 billion by 2050, posing a significant challenge for the agriculture industry. Despite grappling with obstacles such as extreme weather conditions, escalating climate change, and the environmental impact of farming, the imperative to fulfill the growing demand for food remains. In order to address these escalating needs, the agricultural sector must embrace innovative technologies. Novel smart farming applications driven by IoT (Internet of Things) advancements offer the potential to minimize waste and enhance productivity, encompassing improved fertilizer utilization and optimized routes for farm vehicles.

To cater to the ever-increasing food requirements and counterbalance diminishing supplies, rapid advancements in food production technology are indispensable. Agriculture stands as the principal source of achieving this objective. This dynamic is vital for human societies, given the mounting demand for food production. Agriculture serves as a pivotal catalyst for economic development, particularly in nations like India. The scarcity of water and limited availability of arable land has led to decreased water reserves on Earth, prompting farmers to resort to irrigation. Irrigation can be defined as the controlled application of water to soil, tailored to the specific needs of different plant species based on soil characteristics.

Smart farming represents a sophisticated and technology-driven approach to cultivating food sustainably and efficiently. It involves integrating modern Information and Communication Technologies (ICT) into agricultural practices. In IoT-based intelligent farming, a comprehensive system is established to monitor crop fields using an array of sensors (light, humidity, temperature, soil moisture, etc.) while automating irrigation processes. Farmers gain the ability to remotely monitor field conditions from anywhere. The IoT-driven smart farming paradigm proves significantly more efficient when compared to traditional methods. The applications of IoT-enabled smart farming extend beyond routine large-scale agricultural operations. They also have the potential to propel emerging or conventional trends in agriculture, such as organic farming and family farming with specialized livestock or crops, preservation of specific or heirloom varieties, and the facilitation of transparent farming practices.

In the past, technological advancements were limited, necessitating manual tasks such as seeding, plowing, and plant cutting. However, contemporary technology has evolved significantly, rendering manual seeding under sunlight unnecessary. Through the implementation of robotic technology, one can now remotely supervise seeding processes from a comfortable location, eliminating the need for direct involvement.

The modern agricultural landscape demands innovative approaches to enhance overall efficiency and performance. Farmers encounter challenges spanning seed sowing, plowing, plant waste reduction, and weeding. The equipment traditionally employed for these tasks is cumbersome to transport and operate. The migration of rural populations to urban areas has exacerbated these issues.

The research focuses on addressing labor-intensive and time-consuming seed sowing processes in agriculture through the development of an efficient mechanized system. This system, utilizing existing mass-produced components, aims to reduce energy consumption compared to traditional tractors. Emphasizing the pivotal role of agriculture in sustaining nations, the study highlights the need for water conservation and modern technologies in the face of a growing global population. The proposed solution involves an ESP8266 IoT automatic irrigation system, combining sensors and solar-powered robot machines for autonomous seed sowing. The broader objective is to provide farmers with reliable data through a wireless sensor network enabled by IoT technology, addressing challenges such as poor water management, excessive fertilizer use, and labor-intensive practices. Additionally, the implementation of an NPK sensor is proposed to measure soil nutrient levels, offering insights into optimal fertilizer usage. The use of soil moisture and temperature sensors further contributes to judicious water management and provides crucial information for irrigation and fertilization decisions. The research ultimately aims to empower farmers with precision farming techniques, promoting sustainable and efficient agricultural practices.

Presently, robotics technology plays a substantial role across diverse sectors including medicine, industries, and various other domains. Its application extends to agricultural operations, where robots are capable of executing a range of tasks. This technology can be seamlessly integrated into farming systems to alleviate the labor burden on farmers, concurrently saving time, energy, and costs [1–3].

2 Literature Survey

This article highlights how Internet of Things (IoT) technology has significantly transformed various aspects of daily life, introducing intelligence and efficiency to numerous domains. IoT establishes a network of interconnected devices that form a self-configuring network. In the realm of agriculture, the emergence of intelligent smart farming devices powered by IoT is revolutionizing the agricultural landscape. This technology not only enhances agricultural production but also contributes to cost-effectiveness and waste reduction. The primary aim of this study is to introduce a novel smart IoT-based agriculture system that provides real-time data, such as temperature and soil moisture, enabling farmers to engage in smart farming practices, enhance yield, and improve product quality.

The paper outlines a comprehensive framework that facilitates robust field operations and is farmer-friendly. One of the domains where IoT-based research is rapidly advancing and generating new products is the field of "agriculture." The global significance of the agriculture sector, particularly in ensuring food security, cannot be overstated. In India, farmers face multiple challenges, including limited farm size, technological constraints, trade barriers, government policies, and varying climate conditions.

Leveraging IoT in agriculture holds immense potential not only for efficient crop monitoring but also for optimizing crop yield. IoT's influence extends broadly across various industries, including smart homes, healthcare, urban planning, fitness, and industrial sectors. Agriculture is no exception to this trend. IoT and connected devices can revolutionize farming practices, eliminating the need for traditional manual labor and equine assistance. While IoT is widely recognized for its consumer-oriented applications, its implementation in agriculture presents a dynamic landscape.

The role of IoT in agriculture is instrumental in enhancing production rates, reducing costs, and ensuring better resource utilization. Approximately 75% of crop production contributes to farmers' profits through the integration of technology. Incorporating IoT technology into agricultural operations streamlines tasks, automates machinery control, facilitates remote and real-time monitoring, and enhances resource management through predictive maintenance and environmental forecasting. These advancements are poised to bolster revenues in the agriculture sector, enabling farmers to manage larger areas of land efficiently.

Smart agricultural technology empowers farmers with greater control over crop cultivation and livestock rearing. It offers economies of scale, cost reductions, and the conservation of precious resources like water. Waste reduction and increased productivity extend to various aspects of farming, spanning fertilizer usage and the frequency of farm vehicle trips. When implementing an IoT solution for agricultural purposes, sensor selection becomes pivotal. Optimal sensor choice depends on the data you aim to collect and the intended applications. Sensor quality is paramount, ensuring the accuracy and reliability of collected data, ultimately determining the success of the IoT solution [4–6].

2.1 System Overview

The smart agriculture system comprises distinct nodes labeled as node1, node2, node3, and node4. These nodes function as monitoring units, supplying accurate crop information to farmers. In the present architecture, each hub is equipped with a diverse array of sensors and devices, and it establishes a connection with a central server via remote communication modules. Data transmission between the server and client endpoints occurs through the Internet. The Arduino integrated development environment (IDE) serves as an open-source software platform enabling users to write and upload real-time code to the hardware board. This code can also be stored in the cloud, offering an added layer of data redundancy. The system's programming environment is Java-based and built upon processing and other open-source tools. Furthermore, the system ensures seamless compatibility with any Arduino hardware board [7].

2.2 Analysis of Smart Agriculture

To enhance future developments, this system holds potential for expansion to cover larger expanses of land. Furthermore, there is room for integration to assess soil quality and crop growth across varying soil types. The successful interfacing of sensors and microcontrollers has facilitated wireless communication among different nodes. Subsequent endeavors may focus on augmenting sensor capabilities within this device, particularly in areas like pest control. Additionally, the integration of GPS modules can transform this agriculture IoT stick into a comprehensive agriculture precision solution.

Currently, numerous companies are actively contributing to the advancement of IoT (Internet of Things) initiatives. This burgeoning landscape introduces novel concepts and solutions that address business challenges through machine to machine (M2M) communications. A recent discussion highlighted several notable industrial applications of IoT. Precisely measuring soil moisture plays a crucial role in agriculture, enabling farmers to optimize their irrigation practices. Effective management of soil moisture not only conserves water resources but also enhances crop yields and quality during vital stages of plant growth [8].

2.3 Related Models to Smart Agriculture System

Mini greenhouse 1.20 × 0.60 × 0.60: The initiative stemmed from a simple student idea to cultivate flowers in our school's flower pots despite the extreme local climate conditions. To address this, we conceptualized a compact greenhouse boasting professional features. We approached our municipality with a proposal to finance the construction of a larger model based on our prototype.

Farm assistance: Autonomous plant disease detection robot inspired by the work showcased on plantvillage.psu.edu and iita.org, our goal was to employ the Donkey Car platform in crafting an independent robot capable of navigating a farm environment without causing harm to existing plants or soil. This innovative robot employs object detection to identify and mark diseased crops with an eco-friendly color. The conventional method of manual inspection, often involving smartphones for marking crops, is labor-intensive and time-consuming. The introduction of a uniform robotic platform circumvents these issues, significantly expediting the marking process. This approach also enhances platform sharing among multiple farms.

Key challenges: Maintaining the robot's size and weight to prevent crop damage, ensuring navigation without harming existing crops, devising a safe method for marking diseased crops, and identifying a suitable dataset and farm for potential platform testing.

Smart Agro: Agricultural land is a valuable and indispensable resource, vital for modern society's sustenance. Overcoming challenges in farming, such as regulating temperature, humidity, sunlight, and safeguarding crops against threats, is paramount.

The concept of smart farming has evolved to remotely monitor crop conditions. However, current implementations are often cost-prohibitive for small farms and may not adequately cover large agricultural expanses.

Smart Agro aims to establish an accessible network of user-friendly devices that can be strategically placed within monitored fields. These devices gather data pertaining to atmospheric conditions, soil parameters, and light intensity, all transmitted to a cloud-based dashboard for presentation. This transformative approach empowers farms of any scale to embrace the modern era of agriculture. By deploying a scalable fleet of monitoring devices across their land, farms can receive timely alerts, such as impending drought conditions, fostering an evolution in agricultural practices.

2.4 Hardware Requirement

- Node MCU ESP8266: The Node MCU ESP8266 development board is equipped with the ESP-12E module housing the ESP8266 chip, featuring a Tensilica Xtensa 32-bit LX106 RISC microprocessor. This microprocessor is compatible with RTOS and functions within a clock frequency range of 80–160 MHz, which can be adjusted. The NodeMCU is endowed with 128 KB of RAM and possesses 4 MB of flash memory designated for data storage and program execution. Its notable processing capabilities, complemented by built-in Wi-Fi and Bluetooth functionalities, along with its energy-efficient Deep Sleep Operating mode, render it an excellent choice for Internet of Things (IoT) projects.
- Soil sensor: The Soil Moisture Sensor Module functions by gaging the level of soil moisture, achieved through assessing the resistance between two metallic probes when they are inserted into the target soil for monitoring purposes. This module proves valuable for applications like automated plant watering systems or issuing alerts when plants require watering. The sensors operate based on the measurement of resistance between the two forked probes once they are inserted into the soil. The resistance primarily correlates with the moisture content in the soil. This resistance has an impact on a voltage divider, ultimately producing an analog voltage output. This output can be read by an analog input on a microcontroller (uC), providing an approximate indication of soil moisture. Elevated moisture levels translate to reduced resistance, yielding a lower analog voltage reading. Conversely, as the soil experiences dehydration, resistance escalates, leading to higher voltage values.
- Rain sensor: A rain sensor, also known as a rain switch, serves as an activation device triggered by precipitation. The utility of rain sensors spans two primary domains. Firstly, they find application as water conservation mechanisms within automatic irrigation systems. These sensors prompt the system to deactivate when rainfall occurs, preventing unnecessary watering. Secondly, rain sensors are employed to safeguard the interior of vehicles from rain and to facilitate the automatic functioning of windshield wipers. In the context of professional satellite

communication antennas, rain sensors are employed to initiate rain blowers situated on the aperture of the antenna feed. This action removes water droplets from the mylar cover that maintains a sealed and moisture-free environment within the wave-guides.

- Moisture sensor module (LM393): This module generates an output voltage based on the probe's resistance, which is accessible through an analog output (AO) pin. The identical signal is channeled to an LM393 high-precision comparator for digitization and is subsequently presented at a digital output (DO) pin. Comprising a moisture sensor, resistors, capacitor, potentiometer, comparator LM393 IC, as well as power and status LEDs, this integrated circuit forms the core of the moisture sensor module.
- Temperature and humidity sensor (DHT11): The DHT11 represents a fundamental, exceptionally affordable digital sensor for temperature and humidity. It incorporates a capacitive humidity sensor and a thermistor to gage the ambient air conditions, producing a digital signal on the data pin without necessitating analog input pins. Its usage is straightforward, though accurate data acquisition requires precise timing. A notable limitation of this sensor is that fresh data can only be retrieved at two-second intervals. Consequently, with our library, the recorded sensor readings could be up to two seconds behind real-time.
- Light-dependent resistor (LDR): An alternative term for a light-dependent resistor (LDR) is a photoresistor or cadmium sulfide (CdS) cell. It is also recognized as a photoconductor, functioning based on the principle of photoconductivity. Essentially, this passive component acts as a resistor, its resistance diminishing as light intensity declines. Frequently employed in circuits for sensing light variation, the optoelectronic nature of this device renders it particularly suitable for applications such as light-sensitive circuits and light–dark activated switching systems.
- Grow light LED: Grow lights have two primary objectives: replicating the sun's light spectrum or tailoring a spectrum to suit the specific needs of cultivated plants. To mimic outdoor conditions, the grow light emits diverse colors, temperatures, and spectral outputs, alongside adjustable lamp intensities. The optimal spectrum, luminous efficiency, and color temperature vary depending on factors like plant type, growth stage (germination/vegetative or flowering/fruiting), and required photoperiod. This demand for specific characteristics aligns with the plant's unique requirements during distinct time periods [1, 3, 6, 7, 9, 10].

3 Applications

The incorporation of modern sensors and IoT methodologies in agricultural practices represents a profound shift in the farming landscape. Smart farming, characterized by the harmonious merger of wireless sensors and the Internet of Things, holds the potential to revolutionize agricultural techniques beyond imagination. Embracing smart farming principles allows the Internet to offer enhanced resolutions for long-standing

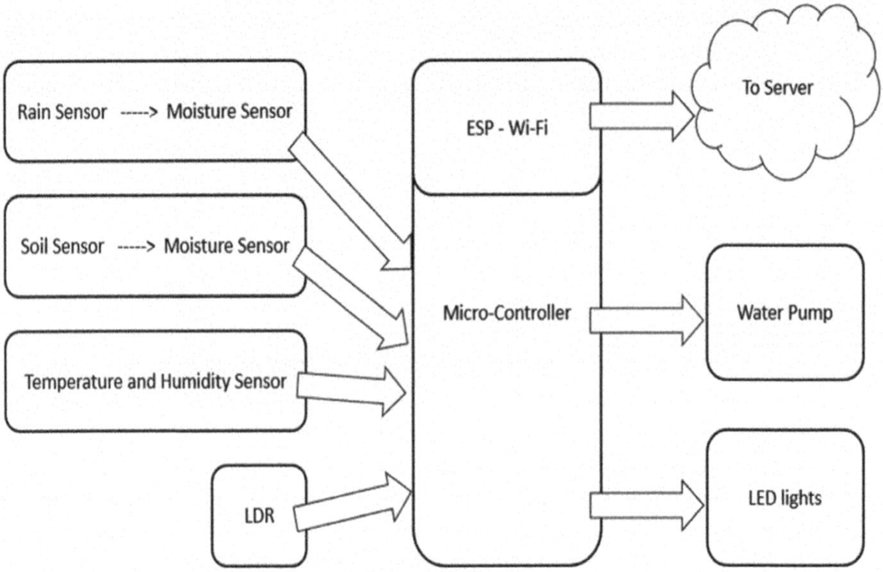

Fig. 1 Block diagram of the system

agricultural challenges, spanning yield enhancement, drought management, soil suitability, irrigation, and pest control. The schematic in Fig. 1 outlines the hierarchical structure encompassing key applications, services, and wireless sensors employed within smart farming endeavors. Noteworthy instances demonstrate how advanced methodologies at various stages contribute to a comprehensive enhancement of overall agricultural efficiency.

3.1 Soil Sampling and Mapping

Examining soil plays a pivotal role in obtaining site-specific insights that inform decision-making at various stages of agricultural practices. The primary goal of soil analysis revolves around assessing the nutrient status of the farmland, influencing subsequent actions based on identified deficiencies. Factors contributing to soil nutrient analysis encompass soil type, crop history, fertilizer usage, irrigation levels, and related considerations. A range of manufacturers offer sensors and tools designed for soil testing, enabling farmers to gage soil quality. Informed by this data, appropriate measures can be implemented to enhance crop growth.

Addressing the critical issue of drought, which significantly impacts crop yield, involves leveraging remote sensors. These sensors are employed to consistently gather soil moisture data, delivering timely SMS and email alerts or website updates to farmers. This real-time information guides irrigation management. By establishing thresholds, once a specific range is reached, an alert prompts action, allowing for timely intervention in water pump operations.

3.2 *Irrigation*

Various irrigation techniques, including drip and sprinkler methods, offer solutions to challenges such as water wastage, which were prevalent in traditional approaches like flood and furrow irrigation. Water scarcity can detrimentally affect crop quality, quantity, and soil nutrient levels, leading to susceptibility to infections harmful to plants. Precisely estimating required water for crops proves challenging due to factors like irrigation method, crop type, soil composition, and soil moisture levels. To address this, wireless sensors integrated into soil and moisture control systems optimize water usage for crops, contributing to improved crop well-being.

The incorporation of IoT techniques further enhances crop efficiency. Notable approaches include crop water stress index (CWSI) irrigation management. Leveraging wireless field sensors, environmental conditions are measured and data transmitted. To calculate CWSI, additional data like weather information and satellite imagery are integrated, resulting in an augmented water usage efficiency that benefits crop growth [6, 10].

4 Advanced Practices in Agriculture

In the past, enhancing crop production centered on seed and fertilizer quality. However, these traditional methods no longer suffice to meet growing demand. Scientists have introduced alternatives, including bioengineered and genetically modified foods, created through genetic engineering that modifies their DNA. Despite these advances, only a few technologies gain public acceptance due to the prevailing preference for organic produce. The agriculture sector undergoes a transformative shift with the advent of IoT, empowering farmers to tackle their challenges. Innovative applications address these issues, leading to heightened crop production, sustainability, and cost-effectiveness.

Numerous years of research have explored the integration of technologies like IoT to enhance conventional agricultural practices. The pivotal role of cutting-edge technology in advancing urban farming is evident, reshaping the landscape of agriculture.

4.1 Greenhouse Farming

Indoor cultivation, recognized as a component of smart agriculture, draws upon historical foundations dating back to the nineteenth century. This approach finds prominence in nations grappling with extreme shifts in climate. Presently, crops cultivated indoors are shielded from the impact of varying environmental conditions, enabling the cultivation of diverse crop types. Facilitating communication within this setup necessitates the integration of sensors and other devices. Factors influencing crop production encompass aspects like ventilation systems, wind control, and monitoring.

In contemporary greenhouse farming, paramount importance is accorded to environmental monitoring. The task involves overseeing multiple measuring points that collectively regulate the indoor climate. This method ensures optimal conditions for crop growth and production.

4.2 Vertical Farming

Meeting the escalating demand for food necessitates an increase in agricultural land, yet pollution and soil erosion contribute to substantial land loss. The adoption of contemporary farming practices, rooted in industrial methods, has adverse effects on soil quality. These challenges are exacerbated by arable land scarcity, posing a significant threat to future food production. Furthermore, the agricultural process heavily relies on fresh water, straining limited water reservoirs. To surmount these hurdles, vertical farming emerges as a viable solution.

Vertical farming predominantly finds application within urban agriculture, creating a controlled environment for cultivating crops. This innovative approach conserves resources effectively. Notably, carbon dioxide levels play a crucial role in this form of cultivation. The use of NDIR (non-dispersive infrared) sensors facilitates precise climate control in vertical farming setups. Gas box sensors contribute to ensuring optimal plant growth conditions. Employing a pseudo dual beam NDIR measurement enhances stability and minimizes optical complexities. IoT-connected farms eliminate the need for manual observation at each growth stage. Various vertical farming applications integrate mint controls, mitigating issues like wastage and incorporating sensor-based solutions.

4.3 Communication

Timely data updates are of paramount importance within the realm of smart agriculture. Regularly reporting information at specified intervals ensures the effectiveness of this approach. Ensuring reliable communication falls within the domain of telecom network operators, pivotal to the process. The communication method employed varies based on the specific applications implemented on the farms.

Wireless technologies such as Bluetooth enable peer-to-peer connections, facilitating seamless data sharing. For shorter distances, Wi-Fi serves as an ideal choice, connecting devices and relaying data to a central server. This data repository allows convenient access to information, enabling users to retrieve details whenever required. Smartphones emerge as valuable tools for crop monitoring, universally accessible. Incorporated GPS functionality in smartphones empowers farmers with location-specific insights [3, 11].

5 Problem Statement

The aim is to establish an efficient decision support system via a wireless sensor network, effectively managing various farm activities and offering pertinent farm-related insights. This encompasses data concerning soil moisture, temperature, and humidity levels. The unpredictability of weather conditions often results in distracting fluctuations in water levels. Farmers encounter these challenges and manage water levels through both automatic and manual means, facilitated by a mobile application. This digital approach enhances convenience for farmers, alleviating time-consuming manual agricultural tasks.

The solution's design prioritizes minimal resource utilization, both in terms of hardware and cost. By doing so, it streamlines the monitoring and maintenance processes of agricultural farms, catering to both automatic and manual operational modes. Crucially, the system should possess the capability to quantify fluctuations in water levels and soil moisture, ultimately contributing to improved agricultural practices [1, 12].

6 Proposed Work

In contemporary times, seed sowing entails labor-intensive processes, either conducted manually or with the aid of tractors. Manual approaches encompass methods such as seed broadcasting or dibbling, and sometimes involve animals for leveling and seed dropping. To address these challenges, a more efficient system is being developed. Seed sowing is a fundamental operation in agriculture, integral to various farming methods like tilling, cultivating, and harvesting. The introduction of mechanized mechanisms eliminates the need for extensive human labor [7]. This smaller machine, comprising existing mass-produced components, requires less energy compared to tractors or other agricultural equipment. Despite the prevalent use of tractors for seeding preparation, it consumes considerable time and often faces labor shortages.

Agriculture plays a pivotal role in the well-being of nations, with water and soil-rich regions capable of producing a diverse array of food items. The increasing global population and its corresponding water demands, coupled with environmental factors affecting freshwater reserves, underscore the urgency of conserving water through modern techniques and advanced technologies. Addressing agricultural challenges through smart agriculture, achieved via automation and IoT technologies, offers a promising solution. The IoT facilitates applications like crop growth monitoring, automatic irrigation decisions, and more.

The proposal involves an ESP8266 IoT automatic irrigation system aimed at modernizing and enhancing crop productivity. A prototype module incorporating sensors feeds data to the ESP8266, which communicates with a robot machine powered by solar energy. The robot autonomously navigates to sow seeds as per user instructions. Data storage in the cloud streamlines further processing. The broader objective is to provide farmers with reliable data, facilitating wireless sensor network-enabled agricultural processes through IoT technology. Current challenges faced by farmers include poor water management, excessive fertilizer use, remote farm monitoring, and labor-intensive practices. The ultimate goal is to provide farmers with effective solutions rooted in precision farming techniques [1, 12].

Components: NPK sensor, soil moisture sensor, rain drop sensor, temperature sensor, PH sensor, perimeter security sensor, Arduino module. Fertilizers play a crucial role in plant growth, and India has witnessed significant utilization of these chemicals. Based on a survey by the National Institute of Agricultural Extension Management, the country consumes approximately 25.6 million tons of fertilizers, with nitrogen (17 MT), phosphorous (6 MT), and potassium (2.5 MT) being the most utilized components. Overuse of fertilizers has led to issues like oxygen depletion, weed growth, algae blooms, and even health concerns like cancers, as highlighted by various studies. To address these challenges, the implementation of an NPK sensor is proposed.

NPK sensor: This sensor is designed to measure the nitrogen, phosphorous, and potassium levels in the soil, which are essential for healthy plant growth. Different crops require varying amounts of NPK. For instance, tomato plantation demands N-P-K values from 6-24-24 to 8-32-16. While soil naturally contains NPK, it often requires supplementation to reach optimal levels. Excessive amounts can be detrimental, making the NPK sensor a valuable tool for farmers to determine the right quantities of NPK to add for optimal crop growth. As per data from the Government of India, groundwater levels have been steadily declining over the past two decades, with agriculture being a major contributor to water wastage. The utilization of soil moisture sensors aids in judicious water management, promoting optimal water usage.

Soil moisture sensor: The soil moisture sensor serves to determine the moisture levels within the soil. When integrated with a water pump, it can activate the pump when the soil's moisture content falls below a predefined threshold, and deactivate it when the moisture content surpasses the threshold. This automated process contributes to tackling the issue of excessive groundwater consumption while concurrently enhancing environmental conditions.

Temperature, a pivotal parameter in agricultural practices, holds a direct influence on crop growth. The growth of most crops thrives within a moderate temperature range, typically spanning from 15 to 32 °C.

Gaining dependable insights into this aspect empowers farmers with the ability to address crucial questions like: When is the appropriate time for watering? What is the optimal quantity of water to apply? When should the application of manure and fertilizers take place?

Temperature sensor: This device is employed to gage the ambient temperature within the vicinity, thus equipping the farmer with valuable information regarding the ideal timing for irrigation, the appropriate volume of water to apply, and the suitable intervals for the incorporation of fertilizers [3, 6–8, 13, 14].

6.1 Advantages

- Large farms can be covered by a limited number of farmers.
- It is time saving, monitoring and providing the conditions which are required for plants is easy by this process.
- The conventional farming practices face certain challenges, which can be addressed through this approach. The irrigation system ensures optimal water supply to plants, preventing overwatering that could hinder their growth. Moreover, natural rainfall might not always provide an adequate water source for the plants.
- Based on the data regarding moisture and temperature received by the cloud, we can adjust the parameters to the desired levels by controlling fans and pumps.
- Efficiency: IoT-enabled agriculture empowers farmers to monitor their crops and conditions in real-time, providing rapid insights. This enables them to anticipate potential problems, make proactive decisions, and take preventive measures. Furthermore, IoT solutions in agriculture introduce automation, such as demand-based irrigation, fertilization, and robotic harvesting.
- Expansion: As the global population approaches 9 billion, with 70% residing in urban areas, the potential for IoT-based greenhouses and hydroponic systems becomes crucial. These innovative approaches facilitate short food supply chains, offering the ability to feed a substantial portion of the population. Smart closed-cycle agricultural systems further extend the possibilities of food cultivation, making it feasible to grow produce in diverse environments such as supermarkets, vertical gardens on skyscrapers, rooftops, shipping containers, and even within the confines of individual homes.
- Reduced resources: Many IoT solutions are dedicated to enhancing resource utilization, encompassing water, energy, and land. IoT-enabled precision farming hinges on data derived from a variety of field sensors, empowering farmers to precisely allocate optimal amounts of resources to individual plants.
- Agility: The incorporation of IoT in agriculture brings about heightened process agility. With real-time monitoring and predictive systems at their disposal, farmers

gain the ability to swiftly react to abrupt shifts in weather, humidity, air quality, and the well-being of crops and soil. In scenarios marked by drastic weather fluctuations, these newfound capabilities empower agricultural practitioners to salvage crops effectively.

- Enhanced product quality: Data-centric agriculture not only increases yield but also enhances product quality. Through the utilization of soil and crop sensors, aerial drone surveillance, and farm mapping, farmers gain deeper insights into the intricate relationships between conditions and crop quality. By leveraging interconnected systems, they can replicate optimal conditions, thereby elevating the nutritional value of their products.

6.2 Disadvantages

- There could be a wrong analysis of weather conditions.
- Device customization to suit farmers' needs may entail costly equipment modifications.
- Malfunctioning data processing equipment or sensors can result in incorrect decision-making.

7 Implementation

- Integration of IoT in the realm of intelligent agriculture: The global population is projected to reach 9.6 billion by 2050—a significant challenge for the agricultural sector. Despite grappling with obstacles like extreme weather, escalating climate change, and the ecological impact of farming, the demand for increased food production remains. To address these mounting requirements, agriculture must embrace innovative technology. Emerging smart farming solutions hinged on IoT advancements hold the promise of minimizing wastage and amplifying productivity. This entails incorporating modern information and communication technologies (ICT) into agriculture. Within the framework of IoT-driven smart farming, a comprehensive system is devised to monitor crop fields, leveraging sensors for metrics like light, humidity, temperature, and soil moisture. This setup empowers farmers to remotely monitor field conditions from any location.
- Integration of soil moisture sensor in intelligent agriculture: Soil moisture sensors gages the volumetric water content present within the soil. The interaction of reflected microwave radiation with soil moisture is employed for remote sensing applications in hydrology and agriculture. Farmers and gardeners can utilize portable probe instruments. These sensors play a pivotal role in efficient irrigation management, leading to improved crop yields, reduced resource utilization, and heightened profitability. By providing insights into the root zone conditions of

crops, soil moisture sensors empower irrigators with a better understanding of the irrigation process.

- Integration of water level sensor in smart agriculture: Adequate water supply is crucial for agricultural and farm production, and it significantly impacts our overall quality of life. Monitoring the water levels in sources like water tanks or bore wells holds substantial importance in agricultural operations. Efficient water management relies on the accurate assessment of water levels in these sources. Maintaining a record of water levels helps conserve water resources and facilitates the study of water consumption patterns. Consequently, water level monitoring emerges as a pivotal activity in the realm of agriculture. For the prototype implementation of this concept, an Arduino UNO board is employed in conjunction with an Ethernet shield to enable Internet connectivity. It is important to note that the water level sensor used in this prototype serves solely for demonstration purposes.

8 Conclusion and Future Enhancement

Our project introduces innovative technological advancements to the farming community, offering them opportunities to enhance their income. The primary objective of this initiative is to transform labor-intensive agricultural practices into more efficient, automated methods. Through the integration of IoT technology, farmers gain the ability to remotely manage and monitor their farms from any location. A significant aspect of this project involves the systematic collection of farm-related data, which can be harnessed and analyzed by both farmers and governmental bodies for improved governance and decision-making. Furthermore, this project has the potential for continuous development by incorporating various new sensors and refining communication protocols. Our automated smart agriculture system effectively reduces the time and resources required for manual operations. Leveraging Internet of Things technology, this system also facilitates the measurement of soil moisture and water levels in fields. While the system operates optimally under ideal conditions, there is room for improvement to address challenges such as low illumination or poor lighting scenarios.

This paper introduces the Agro-logger system, which possesses the ability to transmit and receive data from sensors, as well as retrieve updated and valuable information from the cloud. The cloud platform is linked to multiple Agro-loggers, covering a large geographical area. This collected data undergoes analysis for optimal selection criteria, which incorporates input from both farmers and Agro-loggers. Subsequently, the cloud updates the Agro-loggers with the refined data, adjusting threshold values based on the chosen criteria to enhance crop yield and soil fertility. To implement this system, the Arduino Mini Pro board is chosen due to its sufficient IO pins for sensor connectivity, low power consumption suitable for mobile battery usage during power outages. Xively, now known as pachube, serves as a globally trusted IoT platform, facilitating effortless connection of IoT devices and enabling data storage, analysis, and graphical representation. Xively further refines and interprets actionable results for enhanced device performance, offering real-time updates based on live data.

As mentioned earlier, this project utilizes a Python GUI application, although it can be developed in any programming language. The vocabulary booster is not limited to students; it offers benefits to individuals of all age groups due to its engaging tests that foster creative learning. To enhance accessibility, we can extend this concept by creating a mobile application and a Web interface. These additions will help us broaden the program's reach among the intended audience. Furthermore, incorporating image prompts for specific words could provide visual representations, aiding comprehension, especially for younger learners. By incorporating these enhancements, we can make the application appealing to a diverse range of ages and backgrounds. Our future goal involves building an Android application using Kotlin and XML, available for download on Google Play. Additionally, we look forward to conducting further research to enhance the user experience and introduce crossword puzzles for interactive play with family and friends.

References

1. Yang LD (2011) Implementation of a wireless sensor network with EZ430-RF2500 development tools and MSP430FG4618/F2013 experimenter boards from Texas instruments
2. Lad A, Nandre S, Raichurkar K, Zarkhande S, Charles P (2022) A literature survey on smart agriculture monitoring and control system using IOT. Int J Res Appl Sci Eng Technol (IJRASET) 10:1287–1290
3. Srivastava R, Sharma V, Jaiswal V, Raj S (2020) A research paper on smart agriculture using IoT. Int Res J Eng Technol (IRJET) 07(07)
4. Sekaran K, Meqdad MN, Kumar P, Rajan S, Kadry S (2020) Smart agriculture management system using internet of things. Telkomnika 2020(18):1275–1284
5. Singh G, Kalra N, Yadav N, Sharma A, Saini M (2022) Smart agriculture: a review. Siberian J Life Sci Agric 14(6):423–454
6. Vadapalli A, Peravali S, Dadi V (2020) Smart agriculture system using IoT technology. Int J Adv Res Sci Eng 9:58–65
7. Sehrawat D, Gill NS (2019) Smart sensors: analysis of different types of IoT sensors. In: 2019 3rd international conference on trends in electronics and informatics (ICOEI), Tirunelveli, India, 523–528
8. Ranjan V, Mannajji A, Akshatha H, Megha S, Pratiksha PP (2019) Smart agriculture system. Int J Sci Res Dev (IJSRD) 7(04)
9. Lozano C, Rodriguez O (2011) Design of forest fire early detection system using wireless sensor networks. Electron Electr Eng 3(2):402–405
10. Dhanaraju M, Chenniappan P, Ramalingam K, Pazhanivelan S, Kaliaperumal R (2022) Smart farming: internet of things (IoT)-based sustainable agriculture. Agriculture 12:1745
11. Rehman A, Jingdong L, Khatoon R, Hussain I (2016) Modern agricultural technology adoption its importance, role and usage for the improvement of agriculture. Am-Eurasian J Agric Environ Sci 16(2):284–288
12. Tubaishat M, Madria S (2003) Sensor networks: an overview. IEEE Potentials 22(2):20–23

13. Nakamura FP, Quintão FP, Menezes GC, Mateus GR (2005) An optimal node scheduling for flat wireless sensor networks. In: International conference on networking. Springer, Berlin, Heidelberg, 475–482
14. Yick J, Mukherjee B, Ghosal D (2008) Wireless sensor network survey. Comput Netw 52(12):2292–2330

A New Document Representation Technique for Hate Speech Detection Using a New Term Weight Measure and Word Embedding Techniques

Raju Rdara, N. Swapna, and Urmila Dara

Abstract Due to the advent of various Internet technologies in social media platforms like Twitter, Facebook, LinkedIn, Instagram, and review sites, people are changing their way of communication with other people. These platforms allow the people to freely share their knowledge, opinions to individual or group of people. Most of them used these platforms for welfare of others, but some of them misused these environments by spreading false information, hateful texts about a person, product, or any other entity. Hate speech (HS) is any expression, whether direct or indirect, that disparages another person or group of people based on their gender, sexual orientation, religion, ethnicity, or disability. The terminology of authors employ in their writings is crucial in determining whether or not a text contains hate speech. The weight of a term in a text also play prominent role to differentiate the hate speech from normal text. In this article, we proposed a new document representation method for hate speech detection. The proposed method represents the documents as vectors by using two varieties of information such as the document representation with the features identified through feature selection algorithm and the document vectors represented by using the word vectors generated through word embedding techniques. We developed a new term weight measure to compute the weight value of a vector and compared the performance of proposed term weight measure with different existing term weight measures. The text document vectors are trained with machine learning algorithm. Support vector machine (SVM) is used for producing the model. The SVM achieved 0.8964 score of accuracy for hate speech detection.

Keywords Hate speech · Term weight measures · Hate speech detection · Word embedding techniques · SVM

R. Rdara (✉) · N. Swapna
Vignana Bharathi Institute of Technology, Hyderabad, Telangana, India
e-mail: rajurdara@gmail.com

U. Dara
Cubic Transportation Systems India Pvt. Limited, Hyderabad, India

B. Annappa et al. (eds.), *High Performance Computing, Smart Devices and Networks*, Lecture Notes in Electrical Engineering 1262,
https://doi.org/10.1007/978-981-97-7794-5_24

1 Introduction

Social media's importance in modern human communication is increasing due to its role in such communication. The increased use of social media has facilitated interactions between people who live apart but has also raised new security risks for both individuals and society. The spreading of HS on social media platforms has grown in importance in recent years. Governments and organizations throughout the world have begun developing strategies to address the issue and the possible harm that arises when online discussions turn into actual hate crimes. A statement that disparages an individual or a group of people based on their sexual orientation, race, gender, nationality, ethnicity, religion, color, or any other feature is commonly referred to as hate speech (HS).

Due to the lack of a universally accepted legal definition for hate speech, other scholars came up with their own definitions. More datasets are now accessible for researchers to propose research works with the increased focus on automatic hate speech identification. However, there is not yet a widely used benchmark dataset in the literature; therefore, it is vital to remember that the datasets that are now accessible come from various sources, annotation standards, and domains. The most popular medium for data collecting is Twitter, which is utilized in a number of datasets on cyberbullying, anti-refugee sentiment, and abusive language and hate speech. For the purpose of detecting HS, Davidson et al. [1] created an English dataset. They initially employed an HS lexicon, which includes words and phrases that Internet users recognize as hate speech. Using the lexicon, they used the Twitter API to look for appropriate individuals who contained the terms from the lexicon, finding 33,458 seed people as a result. After that, they extracted a total of 85.4 million tweets from each seed user's timeline. They chose a random sample of about 25,000 tweets containing terms from the lexicon and had the crowd source employees manually annotate them.

This study presents a novel document representation technique for hate speech detection method by using word embedding methods and term weight measures (TWM). This technique uses an FS algorithm to identify the important terms for experimentation. After identification of a collection of words, TWMs are used to calculate each term's weight in the text's vector representation. To calculate the weight of a term, we developed a new TWM. We then compared the effectiveness of the proposed measure with five existing TWMS. This document representation is merged with the document represented generated through the word vectors generated through word embedding techniques. SVM classifier is used for training the final document vectors.

This article is planned in seven sections. Section 2 analyzes the existing proposed approaches for HS detection. The dataset characteristics are represented in Sect. 3. The performance metrics for evaluating the proposed work are defined in Sect. 4. Section 5 provides an explanation of the proposed methodology. Section 6 presents the experiment's results and an analysis of those results. Section 7 discusses the article's conclusions and future goals.

2 Previous Works

Traditional machine learning (ML) methods including LR ("Logistic Regression"), SVMs, and RF ("Random Forest") were used in earlier work on detection of hate speech automatically [2]. Many of the studies investigating HS detection from more recent times use Deep Learning (DL) models. The Waseem and Hovy dataset was used in the work of Badjatiya et al. to show how DL approaches such as CNNs ("Convolutional Neural Networks"), LSTM ("Long Short-Term Memory"), and FastText in HS detection are able to outperform traditional ML models that are trained on n-gram embeddings and bag of words [3, 4]. In addition to using a single tweet in their model, Qian et al. suggested a model by using inter-user and intra-user representations as well. When trained on a foundational LSTM model, this inclusion raises the classification scores [5]. Additionally, in Zhang et al.'s study [6], they determined popular results in HS detection and employed a GRUs ("Gated Recurrent Units") and CNN architecture combination to capture implicit aspects of hateful content.

In order to assess the efficiency of the proposed model on their developed benchmark dataset, Matthew et al. used well-known models including BiRNN, CNN-GRU, BiRNN, and BERT with attention for explainable HS detection [7]. To address a form of hate speech that is underrepresented, ElSherief et al. [8] provided another dataset that is centered on implicit hate speech. They used the dataset to assess their results against well-known baselines (BERT, GPT, and GPT-2) for generation of implied meanings and intended target tasks, and classification.

Alexandra Coroiu et al., addressed [9] the effectiveness of a generalized ML model that is trained on sexist messages and tested on racist messages. Three distinct SVM classifier kernels including linear, radial basis function, and polynomial were tested in their experiment. A dataset of sexist Twitter posts was used to train the SVM using word count and TFIDF features, and a dataset of racist Twitter messages was used to test it. The results show a low $F1$-score value of 0.3 on the testing set, compared to a high value on the training set $F1$-score value of 0.3. In order to automatically detect hate speech, György Kovács et al. suggested [10] a deep learning model based on NLP that combines convolutional and recurrent layers. They used the developed model on the HASOC2019 corpus. Using the HASOC test dataset, the developed model achieved a macro $F1$-score of 0.63 for HS detection.

Sandip Modha et al. devised a method [11] for identifying online aggression which is a variety of HS detection in social media. Based on a Web browser plugin, they created an interface for Twitter and Facebook that shows the aggression comments which are posted on the user's timelines in social media. This plugin's interface provides a tool for citizens who work in major organizations, as well as assistance to the security agency in monitoring social media activity. They conducted experiment with a variety of datasets, including the TRAC 2018 English dataset, the code-mixed Hindi–English dataset, and the dataset that was gathered from Facebook and Twitter via suggested plugins.

People are using social media platforms to communicate hate speech and false or fake information in messages to harm someone's reputation in our society because

Table 1 Data distribution of training dataset

Language/classes	HOF	NOT	Total
English	2261	3591	5852

there are no measures in place to prevent abusive comments. These platforms and governments are responsible for monitoring such abusive language texts before disseminating them to the public. Many issues raised in the detection of hate speech automatically by non-standard variations in word spelling and grammar in communications. This is particularly troublesome in nations where there are speakers of various languages and hate speech produced by these speakers uses code-mixed versions of multilingual terms. A machine learning model for detection of HS was developed by Sreelakshmi et al. [12] using the Hindi–English dataset, which is a kind of code-mixed dataset found in social media. The proposed method classified data samples of 10,000 collected from different sources as hate or non-hate by using FastText, which is a pre-trained library for producing word embeddings from Facebook. The effectiveness of the proposed approach is evaluated against word2vec and doc2vec feature approaches, and it is found that best representation was obtained for features when FastText is paired with (SVM)-radial basis function (RBF). Additionally, they noticed that for code-mixed datasets, character-based features performed the best.

3 Dataset Characteristics

The English dataset used in this experiment was taken from the 2019 FIRE competition's subtask-A [13]. The purpose of the subtask is to determine if the text uses any offensive or hateful language. Table 1 displays the description of the dataset.

Two classes, HOF ("Hate and Offensive") and NOT ("Non-Hate and Offensive"), are included in the dataset. There is profane (PRFN), hate (HATE), and offensive (OFFN) content in the posts of HOF class. There is no hate speech or offensive content in the posts of the NOT class.

4 Evaluation Measures

A generally utilized set of performance criteria has been designed to facilitate solution comparison and assess the efficacy of various machine learning models and HS detection approaches. The fundamental metrics for binary classification are TP, TN, FP, and FN. True Positive, or TP, is the number of documents that are accurately categorized as HOT. The number of documents accurately labeled as NOT is called TN, or "True Negative." Documents incorrectly categorized as HOT are counted as "False Positives," or FPs. Documents incorrectly categorized as NOT are counted as FN ("False Negative").

The ratio of accurately classified documents to total documents is known as accuracy. The accuracy metric is represented by Eq. (1).

$$\text{Accuracy} = \frac{\text{TP} + \text{TN}}{\text{TP} + \text{FP} + \text{TN} + \text{FN}} \tag{1}$$

Precision is a measure of the confidence in the detection of hate speech and is defined as the ratio of positive documents that are correctly predicted to total predicted positive documents. The precision metric is represented by Eq. (2).

$$\text{Precision} = \frac{\text{TP}}{\text{TP} + \text{FP}} \tag{2}$$

The percentage of positive documents that are correctly predicted to all positive documents is known as the recall or DR ("Detection Rate"). This measure expresses the probability that the suggested method will identify hate speech. The recall metric is represented by Eq. (3).

$$\text{Recall} = \frac{\text{TP}}{\text{TP} + \text{FN}} \tag{3}$$

The $F1$-score is a different way to describe how accurate a statistical analysis is. It is calculated by using the harmonic average of recall and precision scores. The $F1$-score metric is represented by Eq. (4).

$$F1\text{-Score} = \frac{2 \times (\text{Precision} * \text{Recall})}{(\text{Precision} + \text{Recall})} \tag{4}$$

In this work, the accuracy measure is used to represent the experimental outcomes.

5 Proposed Document Representation Method

The methodology followed in proposed document representation method is displayed in Fig. 1. In this approach, first, implement various appropriate preprocessing methods on the data in the training data to remove inadequate content from text. Next, extract all terms from the cleaned dataset. These extracted terms are given to two modules such as word embedding techniques and feature selection algorithms. The word embedding techniques convert words as word vectors. These word vectors are used to convert all documents into vector representation by using Doc2Vec model. The feature selection algorithm identifies significant terms from all terms based on the scores of algorithm. Based on the terms scores, select top scored terms as features for the experiment. In this work, we recognized top scored terms of 8000 for HS detection. The identified features are considered for document vector representation and the vector value is computed by using TWM. In the proposed method, different term weight measures and a new TWM is proposed for computing the value of a vector in document vector representation. In this approach, the experiment performed with five existing TWMs such as TF, TFIDF, TFRF, TF-PROB, and TFIEF and one proposed TWM. The chi-square FS algorithm is used to extract primary features. The final document vector is a combination of two representations such as document representation through Doc2Vec model and document representation through terms identified through feature selection algorithm and vector value computed through proposed term weight measure. These document vectors are trained with SVM classifier for generating model. The model predicts the performance of hate speech detection.

5.1 Chi-Square (CHI2) Measure

A popular and straightforward FS technique for determining the important features is chi-square ($\chi 2$), which focuses on the dependencies between various classes and features [14]. This metric calculates a feature's score that is unique to a class. Features with values between 0 and 1 are assigned by using the CHI2 algorithm. A score of 0 implies that there is no dependency between the feature and the class, whereas a score of 1 represents that the feature is more pertinent to the class. Equation (5) represents the CHI2 measure to determine the degree of relevancy between a feature Ti and a class Cj.

$$\chi^2(Ti, Cj) = \frac{N \times (ai\ di - bi\ ci)^2}{(ai + bi) \times (ai + ci) \times (bi + di) \times (ci + di)} \tag{5}$$

5.2 Term Weight Measures

The TWM determines the significance of a term unique to a document by taking into account a number of different types of data. Whether a term is a good representative of a negative or positive class depends on its weight. TWM's are divided into two groups, namely UTWM ("Unsupervised TWM") and STWM ("Supervised TWM") based on the data used in the TWM. While STWMs used the class membership information of the document, UTWMs do not use such information. When compared to UTWM, STWMs are more effective at determining the significance of the term. This article describes an experiment that was carried out by using both supervised and unsupervised TWMs.

The weight of a term in a document is calculated by using TWMs. Distinct TWMs provide different weight values based on how a term is distributed within and between classes of documents. Table 2 presents the existing TWMs that are used in this experiment.

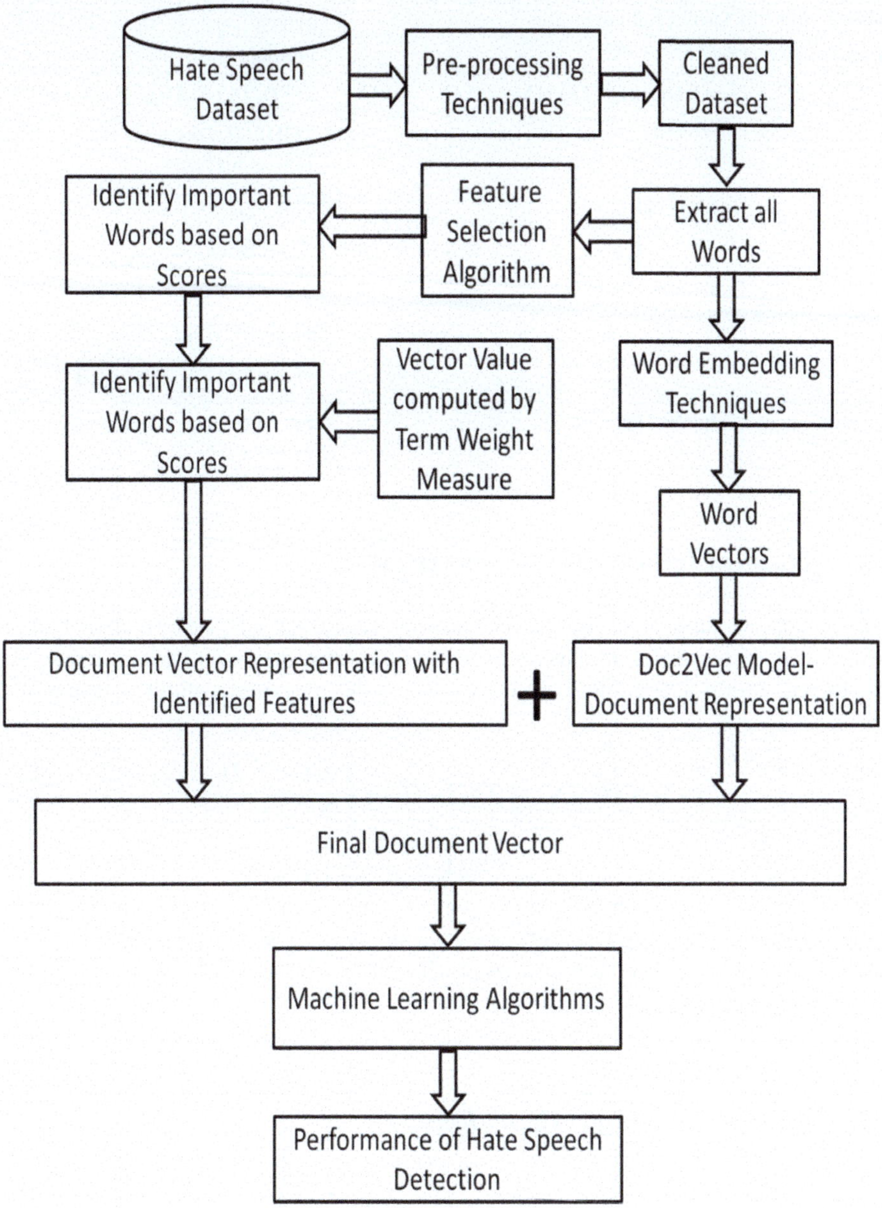

Fig. 1 Proposed document representation model

Table 2 Existing term weight measures

S. No.	TWM	Formula
1	Term frequency (TF) [15]	$\mathrm{TF}(T_i, D_k) = \mathrm{TF}_{ik}$
2	TFIDF [15]	$\mathrm{TFIDF}(Ti, Dk) = \mathrm{TF}ik \times \log\left(\frac{N}{\mathrm{DFi}}\right)$
3	TFIEF [15]	$\mathrm{TFIEF}(T_i, D_k) = \frac{\mathrm{TF}_{ik} \times e^{\frac{-\mathrm{DF}_i}{N}}}{\sqrt{\sum_{i=1}^{n}\left(\mathrm{TF}_{ik} \times e^{\frac{-\mathrm{DF}_i}{N}}\right)^2}}$
4	TFRF [16] ("term frequency-relevance frequency") measure	$\mathrm{TFRF}(T_i, D_k) = \mathrm{TF}_{ik} * \log\left(2 + \frac{A}{\mathrm{MAX}(1,B)}\right)$
5	TF-PROB [17]	$\mathrm{TF} - \mathrm{PROB}(T_i, D_k) = \mathrm{TF}_{ik} \times \log\left(1 + \frac{A}{B} * \frac{A}{C}\right)$

5.2.1 Proposed Term Weight Measure (PTWM)

In this study, we suggested a novel TWM by taking into account the term distribution information across different classes as well as the information about how terms are distributed within a document. Equation (6) is the proposed weight measure for terms.

$$\mathrm{PTWM}(T_i, D_k) = \mathrm{TF}(T_i, D_k) * \frac{\mathrm{CNT}(\mathrm{UT}, D_k)}{\mathrm{CNT}(\mathrm{TT}, D_k)} * \log\left(\frac{\max(1, A)}{\max(1, C)}\right) * \frac{A + D}{B + C} \tag{6}$$

where CNT(UT, D_k) and CNT(UT, D_k) are the counts of unique terms and total number of terms in document D_k, respectively. TF(T_i, D_k) denotes the count of term T_i in the kth document D_k.

5.3 *Machine Learning Algorithms*

Based on data pairings of input–output, the ML algorithm created a model that explains the behavior of the model. In addition to learning data pairs of input–output, the algorithms also memorize problem patterns and structures. Training process is generally divided into two phases, such as the training phase and the validation phase. During the training phase, the algorithm analyzes the provided input data to create a model that is utilized to predict the output. During the validation phase, the model analyzes data that was not utilized during training and determines how well the proposed approach predicts the class label. To create models, researchers employed a variety of well-known algorithms. We classified the data in this work by using the SVM classification technique.

The concept of support vector machines (SVM) was developed by Vapnik et al. in 1992 [18] and gained popularity for both classification and regression problems. The primary goal of this classification technique is to identify the hyperplanes that have the maximum distance between positive and negative examples. Maximal hyperplanes are another name for these hyperplanes. Support vectors are the instances that are located on the hyperplane. SVM is more frequently used to solve problems with two classes that can be separated linearly. The scenario is nonlinearly separable of classes when there are more than two classes in the problem. When the problem is nonlinearly separable, SVM is unable to immediately find maximal hyperplanes and support vectors. The kernel concept is introduced by SVM to address this nonlinearity. The kernel concept converts the nonlinearity feature space into a higher dimensionality feature space, and uses these higher dimensions to separate the instances in a linear fashion. Three different kernel types are supported by SVM such as polynomial, linear, and RBF ("Radial Basis Function") [19].

6 Experimental Results

In this section, the experimented results of TWMs and the experimental results of proposed document representation method are displayed.

6.1 Experimental Results of TWMs

To find the term value in the document vectors, the experiment is run using five distinct TWMs and one proposed TWM. In the experiment, the top scored terms of 8000 are determined. Every iteration of the experiment added 1000 terms to the terms of previous iteration. The classification model is trained by using the SVM classifier. Table 3 presents the accuracies of hate speech detection when various term weight metrics and SVM classifier.

When compared to other TWMs, the proposed TWM achieved 0.855 of accuracy for hate speech detection. It was found that increasing the number of words for the document vectors representation improved the accuracy. Additionally, we found that using more than 8000 terms for document vector representation in the experiment reduced the accuracy of hate speech recognition.

Table 3 Accuracies of proposed model for HS detection with various TWM's and SVM classifier

TWMs/number of features	TF	TFIDF	TFIEF	TFRF	TF-PROB	PTWM
1000	0.59	0.671	0.695	0.753	0.761	0.779
2000	0.603	0.685	0.707	0.764	0.773	0.787
3000	0.615	0.689	0.725	0.77	0.795	0.806
4000	0.63	0.696	0.738	0.778	0.802	0.811
5000	0.647	0.702	0.749	0.791	0.809	0.818
6000	0.659	0.716	0.752	0.817	0.824	0.834
7000	0.671	0.727	0.758	0.821	0.831	0.839
8000	0.663	0.733	0.769	0.824	0.846	0.855

6.2 Experimental Results of Proposed Document Representation Method

In this experiment, the document vector represented with combination of 8000 terms representation and document vector representation through word embeddings. Four word embedding techniques such as Word2Vec [20], Glove [21], FastText [22], and BERT [23] are used for generating word embeddings. Table 4 presents the accuracy of proposed document representation method for HS detection.

In Table 4, the combination of 8000 terms document representation and BERT embeddings based document vector representation accomplished best accuracy of 0.8964 for hate speech detection when compared with other representations.

Table 4 RF classifier accuracy of proposed document representation method for hate speech detection

Features	Accuracy
8000 terms document representation + Word2Vec embeddings based document representation	0.8634
8000 terms document representation + Glove embeddings based document representation	0.8721
8000 terms document representation + FastText embeddings based document representation	0.8789
8000 terms document representation + BERT embeddings based document representation	0.8964

7 Conclusions and Future Scope

Hate speech was targeting a particular individual person or group of people, which can cause cyber bullying, personal trauma, discrimination and panic in the society. In this work, a new document representation method was proposed for predicting the hate speech messages. The proposed document vector representation method represents the document vector with combination of word embedding techniques based document vector and features based document vector representation. The proposed method attained an accuracy of 0.8964 for hate speech detection when SVM classifier is used.

In future, we have a plan to implement Deep Learning (DL) techniques of Gated Recurrent Unit and BERT classifier for classification because it was observed that the ML techniques are not able to identify huge amount of discriminatory messages.

References

1. Davidson T, Warmsley D, Macy MW, Weber I (2017) Automated hate speech detection and the problem of offensive language. In: Proceedings of the eleventh international conference on web and social media, ICWSM 2017, Montréal, Québec, Canada, 15–18 May 2017. AAAI Press, pp 512–515
2. Fortuna P, Nunes S (2018) A survey on automatic detection of hate speech in text. ACM Comput Surv 51(4)
3. Badjatiya P, Gupta S, Gupta M, Varma V (2017) Deep learning for hate speech detection in tweet. In: Proceedings of the 26th international conference on World Wide Web companion, ser. WWW'17 Companion, Perth, Australia: International World Wide Web Conferences Steering Committee, pp 759–760. ISBN: 9781450349147
4. Waseem Z, Hovy D (2016) Hateful symbols or hateful people? Predictive features for hate speech detection on Twitter. In: Proceedings of the NAACL student research workshop. Association for Computational Linguistics, San Diego, California, June 2016, pp 88–93
5. Qian J, ElSherief M, Belding E, Wang WY (2018) Leveraging intra-user and inter user representation learning for automated hate speech detection. In: Proceedings of the 2018 conference of the North American chapter of the Association for Computational Linguistics: human language technologies, vol 2 (short papers). Association for Computational Linguistics, New Orleans, Louisiana, June 2018, pp 118–123
6. Zhang Z, Luo L (2019) Hate speech detection: a solved problem? The challenging case of long tail on twitter. Semantic Web 10:925–945
7. Mathew B, Saha P, Yimam SM, Biemann C, Goyal P, Mukherjee A (2010) Hatexplain: a benchmark dataset for explainable hate speech detection. arXiv:2012.10289; Banks J (2010) Regulating hate speech online. Int Rev Law Comput Technol 24(3):233–239
8. ElSherief M et al (2021) Latent hatred: a benchmark for understanding implicit hate speech. In: Proceedings of the 2021 conference on empirical methods in natural language processing, online. Association for Computational Linguistics, Punta Cana, Dominican Republic, Nov 2021, pp 345–363
9. Coroiu A (2019) The generalization performance of hate speech detection using machine learning. In: 31th twenty student conference on IT, 5th July 2019, Enschede
10. Kovács G, Alonso P, Saini R (2021) Challenges of hate speech detection in social media. SN Comput Sci 2:95

11. Modha S, Majumder P, Mandl T, Mandalia C (2020) Detecting and visualizing hate speech in social media: a cyber watchdog for surveillance. Expert Syst Appl 161:113725
12. Sreelakshmi K, Premjith B, Soman KP (2020) Detection of hate speech text in Hindi-English code-mixed data. Procedia Comput Sci 171:737–744
13. https://hasocfire.github.io/hasoc/2019/dataset.html
14. Yang Y, Pedersen J (1997) A comparative study on feature selection in text categorization. In: Proceedings of the 14th international conference on machine learning, Nashville, pp 412–420
15. Tang Z, Li W, Li Y (2019) An improved term weighting scheme for text classification. Concurr Computat Pract Exper 2019
16. Raghunadha Reddy T, Vishnu Vardhan B, Vijayapal Reddy P (2016) Profile specific document weighted approach using a new term weighting measure for author profiling. Int J Intell Eng Syst 9(4):136–146
17. Raghunadha Reddy T, Vishnu Vardhan B, Vijayapal Reddy P (2017) A document weighted approach for gender and age prediction. Int J Eng Trans B Appl 30(5):647–653
18. Boser BE, Guyon IM, Vapnik VN (1992) A training algorithm for optimal margin classifiers. In: Proceedings of the fifth annual workshop on computational learning theory, COLT'92. ACM, New York, NY, USA, pp 144–152
19. Hofmann T, Schölkopf B, Smola AJ (2008) Kernel methods in machine learning. Ann Statist 36(3):1171–1220
20. Rong X (2014) word2vec parameter learning explained. arXiv:1411.2738
21. Pennington J, Socher R, Manning C (2014) Glove: global vectors for word representation. In: Proceedings of the 2014 conference on empirical methods in natural language processing (EMNLP), pp 1532–1543
22. Joulin A, Grave E, Bojanowski P, Mikolov T (2016) Bag of tricks for efficient text classification. arXiv:1607.01759
23. Devlin J, Chang M-W, Lee K, Toutanova K (2018) Bert: pre-training of deep bidirectional transformers for language understanding. arXiv:1810.04805